T0235248

Systemic and Systematic Risk Management

Systemic and Systematic Risk Management

Joseph Eli Kasser

CRC Press
Taylor & Francis Group
Boca Raton London New York

CRC Press is an imprint of the
Taylor & Francis Group, an **informa** business

First edition published 2020
by CRC Press
6000 Broken Sound Parkway NW, Suite 300, Boca Raton, FL 33487-2742

and by CRC Press
2 Park Square, Milton Park, Abingdon, Oxon, OX14 4RN

Library of Congress Cataloging-in-Publication Data
Names: Kasser, Joe, author.
Title: Systemic and systematic risk management / Joseph E. Kasser.
Description: First edition. | Boca Raton, FL : CRC Press, 2020. |
Includes bibliographical references and index. |
Summary: "This book discusses risk management as it applies to the
problem-solving process for simple, complex, and wicked problems. The book does
this in the context of policies. When applying systems thinking to risk management,
it can be seen that risk management applies to everything in daily life, from crossing
the road, to preventing problems from arising in project management, as well
as in major systems development and social systems. The book will provide ways
of closing the loop and providing the policy makers with the tools to monitor
the policy implementation"—Provided by publisher.
Identifiers: LCCN 2020000359 (print) | LCCN 2020000360 (ebook) |
ISBN 9780367112219 (hardback) | ISBN 9780429025389 (ebook)
Subjects: LCSH: Risk management.
Classification: LCC HD61 .K367 2020 (print) | LCC HD61 (ebook) |
DDC 658.15/5—dc23
LC record available at https://lccn.loc.gov/2020000359
LC ebook record available at https://lccn.loc.gov/2020000360

ISBN: 978-0-367-11221-9 (hbk)
ISBN: 978-0-429-02538-9 (ebk)

Typeset in Times
by codeMantra

Dedication

To my wife, Lily, always caring, loving and supportive

Contents

Preface

Every action taken involves risk: be it making a cup of coffee, slicing bread, playing golf, using public or private transportation, initiating a public or private organization policy, commencing a large engineering project (LEP) or exploring the ocean depths or outer space.

The traditional approach to risk management has been to focus on the risk specific to the activity treating each activity as a special case, then following a process of identifying, analysing and mitigating the associated risks by applying information from the appropriate standards. This book:

- Is different; it employs the systems thinking approach.
- Contains hardly any mathematics.
- Is meant for project managers and policy personnel.
- Treats the risk management process as an instance of the problem-solving process.
- Shows that when plotted in a framework, any type of risk in any type of activity shares properties with a readily identifiable set of risks.
- Shows how systemic and systematic project management and systems engineering can mitigate or prevent most identifiable risks and build robustness into a system to allow it to cope with some types of unforeseen types of risks.
- Provides you with tools to perform risk management in your personal as well as professional life.
- Will not make you an expert in the topics covered in each chapter; it is not intended to. It is intended to teach you enough to communicate with the relevant subject matter (domain) experts to:
    - Make an informed decision on the advice (information) they provide.
    - Detect when they are more ignorant than you which will enable you to ignore their advice and find a real expert.

This book is the fourth in a set of books applying systems thinking to project management and systems engineering. Accordingly, as the focus of the book is on risk management the chapters on risks in project management and systems engineering, are not only a useful accompaniment to books on project management and systems engineering, they also show how the two disciplines are interdependent and related.

As in the previous volumes, this book introduces new concepts based on the systems approach such as the risk framework in Section 1.2 and the firm fixed price contract with penalties and bonuses in Section 8.4.2.1.

Acknowledgements

This book would not have been possible without the co-authors of the papers upon which some of these chapters are based, and colleagues and friends who helped review the manuscripts:

Eileen P. Arnold
Professor Derek K. Hitchins
Victoria R. Williams
Associate Professor Yang Yang Zhao

Author

Joseph Eli Kasser has been a practicing systems engineer for 50 years, a project manager for more than 35 years and an academic for 20 years. He is a Fellow of the Institution of Engineering and Technology (IET), a Fellow of the Institution of Engineers (Singapore) and the author of *'Systems Engineering a Systemic and Systematic Methodology for Solving Complex Problems'*, *'Systemic and Systematic Project Management'*, *'The Systems Thinkers Toolbox: Tools for Managing Complexity'*, *'Perceptions of Systems Engineering'*, *'Holistic Thinking: Creating Innovative Solutions to Complex Problems'*, *'A Framework for Understanding Systems Engineering'* and *'Applying Total Quality Management to Systems Engineering'*, two books on amateur radio and many International Council on Systems Engineering (INCOSE) symposia and other conference and journal papers.

He received the National Aeronautics and Space Administration's (NASA) Manned Space Flight Awareness Award (Silver Snoopy) for quality and technical excellence for performing and directing systems engineering. He holds a Doctor of Science in Engineering Management from the George Washington University. He is a Certified Manager, a Chartered Engineer in both the UK and Singapore and holds a Certified Membership of the Association for Learning Technology. He has been a project manager in Israel and Australia, and performed and directed systems engineering in the US, Israel and Australia. He gave up his positions as a Deputy Director and Defence Science and Technology Organization (DSTO) Associate Research Professor at the Systems Engineering and Evaluation Centre at the University of South Australia in early 2007 to move to the UK to develop the world's first immersion course in systems engineering as a Leverhulme Visiting Professor at Cranfield University. He spent 2008–2016 as a Visiting Associate Professor at the National University of Singapore where he taught and researched the nature of systems engineering, systems thinking and how to improve the effectiveness of teaching and learning in postgraduate and continuing education. He is currently based in Adelaide, Australia. His many awards include:

- National University of Singapore, 2008–2009, Division of Engineering and Technology Management, Faculty of Engineering Innovative Teaching Award for use of magic in class to enrich the student experience.
- Best Paper, Systems Engineering Technical Processes track, at the 16th Annual Symposium of the INCOSE, 2006, and at the 17th Annual Symposium of the INCOSE, 2007.
- United States Air Force (USAF) Office of Scientific Research Window on Science program visitor, 2004.
- Inaugural SEEC 'Bust a Gut' Award, SEEC, 2004.
- Employee of the Year, SEEC, 2000.
- Distance Education Fellow, University System of Maryland, 1998–2000.

- Outstanding Paper Presentation, Systems Engineering Management track, at the 6th Annual Symposium of the INCOSE, 1996.
- Distinguished Service Award, Institute of Certified Professional Managers (ICPM), 1993.
- Manned Space Flight Awareness Award (Silver Snoopy) for quality and technical excellence, for performing and directing systems engineering, NASA, 1991.
- NASSA Goddard Space Flight Center Community Service Award, 1990.
- The E3 award for Excellence, Endurance and Effort, Radio Amateur Satellite Corporation (AMSAT), 1981, and three subsequent awards for outstanding performance.
- Letters of commendation and certificates of appreciation from employers and satisfied customers including the:
  - American Radio Relay League (ARRL).
  - American Society for Quality (ASQ).
  - Association for Quality and Participation (AQP).
  - Communications Satellite Corporation (Comsat).
  - Computer Sciences Corporation (CSC).
  - Defence Materiel Organisation (Australia).
  - Institution of Engineers (Singapore).
  - IET Singapore Network.
  - Loral Corporation.
  - Luz Industries, Israel.
  - Systems Engineering Society of Australia (SESA).
  - University of South Australia.
  - United States Office of Personnel Management (OPM).
  - University System of Maryland.
  - Wireless Institute of Australia.

When not writing and lollygagging, he provides consulting services and training.

1 Introduction

This book is the fourth book in a set of books on the systems approach to problem-solving in project management and systems engineering and:

- Introduces new concepts based on the systems approach such as the risk framework (Section 1.2) and the firm fixed price (FFP) contract with penalties and bonuses (Section 8.2.4.1).
- Complements as well as incorporates material from the following companion volumes by adding new material from a risk perspective using the appropriate citations:
  - *Systems Thinker's Toolbox: Tools for Managing Complexity* (Kasser 2018) which provides more than 100 conceptual tools for systems thinkers.
  - *Systemic and Systematic Project Management* (Kasser 2019a) which discusses the systems approach to project management and how some of the tools in the toolbox are applied in project management.
  - *Systems Engineering: A Systemic and Systematic Methodology for Solving Complex Problems* (Kasser 2019b) which discusses the systems approach to systems engineering and managing the complex, well-structured, ill-structured and wicked problems that are in vital need of risk management as well as how some of the tools in the toolbox are applied in project management.
- Integrates project management with systems engineering rather than treating them separately as is done in most other books.
- Is based on more than 20 years of research and 50 years of experience, much of which has been previously published in peer-reviewed conference papers and journals. Accordingly, citations are provided where appropriate.
- Shows how to proactively incorporate prevention into planning in order to prevent risks, as well as how to mitigate them when they occur.
- Discusses the flow of a policy from creating a policy to implementing one through large engineering projects (LEPs) (Chapter 8) and simpler projects and where the risks arise and should be dealt with.
- Presents the risks in the relationship between policy creation, implementation and project management.
- Discusses risks throughout the policy implementation process and shows how the nature of risks changes from political to financial to technological as implementation proceeds.
- Points out that in most instances the traditionally ignored major risk is that of poor performance by personnel.

1.1   HOW TO READ AND USE THIS BOOK

This is a reference book as well as a text book, so if you are not using it as a textbook, don't read the book sequentially in a linear manner, but prepare for several passes through it. This book is non-fiction. Non-fiction books are different to fiction: stories, novels and thrillers are designed to be read in a linear manner from start to finish.

This book contains a lot of cross-references in the form of (Section n)* to help you navigate through the information. While the majority of cross-references are to material in earlier chapters, there are some forward references to material in later chapters, particularly in this chapter, because it is more logical to discuss the material in the context of the later chapter.

This book is designed to help you learn and use the content in the following manner:

1. *Skim the book*: flip through the pages; if anything catches your eye and interests you, stop, glance at it and then continue flipping through the pages. Notice how the pages have been formatted with dot points (bulleted lists) rather than in paragraphs to make skimming and reading easier.
2. For each chapter:
   - Read the introduction and summary.
   - Skim the contents. Stop and read topics of interest.
   - Look at the drawings.
   - Go on to the next chapter.
3. If you don't understand something, skip it on the first and second readings: don't get bogged down in the details.
4. Work though the book slowly so that you understand the message in each section of each chapter. If you don't understand the details of the example, don't worry about it as long as you understand the point that the example is demonstrating. Refer to the list of acronyms in Table 1.1 as necessary.
5. Successfully manage your risks and all subsequent ones.

Step 1 should give you something you can use immediately. Steps 2 and 3 should give you something you can use in the coming months. Step 4 should give you something you can use for the rest of your life. Step 5 is the rest of your life.

If you are using this book as a textbook, you will note forward references as well as backward references. This is because topics are considered from a number of perspectives, and since topics repeat, topics are discussed in the context in which they most often generally occur.

---

* The print book version of hotlinks.

TABLE 1.1
Acronyms Used in This Book

ADS	Air Defence System
AoA	Analysis of Alternatives
ATR	Acceptance Test Review
BPR	Business Process Reengineering
CAIV	Cost as an Independent Variable
CATWOE	Customers, Actors, Transformation Process, Weltanschauung, Owners, Environmental Constraints
CDR	Critical Design Review
CM	Change Management
CONOPS	Concept of Operations
COTS	Commercial Off the Shelf
CPAF	Cost-Plus Award Fee
CPFF	Cost-Plus Fixed Fee
CPIF	Cost-Plus Incentive Fee
CRIP	Categorized Requirements in Process
DMSMS	Diminishing Manufacturing Sources and Material Shortages
DoD	Department of Defense
DODAF	Department of Defense Architecture Framework
dTRL	Dynamic TRL
DRR	Delivery Readiness Review
ENU	Engaporean National University
ETL	Enhanced Traffic Light
EVA	Earned Value Analysis
FCFDS	Feasible Conceptual Future Desirable Situation
FDD	Feature-Driven Development
FFP	Firm Fixed Price
HKMF	Hitchins-Kasser-Massie Framework
HSI	Human Systems Integration
HTP	Holistic Thinking Perspective
ICBM	Intercontinental Ballistic Missile
ICD	Interface Control Document
ID	Identification
IEEE	Institute of Electrical and Electronics Engineers
INCOSE	International Council on Systems Engineering
IRR	Integration Readiness Review
IST	Idea Storage Template
IV&V	Independent Verification and Validation
JIT	Just-In-Time
LCCE	Life Cycle Cost Estimate
LEO	Low Earth Orbit

(*Continued*)

TABLE 1.1 (*Continued*)
Acronyms Used in This Book

LEP	Large Engineering Project
MBE	Management by Exception
MBO	Management by Objectives
MBSE	Model Based Systems Engineering
MBUM	Micromanagement by Upper Management
MOE	Measures of Effectiveness
MOP	Measures of Performance
MOS	Measures of Suitability
MT	Mission Tasks
MVA	Multi-Attribute Variable Analysis
NASA	National Aeronautics and Space Administration
O&M	Operations and Maintenance
OCD	Operations Concept Document
OCR	Operations Concept Review
OMB	Office of Management and Budget
OODA	Observe–Orient–Decide–Act
OT&E	Operational Test and Evaluation
PAM	Product-Activity-Milestone
PDR	Preliminary Design Review
PERT	Program Evaluation Review Technique
PID	Project Initiation Document
RFP	Request For Proposal
RMA	Reliability Maintainability and Availability
ROI	Return on Investment
SAGE	Semiautomatic Ground Environment
SDP	System Development Process
SEMP	Systems Engineering Management Plan
SETA	Systems Engineering – The Activity
SETR	Systems Engineering – The Role
SLC	System Lifecycle
SRR	System Requirement Review
SSM	Soft Systems Methodology
STALL	Stay Calm, Think, Analyse, Listen, Listen
T&E	Test and Evaluation
TAWOO	Technology Availability Window of Opportunity
TEMP	Test and Evaluation Master Plan
TPM	Technical Performance Measures
TRL	Technology Readiness Level
TRR	Test Readiness Review
WBS	Work Breakdown Structure
WP	Work Package

1.2 THE TRADITIONAL AND SYSTEMS APPROACHES TO RISK MANAGEMENT

Anyone who studies the literature on risk management in different domains such as engineering, finance and management will notice:

1. Risks are generally treated as unique to the situation; for example there are books on financial risks, project management risks, systems engineering risks, logistics risks and failures and reliability.
2. A great deal of similarity in what is being described in the different situations even though the words used (terminology) might differ.
3. Lists of risks presumably based on the author's experience, and those the author has read about in the literature (other people's experience).

So, they could read the books and extract a list of risks, put them in a database and then sort and index them and keep them handy. Then the next time they have to create a risk management plan, open up the risk database and start from the top, determine if each risk listed is applicable to their situation, and if it is, list it and plan how to mitigate or prevent it.* If they are smart, they will also capture how the risk was mitigated or prevented in their database; that way they won't have to think too much when creating the risk management plan, because they can copy the risk and the mitigation/prevention information from the database into their plan. This process is the basic pattern matching process for what can turn out to be a long list of risks.

Perceptions from the *Generic* and *Temporal* holistic thinking perspectives (HTPs) (Section 2.5.1) note these approaches of building lists seem to be paralleling the development of theories of motivation in psychology. For example, Murray identified separate kinds of behaviour and developed an exhaustive list of 39 psychogenic or social needs (Murray 1938). However, the list is so long that there is almost a separate need for each kind of behaviour that people demonstrate (Hall and Lindzey 1957). While Murray's list was very influential in the field of psychology, it has not been applied directly to the study of motivation in organizations. This is probably because the length of the list makes it impractical to use. Lists of risks in the literature can be even longer and more impractical. Inferences from the *Scientific* HTP note that one could expect that similar lengthy lists of risks would only be useful for checking off when pattern matching. If a risk on the list is found, it might be mitigated or prevented. However, if there is a risk that is not in the list, it will be ignored by the pattern matchers until it happens and then reacted to and hopefully added to the list for future activities.

Maslow's hierarchical classification of needs (Maslow 1954, 1968, 1970) has been by far the most widely used classification system in the study of motivation in organizations. Maslow differs from Murray in two important ways, namely:

---

* Someone who is lazy or thinks that they want to be complete could just incorporate the entire list in the database and impress people who haven't the faintest idea as to what risk management is about.

TABLE 1.2
The Risk Framework

		Lifecycle State		
Layer of Complexity	**Initialization**	**Planning**	**Performance**	**Closeout**
5	Social			
4	Supply chain			
3	Business			
2	System			
1	Product			
0	Component			

- *Short*: only containing five categories.
- *Hierarchical*: commonly drawn as a pyramid.

The traditional approach to risk management (Chapter 3) also assigns a probability of occurrence (Section 3.8.3.1) and severity of impact should the risk materialize (Section 3.8.3.2) to each identified risk pertaining to the project and combines them into a risk rectangle (Section 3.8.4). The probability of occurrence and severity of impact are multiplied together, and the risks with the highest values are mitigated (Section 3.8.4).

The systems approach (Section 1.4) is based on the problem-solving process and the risk framework shown in Table 1.2.

The risk framework is a two-dimensional framework. The vertical dimension uses the principle of hierarchies (Section 6.3.1) to aggregate risks in six layers being based on Hitchins' five layers of systems engineering (Hitchins 2000). The layers in order of objective complexity (Section 5.2.2.2.2) from highest to lowest are as follows:

5. *Social*: the business of government. Social projects tend to develop and implement policies and include one or more of the lower layers in order to implement the policy.
4. *Supply chain*: a number of businesses. Supply chain projects may include business and systems projects pending on the complexity of the supply chain.
3. *Business*: a number of systems working interdependently. Business projects often include system projects to develop the systems that are used in the business.
2. *System*: a number of products integrated into a system. System projects system projects tend to develop new systems or upgrade existing systems and often include a number of product projects to develop the subsystems for use in the system.

1. *Product*: a number of components integrated into a product. Product projects tend to develop new products or investigate failures in existing products and ways to mitigate or prevent those failures in the future. Product projects may include a number of component projects to develop new technology or identify alternative technologies for use in the products.
0. *Component*: technological, biological, information, etc. Component projects tend to develop parts or subsystems that are used in products; some products are technology-based ahd others are based on people often called manually intensive, while other products retain various mixtures of people and technology.

The horizontal time dimension contains the four states in the project lifecycle:

1. The project initialization state discussed in Section 7.4.1.
2. The project planning state discussed in Section 7.4.2.
3. The project performance state discussed in Section 7.4.3.
4. The project closeout state discussed in Section 7.4.4.

The risk framework can be used to track actions through the layers and states of their lifecycles. For example, when a public policy is initiated in layer 5, it becomes a project or even a series of projects following the policy lifecycle (Section 9.5.1). When the policy reaches the performance state (Section 9.5.1.4), if, for example, meeting the goal of the policy means that an airport, a transportation system, a new spacecraft or other objectively complex system needs to be developed or acquired, it might become a LEP (Chapter 8) in layers 3 or 4. The LEP is then split into a number of medium and smaller projects in layers 3 and 2. These projects take place in series and in parallel, and might generate further projects in layer 1. Moreover, each project has to be managed, and the state of each project needs to be reported to the organization charged with implementing the policy.

The systems approach uses perceptions from the:

1. *Generic* HTP to note similarities in the risks up and down the columns and along the rows of the risk framework.
2. *Continuum* HTP to note differences in the risks up and down the columns and along the rows of the risk framework.

Examples of similarities and differences include the following:

- In the initialization state, the same problems of financing the project, finding competent core personnel occur at each layer, albeit the amount of financing and the type of personnel may differ.
- Finding the right consultants is a problem faced in all areas of the risk framework.
- The problems posed in planning a project are similar in principle; in each layer the difference is in the scope of the plan.

The systems approach allows the unwieldy large numbers of risks in a list to be aggregated into usable categories. Each area in the risk framework contains risks, the remedying of which poses problems. The solution to a problem in one state creates a new problem for the subsequent state. For example, Fred has a problem in that he is thirsty. He decides to have a drink to quench his thirst (the solution). That gives him a new problem: what to drink. And so on, until he has actually decided what to drink, obtained and drunk it. Then he has to verify that the solution has remedied the original problem, namely after having drunk whatever it was, he is no longer thirsty. If he is not thirsty, the problem is remedied. If he is still thirsty, he drinks a little more and perceives how he feels and drinks a little more until he is no longer thirsty.

Using the systems approach means thinking about what can go wrong (the risks) in each state in the project lifecycle such as something that can:

1. *Cause him to select the wrong solution*: e.g. ignorance. He does not know what to do to quench his thirst.
2. *Impede him from realizing the solution*: e.g. there are no liquids available (really bad) or his preferred choice of beverage is not available (annoying).
3. *Prevent him from realizing the solution*: e.g. something that can cause him to spill the liquid before he can drink it or something that causes the liquid to vanish before he can drink it.

The systems approach builds risk management into both the:

1. *Planning process*: by considering these categories of risks and planning how to mitigate or prevent them. It does this by working backwards from the solution, identifying risks in the later states of the process and planning to prevent or minimize them in earlier states.
2. *Realization process which turns the plan into reality*: by monitoring the implementation of the plan and ensuring that risks identified during the planning process do not occur or having mitigation contingency plans ready should such a risk occur.

The risks are not only developed from lists of past experience, but also by thinking about the activities and products in each state of the process using:

- Critical thinking (Section 2.2).
- Systems thinking (Section 2.4) to understand the situation.
- Systems thinking and beyond (Section 2.5) to conceptualize the risks and how to mitigate or prevent them (Section 7.9).

So, in the systems approach we do not work from lists of risks generated by other people to identify and mitigate risks, but rather we think about the situation and use domain knowledge and expertise to identify and mitigate risks. Perceived from the *Generic* HTP, those lists of risks generated by other people can form a useful checklist, but perceptions from the *Continuum* HTP need to be applied to determine if those risks are pertinent to our situation and if the mitigation techniques used in other situations will work in our situation or will need to be modified in some way.

The systems approach to risk management is to ensure that there is no single point of failure in both the process and the product. Single points of failure in the

1. *Process*: are mitigated by:
   - Having multiple sources of supply, or the knowledge that those sources are available.
   - Reliable and/or redundant production equipment and a maintenance concept that provides for repair within specified times.
2. *Product*: are mitigated by designing security, reliability, redundancy and robustness into the system in the early states of the system development process (SDP) (Section 6.1.5.4) or using a maintenance concept that provides for repair within specified times if the operation of the system permits.

1.3 THE CONTENTS OF THIS BOOK

Since it is unlikely that any one person will have the domain knowledge to cover risk management of the entire enterprise:

- Chapter 1 introduces the book and the risk framework and recommends a way to read this book if you are not using it as a textbook. It continues with an introduction to the systems approach and concludes with a summary of the major risks in any activity and some ways to mitigate them.
- Chapter 2 discusses thinking, systems thinking, the nine HTPs and the benefits of going beyond systems thinking. In particular, it discusses judgement and creativity, thinking, critical thinking, systems thinking and going beyond systems thinking. It concludes with examples of using the HTPs to perceive a number of systems.
- Chapter 3 documents thoughts about risks and risk management systemically and systematically from the nine HTPs and suggests generic risk mitigation and prevention activities. Systemic perceptions of risks using the HTPs include the risk management process, the risk rectangle, risk profiles and risk trees.
- Chapter 4 perceives change and change management (CM) from the HTPs because when risks turn into events, change happens, and processes in which risks occur generally cause change. In particular, the chapter discusses some CM models, resistance to change and how to overcome it, and some aspects of stakeholder management.
- The risk management process is riddled with changes, problems and solutions because managing risks poses problems which need to be solved or remedied and change is an outcome, mostly desired but sometimes undesired. A risk mitigation strategy is a solution to the problem of managing the risk but poses a problem to the people who will have to implement the strategy. Accordingly, Chapter 5 discusses problems and problem-solving

from different perspectives because problem-solving is a major component of risk management. In particular, the chapter discusses the problem, solution and implementation domains and their differences, the traditional problem-solving process and the extended and multiple-iteration problem-solving process for applying risk management to managing complex, well-structured, ill-structured and wicked problems. While discussing managing complexity, the chapter specifies the minimum number of elements in a system for it to be defined as, and managed as, complex.

- Chapter 6 discusses systems and systems engineering from different perspectives because projects are systems and create products which are systems using systems engineering.* Moreover, the project takes place in an environment or context which is a system. The chapter provides an overview of systems, the system lifecycle (SLC), systems engineering and the SDP to help the policymaker, implementer and project manager understand the important aspects of risk management in systems and systems engineering. Specifically, this chapter discusses the nature of systems, properties of systems, hierarchies of systems, supply chains as systems, an introduction to systems engineering, modeling and simulation, the nine-system model and risks in systems and systems engineering.
- Policies and projects are systems which are realized using systems engineering and project management working interdependently. Chapter 6 provided an overview of systems, the SLC, systems engineering and the SDP to help the policymaker, implementer and project manager understand the important aspects of risk management in systems and systems engineering and the interdependency and overlap between risk management in systems engineering and project management. Accordingly, Chapter 7 discusses project management from different perspectives because project management is the heart of effecting any change, caused by the implementation of a policy, a LEP or any type of project. In particular, the chapter provides an overview of the systems approach to projects and project management discussing the project lifecycle, project plans, project planning, generic and specific planning, the planning process and using prevention as well as mitigation to manage risks and generic project risks.
- Chapter 8 provides an introduction to LEPs and managing risks in LEPs because a LEP contains a hierarchy of projects and systems. In particular, it discusses risks in LEPs, the LEP lifecycle and risks based on the non-availability of technology, since technology has become ubiquitous in the 21st century and introduces the FFP contract with penalties and bonuses (Section 8.2.4.1).
- Chapter 9 discusses policies and policy management from different HTPs because many policies are implemented by LEPs. In particular, it discusses policies as a system, the different types of policies, the generic policy lifecycle, policy documents and ways of mitigating risks in policies.
- Chapter 10 is an afterword which summarizes the contents of the book.

---

* Even if they don't call it systems engineering.

Each chapter also contains a discussion of pertinent risks and contains exercises. However, this book will not make you an expert in the topics covered in each chapter, the book is not intended to. It is intended to teach you enough to communicate with the relevant subject matter (domain) experts to:

1. Make an informed decision on the advice (information) they provide.
2. Detect when they are more ignorant than you, which will enable you to ignore their advice and find a real expert.
3. Be able to use the tools to perform risk management in your personal as well as your professional life.

1.4 THE SYSTEMS APPROACH

'The systems approach is a technique for the application of a scientific approach to complex problems. It concentrates on the analysis and design of the whole, as distinct from the components or the parts. It insists upon looking at a problem in its entirety, taking into account all the facets and all the variables, and relating the social to the technological aspects'.

(Ramo 1973)

Namely the systems approach is the application in an interdependent systemic and systematic manner of:

1. Critical thinking (Section 2.2).
2. Systems thinking (Section 2.4).
3. Going beyond systems thinking (Section 2.5).
4. Risk management (Chapter 3).

Moreover, the advantages of the systems approach to problem-solving have been known for centuries. The literature contains many examples including:

When people know a number of things, and one of them understands how the things are systematically categorized and related, that person has an advantage over the others who don't have the same understanding.

Luzatto (circa 1735)

People who learn to read situations from different (theoretical) points of view have an advantage over those committed to a fixed position. For they are better able to recognize the limitations of a given perspective. They can see how situations and problems can be framed and reframed in different ways, allowing new kinds of solutions to emerge.

Morgan (1997)

The systems approach lets you see things differently and produces opportunities. Yet the systems approach is not widely used when dealing with problems and issues. It seems that systems thinking which leads to the systems approach

TABLE 1.3
Active Brainstorming Results in Class

Team	Total Number of Ideas after Brainstorming	Total Number of Ideas after Active Brainstorming	Improvement (%)
1	20	40	100
2	9	89	889
3	22	66	200
4	31	64	106
5	39	79	103
6	28	89	218
7	20	'Too many to count'	Large

must be a learned skill because many people would say that human nature has a non-systems approach to thinking about issues. The non-systems thinker views issues from a single viewpoint/perspective or when multiple viewpoints are used, they are individual non-standardized viewpoints in some random order depending on the person.*

C. West Churchman introduced three standardized views in order to first think about the purpose and function of a system and then later think about the physical structure in *The Systems Approach* (Churchman 1968). These views provided three anchor points or viewpoints for viewing and thinking about a system. Twenty-five years later, *The Seven Streams of Systems Thinking* (Richmond 1993) introduced seven standardized viewpoints. Fifteen years later, Richmond's seven streams were modified and adapted into nine systems thinking perspectives (Kasser and Mackley 2008). The nine systems thinking perspectives introduced nine standardized viewpoints which can be used in sequence or in a random order. The nine systems thinking perspectives cover purpose, function, structure and more, and were later renamed as the HTPs (Section 2.5.1). The number of perspectives was limited to nine in accordance with Miller's rule (Miller 1956) (Kasser 2018: Section 3.2.5). The nine perspectives and the active brainstorming tool (Section 2.5.2) improved the way students applied systems thinking in the classroom and seem to provide a good way of teaching applied systems thinking. For example, the first time they were used in a classroom exercise, the students were asked to brainstorm a topic, then after a brief lecture on active brainstorming they were asked to use active brainstorming on the same topic. The result shown in Table 1.3 (Kasser 2018) shows at least a 100% increase in the number of ideas generated by all the teams.

---

* I have often heard people saying, "let's make sure we are on the same page", or engineers saying, "let's make sure we are on the same wavelength" to try to ensure meeting participants were viewing something from the same perspective or viewpoint.

1.5 THE SYSTEMS APPROACH TO PROBLEM-SOLVING

The systems approach to problem-solving as explained in this book is in sequential order:

1. Assume the problem was raised by the need to remedy an undesirable situation (Schön 1991).
2. Gain an understanding of the undesirable situation by viewing it from different perspectives.
3. Identify the cause of the undesirability if possible.
4. Visualize the situation without the undesirability.
5. Visualize the process of removing all or part of the undesirability from the situation.
6. Examine several alternative solutions* for feasibility, affordability, minimal risk and other pertinent factors.
7. Select the best solution.
8. Realize the solution.
9. Verify that the solution has removed the undesirability from the situation.
10. If the situation is still undesirable, then return to Step 2.

1.6 THE MAJOR RISKS IN ANY ACTIVITY

A number of authors provide lists of various types of risks which if combined can produce a very long list. Experience in behavioural psychology has shown that a list of 39 reasons was too long to be practical (Section 1.2).

As this book shows in the later chapters, the systems approach is based on:

- Identifying generic risks in each area of activity and determining their importance and urgency using domain expertise before managing them.
- Aggregating the generic risks into categories so that common mitigation and prevention techniques may be applied.
- Classifying the major risks to any endeavour into the following categories:
  1. *Lack of a common vision of the goal*†: this situation is represented in Figure 1.1. Without that common vision of the purpose and performance of the solution systems among the stakeholders, time and funds will be expended on non-productive activities. This risk may be mitigated by developing that common vision in the appropriate format for the activity such as in the concept of operations (CONOPS) (Section 6.5.9.1) commonly used in systems engineering.
  2. *People*: incompetent, inexperienced and unmotivated people will not do the job properly. They will waste time and funding doing the wrong thing and introducing errors of commission and omission

---

* A combination of the transition process and solution system in operation.
† The goal may change over time. Each step in the problem-solving process has its own goal, and different activities have their own goals. Accordingly, there may be more than one goal at any time: the short-term as well as the long-term goal of the project.

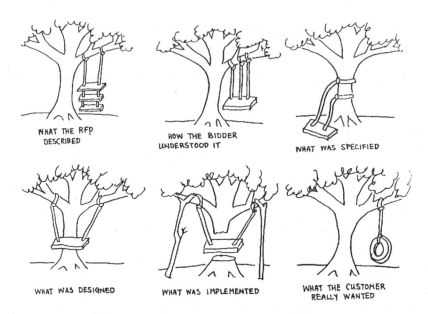

FIGURE 1.1   The consequences of lacking a common vision of the goal. (Republished with permission of ARTECH HOUSE, from *Applying Total Quality Management to Systems Engineering*, Joe Kasser, 1995; permission conveyed through Copyright Clearance Center, Inc.)

(Section 5.1.4.3.1) into what they do even if they follow the SDP (Section 6.1.5.4). This risk may be mitigated by employing competent and experienced people, motivating them by showing them 'what's in it for me*?' (Section 7.3.1.9) and providing just-in-time (JIT) training at appropriate points in the SDP. However, competent people cannot do a proper job with adequate resources and time.

3. *Politics*: people will mess things up for various personal agendas. They can use negative politics to cause failures or modifications that will cause failures. They can withhold resources and fail to meet commitments. They can persuade others to follow their lead with respect to the activity. This risk can often be mitigated by the use of positive politics (Section 7.3.1.10.1).

4. *Funding*: lack of sufficient funding leads to shortcuts, errors and early terminations. This risk can be mitigated by providing adequate funding.

5. *Resources*: insufficient resources lead to failures due to the lack of the resource. This risk can be mitigated by providing adequate resources.

6. *Time*: insufficient time leads to shortcuts and failures. This risk can be mitigated by inserting sufficient time into the schedule.

7. *Technology*: failures happen and need to be minimized and repaired: technology becomes obsolete over time as in diminishing manufacturing sources and material shortages (DMSMS) which results in systems

---

* Me being them.

becoming unmaintainable and needing replacement. This risk may be mitigated by the use of redundancy, reliability and the correct maintenance procedure.

The systems approach to planning a project determines the costs, schedules and resources needed as worst-case values to mitigate the funding, resource and time-related risks. Stakeholders that subsequently reduce those values for political and other reasons then introduce those risks into the endeavour.

1.7   SUMMARY

This chapter introduced the book and the risk framework and recommended a way to read this book if you are not using it as a textbook. The chapter continued with an introduction to the systems approach and concluded with a summary of the major risks in any activity and some ways to mitigate them.

REFERENCES

Churchman, C. West. 1968. *The Systems Approach*. New York: Dell Publishing Co.

Hall, Calvin S., and Gardner Lindzey. 1957. *Theories of Personality*. New York: John Wiley & Sons.

Hitchins, Derek K. 2000. *World class systems engineering – The five layer model* [Website] 2000 [cited 3 November 2006]. Available from http://www.hitchins.net/5layer.html.

Kasser, Joseph Eli. 1995. Applying Total Quality Management to Systems Engineering. Boston: Artech House.

Kasser, Joseph Eli. 2018. *Systems Thinker's Toolbox: Tools for Managing Complexity*. Boca Raton, FL: CRC Press.

Kasser, Joseph Eli. 2019a. *Systemic and Systematic Project Management*. Boca Raton, FL: CRC Press.

Kasser, Joseph Eli. 2019b. *Systems Engineering: A Systemic and Systematic Methodology for Solving Complex Problems*. Boca Raton, FL: CRC Press.

Kasser, Joseph Eli, and Tim Mackley. 2008. Applying systems thinking and aligning it to systems engineering. In *the 18th International Symposium of the INCOSE*. Utrecht, Holland.

Luzatto, Moshe Chaim. circa 1735. *The Way of God*. Translated by Aryeh Kaplan. New York and Jerusalem: Feldheim Publishers, 1999.

Maslow, Abraham Harold. 1954. *A Theory of Human Motivation*. New York: Harper & Row.

Maslow, Abraham Harold. 1968. *Toward a Psychology of Being*. 2nd edition. New York: Van Nostrand.

Maslow, Abraham Harold. 1970. *Motivation and Personality*. New York: Harper & Row.

Miller, George. 1956. The magical number seven, plus or minus two: Some limits on our capacity for processing information. *The Psychological Review* no. 63:81–97.

Morgan, Gareth. 1997. *Images of Organisation*. Thousand Oaks, CA: SAGE Publications.

Murray, Henry A. 1938. *Explorations in Personality*. New York: Oxford University Press.

Ramo, Simon. 1973. The systems approach. In *Systems Concepts*, edited by Ralph F. Miles Jr., 13–32. New York: John Wiley & Son, Inc.

Richmond, Barry. 1993. Systems thinking: Critical thinking skills for the 1990s and beyond. *System Dynamics Review* no. 9 (2):113–133.

Schön, Donald A. 1991. *The Reflective Practitioner*. Aldershot: Ashgate.

2 Thinking

Thinking is a major component of risk management because it is the most important function used in identifying, mitigating and preventing risks. This chapter discusses thinking, systems thinking, the nine holistic thinking perspectives (HTPs) and the benefits of going beyond systems thinking to reduce risk. In particular, the chapter discusses:

1. Judgement and creativity in Section 2.1.
2. Critical thinking in Section 2.2.
3. Systems thinking in Section 2.4.
4. Going beyond systems thinking in Section 2.5.
5. Using the HTPs to perceive systems in Section 2.6.

Thinking is:

- The action that underlies decision-making (Section 5.1.5.1) and problem-solving (Section 5.1.6.1).
- A cognitive act performed by the brain.
- Something that apparently most people do not do.

Cognitive activities (functions) include accessing, processing and storing information. The most widely used cognitive psychology information processing model of the brain likens the human mind to an information processing computer (Atkinson and Shiffrin 1968) cited by Lutz and Huitt (2003). Both the human mind and the computer ingest information, process it to change its form, store it, retrieve it and generate responses to inputs (Woolfolk 1998). These days we can extend our internal memory using paper notes, books and electronic storage. We use our mental capacity to think about something received from a sense (hearing, sight, smell, taste and touch). Perceived from the *Functional* HTP, our mental capacities might be oversimplified in four levels as follows (Osborn 1963: p. 1):

1. *Absorptive*: the ability to observe and to apply attention (lowest).
2. *Retentive*: the ability to memorize and to recall.
3. *Reasoning*: the ability to analyse and to judge.
4. *Creative*: the ability to visualize, to foresee and to generate ideas.

When we sense the world, our brain connects concepts using a process called reasoning or thinking and generally uses a filter to separate the pertinent sensory input from the non-pertinent. This filter is known as a 'cognitive filter' in the behavioural science literature (Wu and Yoshikawa 1998) and as a 'decision frame' in the management literature (Russo and Schoemaker 1989). Cognitive filters and decision frames:

1. Are filters through which we view the world.
2. Include political, organizational, cultural, religious and metaphorical, highlighting relevant parts of the system and hiding (abstracting out) the non-relevant parts.[*]
3. Can also add material that hinders solving the problem.[†] Failure to abstract out the non-relevant issues can make things appear to be more complex and complicated than they are and gives rise to artificial complexity (Section 5.2.2.2.2).

2.1 JUDGEMENT AND CREATIVITY

Our thinking mind is mainly twofold (Osborn 1963: p. 39):

1. *A judicial mind*: which analyses, compares and chooses;
2. *A creative mind*: which visualizes, foresees and generates ideas.

Both judgement and creativity are alike in that they both use analysis and synthesis, but they do so in different ways. In judgement we analyse facts, weigh them, compare them, reject some and keep others, and then create a conclusion. On the other hand, the creative process does much the same but uses imagination to produce ideas instead of judgements.[‡]

2.2 CRITICAL THINKING

The literature on creativity and idea generation generally separates thinking up ideas and applying the ideas. The literature on critical thinking however tends to combine the logic of thinking with applying the ideas using the terms 'smart thinking' and 'critical thinking'. The term 'critical thinking', by the way, comes from the word 'criteria', not from 'criticism'. The diverse definitions of the term critical thinking include:

Disciplined, self-directed thinking displaying a mastery of intellectual skills and abilities-thinking about your thinking while you're thinking to make your thinking better.

Eichhorn (2002)

The art of thinking about thinking while thinking in order to make thinking better. It involves three tightly coupled activities: It analyses thinking; it evaluates thinking; it improves thinking.

Paul and Elder (2006: p. xiii)

---

[*] There is always a risk in defining something as non-relevant, which is one reason for not defining anything as non-relevant which leads to artificial complexity (Section 5.2.2.2.2).
[†] For example, the differences between Catholics and Protestants in Northern Ireland are major to many of the inhabitants of that country, but are hardly noticeable to most of the rest of the world.
[‡] Note that imagination can be used to create innovative judgements!

Judicious reasoning about what to believe and therefore what to do.

Tittle (2011: p. 4)

The process of purposeful, self-regulatory judgement.

Facione (1990) cited in Facione (2011: p. 6)

Purposeful, reflective judgment that manifests itself in giving reasoned and fair-minded consideration to evidence, conceptualizations, methods, contexts, and standards in order to decide what to believe or what to do.

Facione (2011: p. 12)

Depending on the definition, critical thinking covers:

1. The thinking process.
2. The means to evaluate or judge the ideas.

Systems Thinker's Toolbox: Tools for Managing Complexity (Kasser 2018a: Chapter 3) summarizes critical thinking, its application and different methods of assessment.

2.3 BRAINSTORMING

One of the most common ways of thinking about risks is to brainstorm ideas. Brainstorming (Osborn 1963):

* Is often used without mentioning its name.
* Is a tool for generating ideas (ideation) generally used in small groups.
* Is based on a tool called the association of ideas (Kasser 2018a: Section 7.2) which goes at least as far back to the time of Aristotle (384–322 BC) (Osborn 1963: p. 114), where one idea triggers the next.
* Can be used to think about risks to create a risk list applicable to the situation.
* Can be used to think about the effects of a risk should it turn into an event.
* Can be used to estimate the probability of occurrence (Section 3.8.3.1) of a risk.
* Can be used to think about the severity of the impact should it occur (Section 3.8.3.2).
* Is used in a number of variations.

In brainstorming and other ideation techniques, the number of ideas generated will be greater if the participants are notified ahead of time as to the:

1. Purpose of the ideation meeting.
2. Problem to be addressed.
3. Need to bring some ideas with them.

Osborn's four original basics for brainstorming were (Osborn 1963: p. 156):

1. No criticism of ideas.
2. Encourage wild and exaggerated ideas.
3. Go for large quantities of ideas.
4. Build on each other's ideas.

Brainstorming can be performed by an individual or by a team. In team sessions, a group of people get together, are given a problem statement and ideate (produce ideas). The brainstorming team contains:

- A leader/facilitator who guides the session and enforces the rules.
- A scribe who records the ideas.
- A small number of people who will do the ideation.

2.3.1 Requirements for Brainstorming Sessions

Brainstorming sessions shall conform to the following requirements:

1. The facilitator shall introduce the session by stating a specific problem rather than a general situation.*
2. The scribe shall record all of the ideas.
3. The participants shall state their ideas.
4. The participants shall suggest modifications of ideas previously stated to improve the idea or combine two or more ideas into another idea.
5. The participants shall not self-censor their ideas before stating them. This is because other people may like those ideas or see them from a different perspective and subsequently build on those ideas to produce something better.
6. There shall be no criticism of ideas during the session.
7. The ideas shall be evaluated and sorted after the brainstorming session.
8. There shall be no limit on the scope of the ideas. Wild ideas can be tamed down later if appropriate.
9. The only discussion on ideas shall be for the purpose of clarification.
10. The facilitator shall not allow the session to break up into groups.
11. The composition of the participants shall be mixed in age, sex, and experience and domain knowledge.
12. Participants shall ideally be at the same level in the organizational hierarchy to minimize the intimidation effect where people lower in rank in the hierarchy are unwilling to disagree with people higher up.
13. If rank is mixed, lower ranks shall speak first.

While a number of variations of brainstorming have been described in the literature, they all tend to suffer from a number of defects including:

---

* The situation might be summarized to provide the context of the problem.

- Being a generally passive approach because they are based on waiting for the ideas to be generated before writing them down.*
- Team sessions being prone to capture by the most opinionated person in the brainstorming session.
- Being unstructured, while allowing free range of ideas, brainstorming tends to fail to focus on issues pertinent to the session.

Other ideation techniques such as active brainstorming (Section 2.5.2), the nominal group techniques (NGTs) (Delbecq, Van de Ven, and Gustafson 1975) (Kasser 2018a: Section 7.9) and slip writing (Mycoted 2007) (Kasser 2018a: Section 7.10) overcome these defects to various degrees in various manners.

2.4 SYSTEMS THINKING

Systems thinking goes beyond just thinking. It seems to be difficult to learn and difficult to teach. There are books on the history and the philosophy and on what constitutes systems thinking. Similarly, the books on problem-solving tend to describe the problem-solving process. However, they don't generally describe how to do the process. And the few articles that do describe how to use a specific tool published in domain literature and do not get a wide distribution outside that domain, so few people hear about it and even fewer people actually use it.†

2.4.1 THE TWO DISTINCT TYPES OF SYSTEMS THINKING

One reason for the lack of good ways of teaching systems thinking might be that when different people are asked to define systems thinking, they provide different and sometimes conflicting definitions. However, these definitions can be sorted into two classifications:

1. *Systemic thinking*: thinking about a system as a whole to gain an understanding of the system.
2. *Systematic thinking*: employing a methodical step-by-step manner (process) to think about something or to achieve a goal.

Many proponents of systems thinking consider either systemic or systematic thinking to be systems thinking not realizing that each type of thinking seems to be a partial view of an unknown whole, in the manner of the fable of the blind men feeling

---

* There are variations which trigger ideas using 'what' and 'why' questions.
† I found that out the hard way when I started to teach systems thinking at the University of South Australia (UniSA) in 2000. I could describe the benefits and history of systems thinking but no one at the Systems Engineering and Evaluation Centre (SEEC) could teach how to use systems thinking very well. We could teach causal loops in the manner of *The Fifth Discipline* (Senge 1990), but that was all. There weren't any good textbooks that approached systems thinking in a practical manner. So, I ended up moving half way around the world to Cranfield University in the UK to develop the first version of a practical and pragmatic approach to teaching and applying systems thinking to systems engineering under a grant from the Leverhulme Foundation.

parts of an elephant and each identifying a single and different animal (Yen 2008). However, both types of systems thinking are needed (Gharajedaghi 1999). Consider each of them:

2.4.1.1 Systemic Thinking

Systemic thinking has three steps (Ackoff 1991):

1. A thing to be understood is conceptualized as a part of one or more larger wholes, not as a whole to be taken apart.
2. An understanding of the larger system is sought.
3. The system to be understood is explained in terms of its role or function in the containing system.

Proponents of systemic thinking tend to:

- Equate causal loops or feedback loops with systems thinking because they are thinking about relationships within a system (Senge 1990, Sherwood 2002).
- Define systems thinking as looking at relationships rather than unrelated objects, connectedness, process rather than structure, the whole rather than its parts, the patterns rather than the contents and the context of a system (Ackoff, Addison, and Andrew 2010: p. 6).

The benefits of systemic thinking have also been known for a long time. For example:

When people know a number of things, and one of them understands how the things are systematically categorized and related, that person has an advantage over the others who don't have the same understanding.

Luzatto (circa 1735)

People who learn to read situations from different (theoretical) points of view have an advantage over those committed to a fixed position. For they are better able to recognize the limitations of a given perspective. They can see how situations and problems can be framed and reframed in different ways, allowing new kinds of solutions to emerge.

Morgan (1997)

Having a set of standardized perspectives or viewpoints systemizes the use of the viewpoints to understand something and facilitates communications. The number of standardized viewpoints has evolved over the years. For example, C. West Churchman introduced three standardized views in order to first think about the purpose and function of a system and then later think about the physical structure (Churchman 1968). These views provided three anchor points or viewpoints for viewing and thinking about a system. Twenty-five years later, seven standardized viewpoints were introduced (Richmond 1993). Richmond's seven streams are:

1. *Dynamic thinking*: which frames a problem in terms of a pattern of behaviour over time.

2. *System-as-cause thinking*: which places responsibility for a behaviour on internal factors which manage the policies and plumbing of the system.
3. *Forest thinking*: which is believing that to know something you must understand the context of relationships.
4. *Operational thinking*: which concentrates on getting at causality and understanding how behaviour is actually generated.
5. *Closed-loop thinking*: which views causality as an on-going process, not a one-time event. With the effect of feeding back to influence the causes, and the causes affecting each other, namely the causal loop.
6. *Quantitative thinking*: which accepts that you can always quantify something even though you can't always measure it.
7. *Scientific thinking*: which
   - Recognizes that all models are working hypotheses that always have limited applicability.
   - Provides the hypothesis or inference of causes, and potential remedial actions.

2.4.1.1.1 Context Diagrams

A useful tool for perceiving an issue systemically is the context diagram or relationship chart which can (Burgh 2011):

- Help define and agree the scope or boundary of the system of interest.
- Provide a simple high-level picture of the system of interest. All systems operate in an environment; failure to pay attention to that environment will lead to a high risk of failure.
- Help identify the elements in the environment of the system of interest that it interacts with.
- Identify and define the external interfaces the system of interest logically has to have with the outside world. Most system issues or problems occur at these interfaces and a context diagram emphasizes them and encourages their clear definition.
- Allow the whole team to share information and agree at a common understanding when used within a team.

A context diagram, also known as a relationship chart, is a concept map (Kasser 2018a: Section 6.1.2) that shows the relationship between the system or item of interest and external entities. The system is usually located in the centre of the diagram and the other entities surround it in a circle. Lines are drawn between the system and the other entities to represent relationships. A context diagram only shows that a relationship exists, namely there is an interface between the entities. The entities may be enclosed in circles or boxes. When the lines are labelled with the data that flows between the system and the entities, the diagram is known as a data flow diagram. The context diagram should not be confused with a flowchart (Kasser 2018a: Section 2.7) because it does not provide any information about the data flows (timing, sequences, etc.), other than data is flowing across the interface.

2.4.1.1.2   Systematic Thinking
Systematic thinking:

- Is mostly discussed in the literature on problem-solving, systems think-ing and critical thinking. It is often taught as the problem-solving pro-cess (Section 5.1.1.2.4) or the system development process (SDP) (Section 6.1.5.4).
- Provides the process for systemic thinking which helps understand and remedy the problematic situation. The benefits of systematic thinking have been known for a long time, hence the focus on process and controlling the process.
- Is made up of two parts:
  - Top-down (analysis):
    - Is breaking a complicated topic into several smaller topics and thinking about each of the smaller topics.
    - Can be considered as a top-down approach to thinking about some-thing and is associated with René Descartes (1637, 1965).
    - Has been termed 'reductionism' because it is often used to reduce a complicated topic to a number of smaller and simpler topics.
  - Bottom-up (synthesis):
    - Refers to combining two or more entities to form a more compli-cated entity.
    - Can be considered as a bottom-up approach to thinking about something.

When we think about something, we tend to mix and combine top-down (analysis) and bottom-up (synthesis) thinking.

2.4.1.1.3   Combining Analysis and Synthesis
When faced with a complicated problem, we break it up into smaller simpler prob-lems (analysis), then solve each of the simpler problems and hope that the com-bination of solutions to the smaller problems (synthesis) will provide a solution to the large complicated problem. For example, consider the problem of making a cup of instant coffee. We use analysis to identify the components that make up the complete cup of instant coffee. So, the coffee powder, cup, hot water, cream and sugar spring to mind. We then use synthesis to create the cup of instant cof-fee from the ingredients. When we think of the process, we think of mixing the ingredients and so we think of a spoon; when we think of heating the water, we think of a kettle and gas or electricity as the fuel. A typical initial set of ideas for making a cup of instant coffee is shown in Figure 2.1 (Kasser 2018). The spoon is drawn as an assistant to the cup of coffee because it is used during the process of creating the cup of instant coffee and then discarded or recycled for creating a subsequent cup. The kettle and gas/electricity are associated with heating the water and so are shown in a similar manner as assistants to the hot water. However,

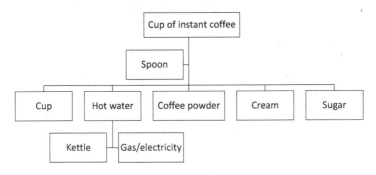

FIGURE 2.1   Initial set of ideas pertaining to a cup of instant coffee.

this arrangement mixes concepts at different levels of the hierarchy and a better arrangement of the ideas is shown in Figure 2.2 (Kasser 2018a). The insertion of an abstract or virtual 'ingredients' concept into the chart clarifies the arrangement of ideas showing which ideas constitute the ingredients and which ideas are associated with other aspects of the cup of instant coffee. However, Figure 2.2 should only be considered as an interim or working drawing and a better drawing is shown in Figure 2.3 (Kasser 2018a), which clearly distinguishes between the items associated with the cup of instant coffee and the aggregation of the spoon and kettle into an abstract concept called 'kitchen items', the constituents of which are used in the process of creating the cup of instant coffee but not in the final product. However, Figure 2.3 contains too many items and should be replaced by a set of drawings starting with Figure 2.4 (Kasser 2018a) at the top level and the drawings for each subsystem: kitchen items, cup and ingredients in accordance with the principle of hierarchies (Section 6.3.1).

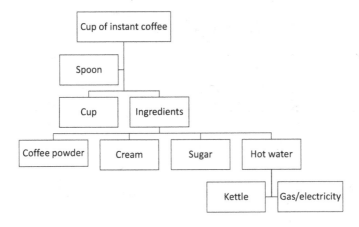

FIGURE 2.2   Hierarchical arrangement of concepts.

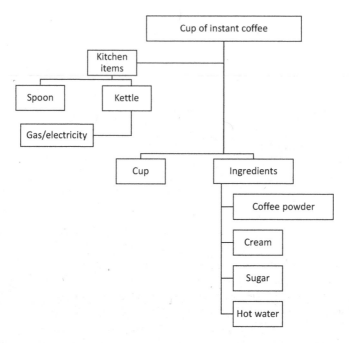

FIGURE 2.3   A better hierarchical arrangement.

2.5   BEYOND SYSTEMS THINKING

The 'beyond' part of 'systems thinking and beyond' is:

- Where the problem definitions and solutions come from.
- Sometimes called holistic thinking.*

FIGURE 2.4   A simpler hierarchical arrangement.

---

* Perceptions from the *Generic* HTP note that there are other definitions of holistic thinking in the literature.

- Emerged from research in 2008 which modified Richmond's seven streams and adapted them into nine systems thinking perspectives (Kasser and Mackley 2008).*
- Changes Ackoff's first step, from
    'A thing to be understood is conceptualized as a part of one or more larger wholes, not as a whole to be taken apart' (Ackoff 1991)
    into
    A thing to be understood is conceptualized as a part of one or more larger wholes, *as well* as a whole to be taken apart depending on the nature of the problem.

For example, if the problem is:

- *An operations research problem*: then the researcher needs to conceptualize the system as a part of one or more larger wholes per Ackoff's definition.
- *To repair a system*: then the repairer will need to understand the structure and function of the system in order to identify the failure and make the repair.
- *To design a system*: then the designer will need to understand what the system is expected to do in its environment as well as how a set of components may be integrated into a system that will meet its requirements.

2.5.1 THE HTPS

The nine systems thinking perspectives introduced nine standardized viewpoints which cover purpose, function, structure and more and were later renamed as the HTPs (Kasser 2018a: Section 10.1). The nine HTPs:

- Are summarized in this section.
- Are widely used in this book.
- Are a systemic tool for:
    - Identifying and managing risks.
    - Gaining an understanding of a problematic situation.
    - Inferring the cause of the undesirability in the situation.
    - Inferring a probable solution to the problems posed by removing the undesirability from the situation.
- Provide a standard set of nine internal, external, progressive and remaining perspectives (anchor points) with which to view a situation.
- Go beyond systems thinking's internal and external views by adding quantitative, temporal, generic and continuum perspectives. This approach:
    - Separates facts from opinion, where:
        - *Facts* are perceived from the eight descriptive HTPs.
        - *Opinion* comes from the insights from the *Scientific* HTP.

---

* The number of perspectives was limited to nine in accordance with Miller's rule (Miller 1956) (Kasser 2018a, Section 3.2.5).

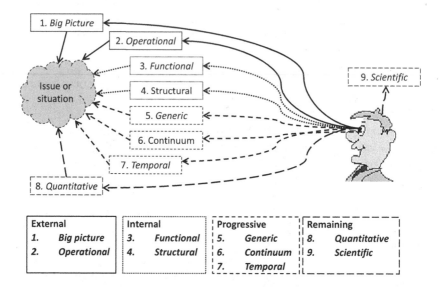

FIGURE 2.5   The nine holistic thinking perspectives.

- Provides an idea storage template (IST) (Kasser 2018a: Section 14.2) for organizing information about situations in case studies and reports in a format that facilitates storage and retrieval of information about situations (Kasser 2018a: Section 10.1.7).

The nine HTPs shown in Figure 2.5 (Kasser 2018a, 2018b) are organized in four groups as follows:

1. *Two external HTPs based on Ackoff's first step are*:
   1. *Big Picture*: includes the context for the system, the environment and assumptions. Tools used to show perceptions from this HTP include context diagrams.
   2. *Operational*: what the system does as described in scenarios; a black box perspective. Tools used to show perceptions from this HTP include timing diagrams and flowcharts.
2. *Two internal HTPs added by the modified first step are*:
   3. *Functional*: what the system does and how it does it; a white box perspective. Tools used to show perceptions from this HTP include function flowcharts and timing diagrams.
   4. *Structural*: how the system is constructed and its elements are organized. Tools used to show perceptions from this HTP include organization charts, work breakdown structures (WBS) and other hierarchical views.
3. *Three progressive HTPs*: where holistic thinking begins to go beyond systems thinking, are the:

5. *Generic*: perceptions of the system as an instance of a class of similar systems; perceptions of similarity. This perspective can identify patterns of behaviour and produces the concept of inheritance. For example:
   - The system inherits from . . .
   - The system behaves like . . .
   - The system looks like . . .
   - The system has the same type of (generic) risks as . . .
6. *Continuum*: perceptions of the system as but one of many alternatives and perceptions of differences. For example, when hearing the phrase, 'she's *not just a pretty face*',[*] the thought may pop up from the *Continuum* HTP changing the phrase to 'she's *not even a pretty face*',[†] which means the reverse. Also, this perspective provides perceptions of what happens when things do not work as intended and risks. Use of the *Continuum* HTP:
   - Is sometimes known as divergent thinking.
   - Leads to a range of solutions rather than a single solution.
   - Helps visualize things in shades of grey rather than in black and white.
   - Indicates that either/or solutions are only two points on a continuum of potential solutions.
7. *Temporal*: perceptions of the past, present and future of the system. Tools used to show perceptions from this HTP include graphs and timelines.
4. *Two Remaining HTPs* are:
8. *Quantitative*: perceptions of the numeric and other quantitative information associated with the other descriptive HTPs such as the numbers in estimates of risk probability and severity. Tools used to show perceptions from this HTP include graphs, tables and spreadsheets.
9. *Scientific*: insights and inferences from the perceptions from the descriptive HTPs leading to the hypothesis or guess about the issue after using critical thinking (Section 2.2).

The first eight HTPs are descriptive, while the ninth (*Scientific*) HTP is prescriptive.

There is no prescribed order for using the HTPs to view something. Big picture thinkers might want to start with the *Big Picture* HTP, whereas detailed thinkers might want to start with the *Functional* HTP or *Quantitative* HTP. When reading each chapter, there is no need to follow the order in which they are written.

2.5.2 Active Brainstorming

Active brainstorming (Kasser 2018a: Section 7.1):

- Is a systemic and systematic thinking tool that increases the number of ideas produced, and overcomes some of the defects, in brainstorming (Section 2.3).

---

[*] Which acknowledges that she is smart.
[†] Which means that not only is she not smart, she is also not pretty.

- Sessions are organized in the same way as brainstorming sessions.
- Should only be used after the initial flow of ideas from brainstorming dry up.
- Produces additional ideas relating to the problem or issue in a systemic and systematic manner. The data in Table 1.3 are from seven teams in the first postgraduate classroom brainstorming/active brainstorming exercise at the National University of Singapore (NUS) showing the increase in the number of ideas. Later exercises provided similar results.[*]
- Achieves these increases in the number of ideas generated by examining the issue from each of the HTPs (Section 2.5.1) and triggering ideas by asking the Kipling questions (Kasser 2018: Section 7.6) 'who', 'what', 'where', 'when', 'why' and 'how' in a systemic and systematic manner.

2.5.2.1 Using Active Brainstorming

While an individual can perform active brainstorming, it is best used in a workshop or session in the context of the following three-stage process:

1. Before the session begins.
2. The session or workshop.
3. Post workshop idea sorting and storing.

2.5.2.1.1 Before the Session Begins

Before the session begins:

1. Identify who needs to be present.
2. Invite them providing a reason for them to attend and tell them what issue or problem the session will be discussing to give them time to think up ideas before the session.
3. Determine who is to be:
   1. The facilitator to pose the questions.
   2. The scribe to write the ideas on the whiteboard or other recording medium.

2.5.2.1.2 The Session or Workshop

The requirements for the session or workshop are the same as the requirements for a brainstorming session (Section 2.3.1). When the session begins, the facilitator should remind the participants as to why the meeting was called and what the initial question was. There will be a natural tendency to generate spontaneous ideas in an unstructured brainstorming manner, particularly in a session containing newcomers to the technique. The ideas will include answers, further questions, names of people to contact for more information and the need for further analysis.

---

[*] The actual number depended on how well the students understood the concept of active brainstorming. The 'too many to count' is probably 'too lazy to count'.

TABLE 2.1

The Active Brainstorming Idea-Triggering Template

	Who?	What?	Where?	When?	Why?	How?
Big Picture						
Operational						
Functional						
Structural						
Generic						
Continuum						
Temporal						
Quantitative						
Scientific						

The facilitator should:

- Not attempt to stem the flow of ideas and ask the participants to wait for the appropriate question.*
- Just make sure the scribe records the ideas in whatever media is being used for the purpose (whiteboard, flip charts, mind mapping software, etc.)

Once the initial flow of ideas stops, the facilitator starts the true active brainstorming process. Using the active brainstorming idea-triggering template shown in Table 2.1 to perceive the situation from a perspectives perimeter (Kasser 2018a: Chapter 10) based on the HTPs (Section 2.5.1), the facilitator begins by posing the Kipling questions (Kasser 2018a: Section 7.6) from the *Big Picture* HTP. The initial cues for the active brainstorming questions can come from the ideas generated from the regular brainstorming session performed just prior to the commencing the active brainstorming session. For example, if one of the ideas produced by regular brainstorming was 'she plays the flute', then active brainstorming can focus on that idea and expand it beginning with questions such as:

- Why is she playing the flute?
- Who is she playing with?
- Where is she playing it?

When starting the questioning sequence for the *Operational* HTP, it often helps to draw a flowchart (Kasser 2018a: Section 2.7), context diagram (Section 2.4.1.1.1) or a similar concept map (Kasser 2018a: Section 6.1.2) to provide a focus for questioning. Once the facilitator has posed a question, the scribe should record the ideas in the same place that the initial flow of ideas was stored. The scribe should not store the responses in the active brainstorming idea-triggering template shown in Table 2.1

---

* Which is one reason the ideas are not documented in the idea-triggering template shown in Table 2.1.

during the session since doing so tends to divert the session into a discussion (dispute) as to the area in which to store the idea and interferes with the flow of ideas.

If no ideas come forth immediately, and sometimes they don't because not all areas are pertinent to every issue, the facilitator should skip to the next question or even the next row or column. However, the scribe should note the question so that it can be assigned as an action item to be answered following the workshop.

At the end of the flow of ideas from the last question in a row, the facilitator moves down to the first column in the subsequent row. Expect a question posed one area of Table 2.1 to sometimes generate ideas that pertain to other areas. Examples of typical questions posed from the HTPs are provided in Section 2.5.2.2. The facilitator should ensure that the discussions triggered by each question are terminated when the flow dries up or starts generating redundant ideas. The workshop is terminated once the ideas stop flowing.

2.5.2.1.3 Post Workshop Idea Sorting and Storing
Once the workshop is complete:

- The ideas are sorted into the ISTs (Kasser 2018a: Section 14.2) or even the HTPs used as an IST.
- Action items are assigned to specific personnel to research answers to the questions that were unanswered during the workshop.
- These findings are then stored in the appropriate IST.

2.5.2.2 Typical Active Brainstorming Questions
The following risk-based questions are intended as a starting point for thoughts and questions. The list is not intended to be complete and not all questions may be appropriate to specific situations. In addition, the questions posed in lateral thinking (de Bono 1973) (Kasser 2018a: Section 7.7), reliability-centered maintenance (Moubray 1997) (Section 3.8.2.1) and the list of typical questions in (Kasser 2018a: Section 7.1.2) may also be used while active brainstorming risks.

2.5.2.2.1 Typical Questions from the Big Picture Perspective
1. What is the purpose of the system?
2. Why is this system there in the first place?
3. What are the assumptions underlying the system?
4. What is the effect of the system not providing all or some of the functions?
5. What could cause the system to stop performing each of its functions?
6. How would a failure manifest itself?
7. How could the system be damaged by sabotage or other external factors?
8. What external factors could damage the system?

2.5.2.2.2 Typical Questions from the Operational Perspective
9. What are the operational scenarios? (open system view).
10. What can happen in a scenario so that the system fails to perform its specified function?

2.5.2.2.3 Typical Questions from the Functional Perspective

11. What activities does/will it perform? (closed system view).
12. How does it perform those activities?

2.5.2.2.4 Typical Questions from the Structural Perspective

13. What happens if a part of the structure fails?
14. What internal factors could damage the system?

2.5.2.2.5 Typical Questions from the Generic Perspective

15. What risks does this system inherit from its class of systems?
16. How are these risks mitigated, prevented and monitored in other instances of the class of system?
17. Did those risks occur?
18. What happened when one of the generic risks occurred?

2.5.2.2.6 Typical Questions from the Continuum Perspective

19. Why is this system different?
20. How is this system different?
21. What are the generic risks in the system?
22. Would the effect of a generic risk be different in this system and why?
23. What are the specific risks in the system?
24. What type of failure would stop a function from performing as specified?

2.5.2.2.7 Typical Questions from the Temporal Perspective

25. How did this situation arise?
26. Is the number of risks changing (decreasing/increasing) and why?
27. How much time will it take to lower the probability of occurrence?
28. How much time will it take to reduce the severity of the impact?

2.5.2.2.8 Typical Questions from the Quantitative Perspective

29. What is the estimated probability of the risk occurring?
30. What is the estimated probability based on (assumptions and facts)?
31. What is estimated severity of the impact should the risk become an event (occur)?
32. What is the estimated severity based on (assumptions and facts)?
33. What will it cost to lower the probability of occurrence?
34. What will it cost to reduce the severity of the impact?
35. What are the cost estimates based on?

2.5.2.2.9 Typical Questions from the Scientific Perspective

36. What could we use ... for?
37. How will we know when the risk no longer exists or is remedied?
38. Could ... prevent the risk of ...?
39. Could we use ... instead of ...?

2.5.2.3 Key Questions

As you gain experience in active brainstorming, you will learn which type of questions from the areas provide the pertinent insight to the various types of issues being discussed and focus on those areas. These questions are known as key questions. The answers to the key questions provide the pertinent information and insight to achieve the goal such as defining the correct problem and identifying the correct feasible solution. Key questions will depend on the situation.

2.5.2.4 Questions to Focus on Problems and Situations

Questions to focus on problems and situations can be framed with the appropriate wording (Ruggiero 2012: p. 129). For example, questions starting with:

- 'How can ...?' tend to produce problems.
- 'Is ...?', 'Does ...?' or 'Should ...?' tend to produce situations.

2.5.2.5 Using Active Brainstorming as an Individual

You can do active brainstorming in a regular brainstorming meeting (Kasser 2018a: Section 7.3), or you can do it on your own when examining a situation. You ask the same questions silently to yourself and act as your own facilitator. If you are in a meeting, then as you answer your own questions, call out the ideas (answers you thought of). If you can't think of an idea to your own question, then call out the question for someone else to answer; if it can't be answered, the question should be noted by the facilitator as an idea to be researched later. Table 2.1 reduced to business card size makes a good memory prompt and doesn't let other people see what you are doing. By doing this, you will quickly develop a reputation for being full of ideas and asking good questions.

2.6 USING THE HTPs TO PERCEIVE SYSTEMS

Consider the following examples of perceptions using the HTPs:

2.6.1 A Camera

When perceived from the HTPs, the perceptions might include:

1. *Big Picture*: where cameras are used and for what purpose.
2. *Operational*: capturing images, transporting safely, viewing images, adjusting settings and charging the battery.
3. *Functional*: capturing images, storing images, retrieving images, deleting images, battery charging functions, etc.
4. *Structural*: camera body, camera case and charger.
5. *Generic*: painting, sketching and other image capture methods/devices.
6. *Continuum*: different types and models of cameras, different materials used to construct camera, what could stop the camera being used (risk and failure analysis).

TABLE 2.2

The Functional/Structural Decomposition of a Camera System

	Understanding How It Works	Capturing Images	Transporting Camera	Recharging Battery
Camera	X	X	X	X
Camera case			X	
Charger				X
Operator		X	X	X

7. *Temporal*: evolution of the image capturing media from photographic plates to film to solid-state memory.
8. *Quantitative*: includes the number of pixels per mm, lens characteristics, etc.
9. *Scientific*: depends on problem or issue

The pertinent HTPs and system boundaries will depend on the nature of the problem or issue. For example:

- *Understanding how a camera works*: perceptions from the *Functional* and *Structural HTPs*. The system bounds the camera as a closed system.
- *Capturing images*: perceptions from the *Operational* HTP. The system bounds the camera and operator.
- *Transporting camera*: perceptions from the *Operational HTP. The* system bounds the camera, operator and camera case.
- *Recharging a camera*: perceptions from the *Operational* HTP. The system bounds the camera, operator and charger.

The functional/structural decomposition drawn in Table 2.2 shows the following:

- An X indicates that the item is used in the scenario. For example, the camera is used in all the scenarios, but the charger is only used when recharging the battery.
- The charger is only used when recharging the battery. The camera design could use an internal charger or an external charger. If there is a need for a low weight or a small volume, the charger may be an external charger since it is only used when recharging the battery.
- The camera case is also only used when transporting the camera. This allows a separate case to be designed if protecting the lenses and other vulnerable parts of the camera is not practical during transportation or would add significantly to the weight of the camera.6

Building the case into the camera lowers the risk of damage to the lens but increases the weight. Not building the case into the camera adds no weight but increases the risk of damage to the lens. The compromise often used is to build a movable lens cover into the camera which protects the lens but not the body of the camera.

2.6.2   A House

When perceived from the HTPs, the perceptions might include:

1. *Big Picture*: location, purpose and assumptions.
2. *Operational*: scenarios showing what the house is used for each weekday morning, afternoon and evening, as well as weekend and holiday activities, e.g. preparing and eating a meal, walking the dog, mowing the lawn, etc.
3. *Functional*: functions performed in the scenarios, e.g. eating, sleeping, reading, talking, accessing the Internet, throwing out the rubbish, etc.
4. *Structural*: electrical, plumbing, heating, cooling, etc.
5. *Generic*: similarity with other houses and buildings and structures serving the same purpose, e.g. tents, apartments, etc.
6. *Continuum*: differences from other houses and buildings and structures serving the same purpose, e.g. tents, apartments, etc., as well as things that could preclude the house from being used for its intended purpose. Risks in the other HTPs, e.g. effects of failures or external threats such as roof leaks.
7. *Temporal*: evolution of houses over time, maintenance and repairs, extensions, etc.
8. *Quantitative*: includes the numbers of rooms, costs, prices, land size, etc.
9. *Scientific*: inferences depend on the risk being considered, the problem or issue.

2.6.3   A Car

Consider a car as the system in the context of home family life. When perceived from the HTPs, the perceptions might include:

1. *Big Picture*: road network, cars drive the economy, etc.
2. *Operational*: going shopping, taking children to school, etc.
3. *Functional*: starting, stopping, turning, accelerating, decelerating, crashing (undesired but possible function), etc.
4. *Structural*: car with doors, chassis, six* wheels and boot.†
5. *Generic*: (four-wheeled land vehicle) trucks, vans, etc.
6. *Continuum*: different types of engines and vehicles (land and non-land), wide choice of manufacturers and models, etc., as well as things that could preclude the car from being used for its intended purpose such as failures and external threats.
7. *Temporal*: examples of the evolution of the car, e.g. Stanley Steamer, Ford Model T, internal combustion, Ford Edsel, hybrid cars, electric cars, etc.
8. *Quantitative*: includes kilometres per hour (kph), engine power, number of passengers, four doors, six wheels, cost, price, information about risk (probability) of part failures, cost of the car, etc.
9. *Scientific*: depends on problem/issue.

---

* Four wheels touch the road, one is a spare and the sixth is the one the driver uses to steer the vehicle.
† Known as a 'trunk' in the US.

The information needed will depend on the issue being examined, and not all information may be pertinent in any given situation.

2.6.4 A Policy

When perceived from the HTPs, the perceptions might include:

1. *Big Picture*: the context or system in which the policy is conceptualized, implemented and validated, the assumptions underlying the policy, etc.
2. *Operational*: the scenarios in the policy cycle.
3. *Functional*: thinking, planning organizing directing controlling staffing, analysing, decision-making, problem-solving, communicating, influencing, etc.
4. *Structural*: policy documents, organizations, etc.
5. *Generic*: problem-solving and decision-making.
6. *Continuum*: range of different types of policies and implementation methods, etc., as well as things that could preclude the policy from being created and implemented.
7. *Temporal*: lessons learned from similar previous polices addressing similar previous undesirability, etc.
8. *Quantitative*: includes cost, return on investment (ROI), metrics to show policy is having an effect, etc.
9. *Scientific*: depends on the specific problem/issue the policy is addressing.

The information needed will depend on the issue being examined, and not all information may be pertinent in any given situation or even available without research and study.

2.6.5 Further Examples of the Use of the HTPs

Further examples of the use of the HTPs to perceive and store information in this book include perceptions of:

1. The undesirable Engaporean National University (ENU) traffic light situation in Section 2.7.1.
2. Risks and risk management in Chapter 3.
3. Change and change management (CM) in Chapter 4.
4. Problem-solving in Section 5.1.
5. The nature of systems in Section 6.1.
6. The perennial problem of poor requirements in Section 6.8.3.
7. Project activities in Section 7.3.1.1.
8. Perspectives of policies in Chapter 9.

2.7   EXERCISES

This section in each chapter provides a number of exercises as opportunities to apply the knowledge discussed in the chapter. The general requirements for each exercise are:

1. Create a compliance matrix (Kasser 2018b: Section 9.5.2) for the exercise that contains all the requirements.
2. Use brainstorming (Section 2.3) and active brainstorming (Section 2.5.2) to pose questions from each HTP.
3. Do a little research on the Internet or in the library (books and journals) to find reasonable answers. Demonstrate compliance to this requirement by citing the sources using the same style as in this book, namely (author, date).
4. Use imagination when making reasonable assumptions, document them as assumptions as opposed to facts.
5. When documenting conceptions of the situation, use lists and charts as appropriate.
6. Discuss any lessons learned from the exercise.
7. Spend no more than 60 minutes on each exercise.

The exercises are set in Engaporia, a fictitious old British colony, with a stable democratic government and a small population. It is a non-aligned, mostly ignored member of the UN located between the sea and the mountains. The exercises are cumulative, so the exercise in each chapter uses information from the previous exercises.

The exercises are opportunities to practice systems thinking and beyond (Section 2.5) and use the HTPs. A partial acceptable solution is provided for the ENU traffic light upgrade (Section 2.7.1). It is up to the reader to use the sample as a template when performing the remaining exercises.

2.7.1   THE ENU TRAFFIC LIGHT UPGRADE

The ENU traffic light upgrade is a project that has technical and political issues requiring an understanding of technology and people's behaviour. Accordingly, it is used as an example in each chapter to illustrate those aspects as pertaining to the material discussed in each chapter.

The Minister of education in the government of Engaporia had heard that students crossing the main road outside the entrance to ENU were complaining about the long delays at the traffic light especially during morning and afternoon rush hours.

In addition to meeting the generic requirements in Section 2.7:

1. Draw a context diagram of the traffic light and the entities associated with the traffic light.
2. Document perceptions of the situation from the descriptive HTPs.
3. Draw at least three causal loop scenarios from the *Operational* HTP.
4. Recommend a solution (one way of preventing future complaints) with rationale (*Scientific* HTP).

TABLE 2.3

Compliance Matrix for the Exercise in Section 2.7.1

ID	Instruction	Deliverable	Section	Completed
1	Brainstorming or active brainstorming	No	All	Yes
2	Research	No	All	Yes
3	Use imagination	No	All	Yes
4	Use lists and charts	No	All	Yes
5	Spend less than 60 minutes	No	All	Yes
6	Preliminary compliance matrix	No[a]	2.7.1.1.2	Yes
7	Assumptions	Yes	2.7.1.1.1	Yes
8	Context diagram	Yes	2.7.1.1.3	Yes
9	Perceptions of the situation	Yes	2.7.1.1.4	Yes
10	At least three causal loops	Yes	2.7.1.1.5	Yes
11	Recommend a solution	Yes	2.7.1.1.6	Yes
12	Lessons learned	Yes	2.7.1.1.7	Yes
13	Updated compliance matrix	Yes	2.7.1.1.2	Yes

[a] The preliminary compliance matrix is not a deliverable. It is the updated compliance matrix which is the deliverable. The preliminary compliance matrix is a useful tool for working through the exercise. Once updated it's a useful tool for checking the work is complete before delivering the product.

2.7.1.1 One Acceptable Solution

A completed exercise could take the form shown in this section. Note it provides one acceptable solution, not the single correct solution as there will be other acceptable solutions (Section 5.1.1.2.2).

2.7.1.1.1 The Assumptions

The assumptions include:

- The Ministry of Education will turn the problem over to the Ministry of Transport for resolution.
- The Ministry of Transport will use its standard operating procedure for the maintenance and upgrade of traffic lights in the traffic light system, namely subcontract it to a member of its pool of contractors.
- The urgency of dealing with the situation is such that it can be inserted into the list of projects with a normal priority.

2.7.1.1.2 The Compliance Matrix

The partially completed compliance matrix is shown in Table 2.3.*

---

* Placing the section number in the completed column not only shows that the requirement has been met but shows where the response can be found, mitigating the risk of the instructor failing to find and grade the response.

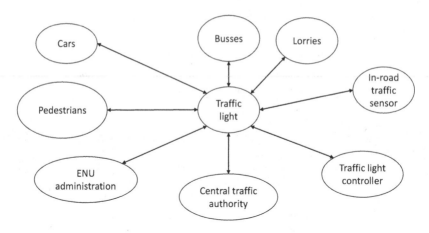

FIGURE 2.6   The ENU traffic light situation context diagram.

2.7.1.1.3   The Traffic Light Situation Context Diagram

The traffic light situation context diagram is shown in Figure 2.6. It clearly and only identifies the entities that interface with the traffic light.*

2.7.1.1.4   Perceptions of the Situation from the Descriptive HTPs

Perceptions of the situation from the different HTPs include:

- *Big Picture*: pedestrian complaints about delays, traffic lights being a part of the road network (system), government involvement in the situation and ENU, and the assumptions in Section 2.7.1.1.1.
- *Operational*: traffic flows, pedestrians crossing the road and various scenarios which included pedestrians crossing the road, and other people driving for various purposes such as commuting to work and ENU.
- *Functional*: pedestrians and vehicles waiting, pedestrians walking and vehicles being driven.
- *Structural*: vehicles, pedestrians, traffic lights, traffic sensors and the traffic light controller.
- *Quantitative*: amount of waiting time, numbers of vehicles, numbers of pedestrians, reaction times of sensors.
- *Generic*: the problem has arisen before in other situations and has been successfully remedied [key observation].
- *Continuum*: differences between the ENU situation and the other situations, differences in the ENU situation at various times of the day.
- *Temporal*: the number of complaints had been slowly increasing over the last year.

---

* And may be incomplete.

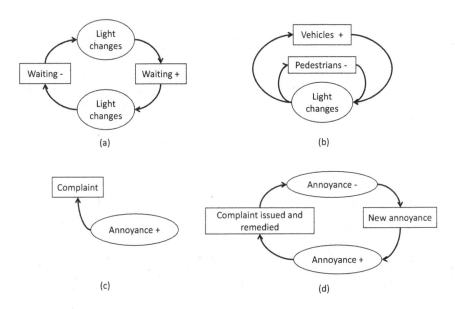

FIGURE 2.7 The causal loops. (a) Change in numbers waiting; (b) change in ratio of numbers of pedestrians and vehicles waiting; (c) increase in annoyance until cause is remedied; (d) change in annoyance as one cause is remedied and another cause arises.

2.7.1.1.5 The Causal Loops

Three causal loops are shown in Figure 2.7. They are:

- The first loop shown in Figure 2.7a applies to both pedestrians and vehicles. It shows that the number of pedestrians or vehicles waiting increases until the light changes. At that point in time, the number waiting decreases as they move through the crossing until the light changes again and the loop begins again.
- The second loop shown in Figure 2.7b shows the inverse relationship between the number of pedestrians and the number of vehicles waiting. As the number of pedestrians decreases because they are crossing the road, the number of vehicles increases because they are stopped by the light. When the light changes, the vehicles can move and the pedestrians must wait.
- The third relationship shown in Figure 2.7c is an open loop showing that pedestrian annoyance increases until it exceeds an undefined threshold and a complaint is generated. Since there is no way of knowing if generating the complaint resets the annoyance level, the loop cannot be closed.
- The fourth loop shown in Figure 2.7d shows that pedestrian annoyance about the waiting time increases until the situation is remedied. The annoyance then decreases until a new annoying issue appears and the pedestrian annoyance increases.

2.7.1.1.6   The Recommended Solution

The recommended solution to the ENU situation is to install countdown timers on the traffic lights counting down the seconds until the lights change.

The recommendation is based on perceptions from the *Generic* HTP. Lessons learned from similar situations (the *Generic* HTP) note that the undesirable situation was not the long wait, but it was the uncertainty of not knowing how much longer the wait would be. The situation had been remedied in similar situations by installing countdown timers on the lights counting down the seconds until the lights were to change.

2.7.1.1.7   Lessons Learned from the Exercise

Lessons learned were:

- The compliance matrix is very handy for keeping track of what needs to be done and making sure that it is being done.
- This example shows that perceptions from the traditional internal and external HTPs contributed to gaining an understanding of the problem at ENU.
- The out-of-the-box solution to the problem came from perceptions from the *Generic* HTP.

2.7.2   THE ENGAPOREAN MAID REDUCTION POLICY

The Engaporean maid reduction policy is a policy that has political issues requiring an understanding of politics and people's behaviour. Accordingly, it is used as an example in each chapter to illustrate those aspects as pertaining to the material discussed in each chapter.

The Minister of the interior in the government of Engaporia would like to reduce the number of foreign workers in the country. She is thinking of reducing the number of maids serving households as one way of reducing the number of foreign workers in the country.

In addition to meeting the generic requirements in Section 2.7:

1. Draw a context diagram of the maid and the entities associated with the maid.
2. Document perceptions of the situation from the descriptive HTPs.
3. Draw at least three causal loop scenarios from the *Operational* HTP.
4. Recommend a solution (one way to reduce the number of maids) with rationale (*Scientific* HTP).

2.7.3   THE ENGAPOREAN DROUGHT RELIEF POLICY

The Engaporean drought relief policy is a policy that has political issues requiring an understanding of ways of modifying existing processes and people's behaviour. Accordingly, it is used as an example in each chapter to illustrate those aspects as pertaining to the material discussed in each chapter.

The country of Engaporia has been suffering from a drought due to a reduction in rainfall over the past few years. The government has noted that the water situation will become critical within three years if nothing is done to alleviate the situation and is considering a number of options. The Minister of tourism in the government of Engaporia would like to instigate a policy to encourage the million tourists arriving by air annually to voluntarily donate at least one bottle of water to the country upon arrival at the airport.

In addition to meeting the generic requirements in Section 2.7:

1. Draw a context diagram of a bottle of water and the entities associated with the bottle.
2. Document perceptions of the policy in action (tourists bringing and not bringing bottles) from the descriptive HTPs.
3. Draw at least three causal loop scenarios from the *Operational* HTP.
4. Recommend a solution (one way to implement the policy) with rationale (*Scientific* HTP).

2.7.4 THE OFF-WORLD MINING POLICY

The off-world mining policy is probably the most complex and complicated policy that could be attempted at this time. It contains the potential for all the technological, management, engineering and political risks that could be imagined as well as being unforeseen in a policy. Accordingly, it is used as an example in each chapter to illustrate those aspects as pertaining to the material discussed in each chapter.

The Minister of science in the government of Engaporia would like to instigate a visionary policy to mine the resources in the asteroid belt and to mine hydrocarbons from the atmosphere of Jupiter. She estimates that it will take at least 50 years for the policy to affect the transition from terrestrial mining to off-word mining. She would also like to benefit Engaporia by constructing a commercial spaceport to handle the anticipated launches and landings in the early days of the policy performance period until the volume reaches a point where more ports will be built in other nations.

In addition to meeting the generic requirements in Section 2.7:

1. Draw a context diagram of a miner and the entities associated with the miner.
2. Document perceptions of the policy in action (the mining is taking place) from the descriptive HTPs. Elaborate perceptions from the *Operational* HTP into:
   1. Transporting miners from the Earth to the asteroids.
   2. Mining the resources in the asteroid belt.
   3. Mining the hydrocarbons from the atmospheres of the Jupiter.
   4. Transporting the mined resources to the Earth.
3. Draw at least three causal loop scenarios from the *Operational* HTP.
4. Recommend a solution (one way to implement the policy) with rationale (*Scientific* HTP).

2.8 SUMMARY

This chapter discussed thinking, systems thinking, the nine HTPs and the benefits of going beyond systems thinking. In particular, it discussed judgement and creativity, critical thinking, systems thinking and going beyond systems thinking and concluded with examples of using the HTPs to perceive a camera, a house, a car and a policy.

REFERENCES

Ackoff, Russel L. 1991. The future of operational research is past. In *Critical Systems Thinking Directed Readings*, edited by Robert L. Flood and Michael C. Jackson. New York: John Wiley & Sons. Original edition, Journal of the Operational Research Society, Volume 30, 1979.

Ackoff, Russsel L., Herber J. Addison, and Carey Andrew. 2010. *Systems Thinking for Curious Managers*. Axminster: Triarchy Press Ltd.

Atkinson, Richard C., and Richard M. Shiffrin. 1968. Human memory: A proposed system and its control processes. In *The Psychology of Learning and Motivation: Advances in Research and Theory*, edited by Kenneth W. Spence and Janet Taylor Spence, Vol. 2, 89–195. New York: Academic Press.

Burgh, Stuart. 2019. *The systems engineering tool box*. Burge Hughes Walsh Limited 2011 [cited 9 May 2019]. Available from https://www.burgehugheswalsh.co.uk/uploaded/1/documents/cd-tool-box-v1.0.pdf.

Churchman, C. West. 1968. *The Systems Approach*. New York: Dell Publishing Co.

de Bono, E. 1973. *Lateral Thinking: Creativity Step by Step*. New York: Harper & Row.

Delbecq, Andre L., Andrew H. Van de Ven, and David H. Gustafson. 1975. Group techniques for program planning: A guide to nominal group and Delphi processes. In *Management Applications Series*, edited by Alan C. Filley. Glenview, IL: Scott, Foresman and Company.

Descartes, René. 1637, 1965. *A Discourse on Method*. Translated by Elizabeth Sanderson Haldane and George Robert Thomson Ross, Part V. New York: Washington Square Press.

Eichhorn, Roy. 2002. *Developing thinking skills: Critical thinking at the army management staff college* 2002 [cited 11 April 11 2008]. Available from http://www.amsc.belvoir.army.mil/roy.html.

Facione, Peter. 1990. *Critical Thinking: A Statement of Expert Consensus for Purposes of Educational Assessment and Instruction*. Newark, DE: American Philosophical Association.

Facione, Peter. 2011. *THINK Critically*. Upper Saddle River, NJ: Pearson Education Inc.

Gharajedaghi, Jamshid. 1999. *System Thinking: Managing Chaos and Complexity*. Boston, MA: Butterworth-Heinemann.

Kasser, Joseph Eli. 2018a. *Systems Thinker's Toolbox: Tools for Managing Complexity*. Boca Raton, FL: CRC Press.

Kasser, Joseph Eli. 2018b. Using the systems thinker's toolbox to tackle complexity (complex problems). In *SSSE Presentation at Roche*. Zurich: Swiss Society of Systems Engineering.

Kasser, Joseph Eli, and Tim Mackley. 2008. Applying systems thinking and aligning it to systems engineering. In *the 18th International Symposium of the INCOSE*. Utrecht, Holland.

Lutz, Stacey T., and William G. Huitt. 2003. *Information processing and memory: Theory and applications*. Valdosta State University 2003 [cited 24 February 2010]. Available from http://www.edpsycinteractive.org/papers/infoproc.pdf.

Luzatto, Moshe Chaim. circa 1735. *The Way of God*. Translated by Aryeh Kaplan. New York and Jerusalem: Feldheim Publishers, 1999.

Miller, George. 1956. The magical number seven, plus or minus two: Some limits on our capacity for processing information. *The Psychological Review* no. 63:81–97.

Morgan, Gareth. 1997. *Images of Organisation*. Thousand Oaks, CA: SAGE Publications.

Moubray, John. 1997. *Reliability-Centered Maintenance*. 2nd edition. New York: Industrial Press Inc.

Mycoted. 2011. *Crawford slip writing*. Mycoted 2007 [cited 18 January 2011]. Available from http://www.mycoted.com/Crawford_Slip_Writing.

Osborn, Alex F. 1963. *Applied Imagination Principles and Procedures of Creative Problem Solving Third Revised Edition*. New York: Charles Scribner's Sons.

Paul, Richard, and Linda Elder. 2006. *Critical Thinking: Learn the Tools the Best Thinkers Use – Concise Edition*. Upper Saddle River, NJ: Pearson Prentice Hall.

Richmond, Barry. 1993. Systems thinking: Critical thinking skills for the 1990s and beyond. *System Dynamics Review* no. 9 (2):113–133.

Ruggiero, Vincent Van. 2012. *The Art of Thinking. A Guide to Critical and Creative Thought*. 10th edition. Boston, MA: Pearson, Education, Inc.

Russo, J. Edward, and Paul H. Schoemaker. 1989. *Decision Traps*. New York: Simon and Schuster.

Senge, Peter M. 1990. *The Fifth Discipline: The Art & Practice of the Learning Organization*. New York: Doubleday.

Sherwood, Dennis. 2002. *Seeing the Forest for the Trees. A Manager's Guide to Applying Systems Thinking*. London: Nicholas Brealey Publishing.

Tittle, Peg. 2011. *Critical Thinking: An Appeal to Reason*. New York: Routledge.

Woolfolk, Anita E. 1998. Chapter 7: Cognitive views of learning. In *Educational Psychology*, 244–283. Boston, MA: Allyn and Bacon.

Wu, Wei, and Hidekazu Yoshikawa. 1998. Study on developing a computerized model of human cognitive behaviors in monitoring and diagnosing plant transients. In *the 1998 IEEE International Conference on Systems, Man, and Cybernetics*. San Diego, CA.

Yen, Duen Hsi. 2008. *The blind men and the elephant* [cited 26 October 2010]. Available from http://www.noogenesis.com/pineapple/blind_men_elephant.html.

3 Risks and Risk Management

This chapter documents thoughts about risks and risk management systemically and systematically from the nine holistic thinking perspectives (HTPs) and suggests generic risk mitigation and prevention activities. Systemic perceptions of risks using the HTPs include the risk management process, the risk rectangle, risk profiles and risk trees. Perceptions of risk management from the various HTPs include the following.

3.1 BIG PICTURE

Perceptions from the *Big Picture* HTP include:

1. Risk management is built into every action we take,* be it getting up in the morning, systems engineering, project management, large engineering projects (LEPs) or initiating a new public policy as discussed in Section 3.1.1.
2. Some risks are well managed; others are not.
3. People who can predict risks that others cannot visualize are often known as prophets.
4. Any action under consideration takes place in an environment in which other actions are taking place. Moreover, those actions can affect the action under consideration.
5. The effect of those other actions in the action under consideration ranges from insignificant to the end of the world as we know it (Section 3.6) and can have a positive or a negative effect.

3.1.1 EVERY ACTION TAKEN INVOLVES RISK

Every action taken involves risk, for example:

- *Making a cup of coffee*: risks include having no water, no way to heat the water and running out of coffee and/or sweetener and creamer.†
- *Slicing bread*: risks include slicing a hand or finger instead of the bread.
- *Using public transportation*: risks include transportation does not show up when scheduled.
- *Playing golf*: risks include being hit by someone else's ball, losing the ball and being hit by a meteorite or a lightning bolt while out on the course.

---

* If it isn't, it should be.
† Unless you don't use sweetener and creamer

- *Using private transportation*: risks include traffic accidents, running out of fuel and being caught speeding.
- *Initiating a public organization policy*: risks include addressing the wrong problem, resistance to change and bureaucratic inertia.
- *Initiating a private organization policy*: risks include addressing the wrong problem and resistance to change.
- *Commencing a LEP*: risks include stakeholders don't like the location, running out of funds, cost escalations, schedule delays, fraud, waste, mismanagement and sabotage.
- *Exploring the ocean depths*: risks include running out of air while under water, getting the bends and being crushed by the pressure.
- *Exploring outer space*: risks include vehicle accident during launch, meteorites holing the vehicle, loss of communications, running out of air while in space and vehicle accident while landing.
- *Deciding what food to have for dinner*: risks include food poisoning.
- *Standing on a log*: risks include falling down and breaking an elbow. When this event occurs, it is known as an accident.

Each of the full set of risks for these and other activities has both:

1. A probability of occurrence (Section 3.8.3.1).
2. A severity of impact or effect should it occur (Section 3.8.3.2).

3.2   OPERATIONAL

Perceptions from the *Operational* HTP include:

1. The risk management process discussed in Section 3.2.1.
2. The system lifecycle in which the projects take place (Section 6.1.5.3).
3. The project lifecycle in which the risks are being managed (Section 7.4).

3.2.1   THE RISK MANAGEMENT PROCESS

In the traditional management paradigm, risk management is documented as separate management and engineering processes and often treated as such. In the systems approach, risk management is built in to the system development process (SDP) during both the planning and performance states.

3.2.1.1   Performing Risk Management in the Traditional Way

The traditional approach to risk management is (DOD 2001):

1. *Risk identification*: reviewing the parts of the system down to the level being considered and identifying all risk events.
2. *Risk analysis*: analysing each risk event to determine the probability of occurrence and consequences/impacts, along with any interdependencies and risk event priorities.

3. *Risk mitigation and contingency planning*: planning mitigation actions and creating contingency plans (Section 7.4.2.2). Translating risk information into decisions and actions (both present and future) and implementing those actions.
4. *Risk tracking*: tracking the risks. Monitoring the risks and actions taken against risks.
5. *Risk controlling*: monitoring them and correcting deviations from planned risk actions.
6. *Risk communicating*: between the team and management. Providing visibility and feedback data internal and external to the project on current and emerging risk activities.

3.2.1.2 Performing Risk Management Using the Systems Approach

The systems approach to risk management is slightly different by adding risk mitigation and prevention planning and focusing on the process as well as the product, namely:

1. *Risk identification*: performed in the planning process[*] by examining each:
    1. *Process step to identify the risks*. In this context, a risk is anything that can negatively impact the cost of the process and the schedule.
    2. *Product being produced by that process*. In this context, a risk is anything that can cause a failure in operation or increase the cost of the product.

        Potential risks can be identified based on the literature and experience which results in a long list of risks. Effective planners with expertise in the solution and implementation domains can think about situations and conceptualize factors that could cause cost and schedule escalations as well as system and subsystem failures using the five questions from reliability-centered maintenance (Moubray 1997) (Section 3.8.2.1) instead of copying lists of risks; mediocre planners can check boxes on lists.
2. *Risk analysis*: consists of the following two parts:
    1. *Risk effect analysis*: performed from two perspectives: the cause and the effect of failures:
        1. *Cause*: starting with a proposed failure, inferring the symptoms that could arise from that failure to identify that cause.
        2. *Effect*: starting with a symptom, deducing what failure could have caused it (root cause).
    2. *Risk impact analysis*: analysing each risk to determine the probability of occurrence and consequences/impacts, along with any interdependencies.
3. *Risk mitigation and prevention planning*: planning mitigation actions and prevention. Translating risk information into decisions and actions (both present and future). This is done in the planning process as part of creating

---

[*] For a project, LEP, policy or any other purpose.

the process steps by inserting the prevention and mitigating activities into the appropriate processes

4. *Risk tracking*: monitoring the risks and actions taken against risks.
5. *Risk controlling*: taking the planned corrective action in the appropriate process steps. Should an unforeseen event occur that will negatively impact the cost and schedule, the occurrence of the event identified the risk post facto. The risk effect and impact analyses are performed and risk mitigation activities identified, costed and tentatively scheduled. A change request (Section 4.9.2) is then issued if accepted, is implemented.
6. *Risk communicating*: between the team and management. Providing visibility and feedback data internal and external to the project on current and emerging risk activities. This can be done using the enhanced traffic light (ETL) chart (Kasser 2018: Section 8.16.2) to summarize those activities.

3.2.2 Communicating with the Public and Other Stakeholders about Risks

Communicating with the public and other stakeholders is one of the ways of:

- Creating a venue where uncertainties can be addressed and questions answered.
- Dealing with conflicting interests and cultures of the various interested and affected parties.
- Dispelling fears about long-term effects from the risk and risk management.
- Improving the transparency and credibility of those implementing the risk management.
- Overcoming the risk of resistance to change by stakeholders who oppose the project or policy.
- Providing information that mitigates the fear of the unknown by transforming the unknown into the known.
- Telling the stakeholders what is in it for them and why they should support the effort.

3.2.2.1 Dealing with Public Fear and Uncertainty

Public fear and uncertainty are a major cause of resistance to change and need to be addressed in any project that affects the public. The prime directive is to remove the uncertainty. There is a belief among change implementers that vagueness helps because it does not allow resistance to a specific issue to coalesce. While this is true and is a positive side of uncertainty, the negative side is much more powerful and will generate resistance among nearly all, if not all, the stakeholders. By removing uncertainty, the stakeholders expected to resist the specific issue can be identified and dealt with (Section 7.10). The best way to remove the uncertainty is to communicate specific, concise and clear information about the benefits to the particular stakeholders.

3.2.2.2 Communicating through the Media

While the media may not be a stakeholder, it is an information conduit that can be used to facilitate communications with stakeholders. Like any other activity, communications needs planning and preparation to get the best out of the communications. Things to consider to minimize the risk of not communicating to the target audience include:

- Avoiding being negative; state facts in positive terms.
- Using different media (e.g. newspapers, radio, television and websites) to reach different target audiences including:
  - The general public.
  - Special interest groups.
  - Individuals.
  - Institutions.
  - The media.
- Thinking about how you'd like to see the message in the medium and prepare it accordingly.
- Focusing on the current situation not the history.
- Creating an honest message. While you may leave out facts,[*] never lie.
- Customizing the message for different target audiences. Use their language and idioms.
- Ways of making it easy for the media to pass on the message to their audiences.

Communications can be one-way or two-way as discussed herein.

3.2.2.2.1 One-Way Communications

One-way communications include:

1. Advertisements in the local media (newspapers, radio and television).
2. Billboard advertisements.
3. Brochures and leaflets.
4. Press releases:
   - Are an abstract of the relevant information.
   - Start with the most important information.
   - End with the details.
   - Should:
     1. Be used to inform the public.
     2. Contain contact information for more details.
   - Should not:
     1. Use overly scientific language.
     2. Contain technical details.

---

[*] Some people call this lying by omission.

3.2.2.2.2   Two-Way Communications

Two-way communications include:

- Social media.
- Citizen panels.
- Emails.
- Focus groups.
- Internet websites with a contact provided for feedback.
- *Press conferences*: press conferences are a way of communicating to several members of the press in a single meeting. Things to consider when organizing press conferences include:
  - Journalists will only attend press conferences if they feel they will get new and newsworthy information, so word the invitation accordingly.
  - Focus on the risk/event generally; do not use scientific language or technical details but have those details available as backup information.
  - If you do not know the answer to a question, the best answer is, 'we do not have enough information to answer the question at this time but are working on getting the information'.
  - Do not speculate; stick to the facts.
  - Make the story personal; give examples of people (to be/have been) affected.
  - Keep the conference short.
  - Never antagonize the press.
- Public presentations and discussions.
- Public exhibitions.
- Social media.
- Surveys.
- Telephone, texting and emails.
- Visits to schools.
- Visits to constituencies.
- Written materials with an opportunity for feedback, e.g. brochures.

Perceived from the *Generic* HTP panels, visits, focus groups and press conference are meetings and presentations and should use the techniques for effective presentations (Kasser 2019a: Section 2.1.6.2.1) and effective meetings (Section 7.3.1.5).

3.3   FUNCTIONAL

Perceptions from the *Functional* HTP include:

1. Thinking (Chapter 2).
2. Using the thinking functions in problem-solving (Section 5.1.5) in the risk management process (Section 3.2.1).

3. Planning organizing directing controlling staffing, analysing, decision-making, communicating, influencing and the other functions of project management (Section 7.1).

3.4 STRUCTURAL

Perceptions from the *Structural* HTP include:

1. The risk framework shown in Table 1.2.
2. Categories of risks discussed in Section 3.4.1.
3. The definitions discussed in Section 3.4.2.
4. The hierarchies in which risks are managed discussed in Section 3.4.3.
5. The Hitchins-Kasser-Massie framework (HKMF) discussed in Section 3.4.4.

3.4.1 CATEGORIES OF RISKS

Perceptions from the *Structural* HTP note that risks can be arranged in various categories. Moreover, a particular risk may fit into more than one category at a time. For example, a risk with a high level of uncertainty may also have the potential for controversy. Categories of risks include:

1. Systemic risks discussed in Section 3.4.1.1.
2. Routine risks discussed in Section 3.4.1.2.
3. Risks with high uncertainty discussed in Section 3.4.1.3.
4. Risks with the potential for controversy discussed in Section 3.4.1.4.
5. Health and safety risks.
6. Security risks.

3.4.1.1 Systemic Risks

Systemic risks are risks that affect the whole system. If the risk occurs and becomes an event, the system fails; for example:

- A single-point failure in a system.
- A single-point failure in a process. For example, if the project does not receive funding there is a risk that the project will be cancelled.
- If the project does not receive support from a specific stakeholder, there is a risk that the project will be cancelled.

Systemic risks may be mitigated or prevented by:

- An alternative design.
- Elimination of parts with low reliability.
- Adding redundancy to cope with the failures in both the product (system) and the process.

- Using a standard operating procedure to preclude the risk from occurring in the system in operation – a solution of last resort because people will bypass the safeties in the process for various reasons.

3.4.1.2 Routine Risks

Routine risks are generally those where the risks and potential consequences are known. These risks are built-in risks adopted because if the risk does not turn into an event, there is nothing else that can perform as well. For example:

- A system occasionally provides incorrect status information due to a known intermittent glitch.
- A project needs to use some technology because when it works, nothing else comes close.
- Side effects can appear n% of the times something is used. For example, in medicinal drugs where drowsiness is a side effect of pain killing and cold and flu symptom-suppressing tablets.
- The risk of being mugged when travelling along a particular road.

Routine risks may be mitigated by avoidance, standard operating procedures, warning labels and paying attention to detail.

3.4.1.3 Risks with High Uncertainty

Risks with high uncertainty are risks whose causes are not well understood, being beyond the boundary of knowledge (Section 5.2.4.2). Attempts to mitigate this category of risk include using the research problem-solving process (Section 5.1.4.4.1) to identify:

- Factors contributing to the uncertainty.
- Ways of evaluating and mitigating each factor separately.

3.4.1.4 Risks with the Potential for Controversy

These risks trigger highly controversial or emotional responses regardless of the certainty of the effect. For example, use of hazardous materials and transportation of nuclear waste.

These risks can be mitigated by communicating with the public (Section 3.2.2) and other methods for risk mitigation and prevention discussed in this book and the literature.

3.4.2 DEFINITIONS

Perceptions from the *Structural* HTP noted the following definitions:

- *Uncertainty*: the lack of complete certainty, i.e. the existence of more than one possibility. The 'true' outcome/state/result/value is not known (Hubbard 2009).

- *Measurement of uncertainty*: a probability assigned to an uncertain outcome or consequence. For example, 'there is a 60% chance of rain this afternoon'.
- *Risk*: a state of uncertainty where the possibilities involve a loss, catastrophe, or other undesirable outcome or consequence. Risks are:
  - Concerned with the future; the past cannot be changed.
  - The result of decisions and actions that cause changes.
  - Unavoidable but can often be prevented or mitigated.
- *Opportunity*: a state of uncertainty where the possibilities involve a desirable outcome or consequence.
- *Event*: a state of certainty. Something expected or unexpected has occurred; what was a risk, opportunity or consequence has occurred and turned into an event. For example, as of this morning there was a probability of 40% that it would rain in the afternoon (risk) and in fact it rained at 4 pm (event).
- *Accident*: an unexpected or unintended event that usually results in an undesirable outcome or consequence such as damage or loss.

Note if an event with a negative effect cannot be predicted, it is not a risk. After it occurs, it becomes a risk of a future occurrence. When a predicted risk occurs, it is not an accident, but it may have occurred due to negligence.

3.4.3   HIERARCHIES IN WHICH RISK ARE MANAGED

Risks like all other human endeavours can be classified according to the principle of hierarchies (Section 6.3.1) as shown in the risk framework in Table 1.2.

3.4.4   THE HKMF

The HKMF (Kasser 2018: Section 5.2)

- Is shown in Figure 3.1.
- Can be used to trace the flow of policies, LEPs and their associated risks though the system development process (SDP) (Section 6.1.5.4) part of the policy and project lifecycles.

The two dimensions of the framework plot the system or product layer of complexity and process (lifecycle) state on different axes where:

1. *The vertical or product axis*: the five layers of complexity (Hitchins 2000) (Section 1.2).
2. *The horizontal or timeline axis*: the nine states of the system lifecycle (SLC) (Section 6.1.5.3).

The out-of-the-box idea for the HKMF came from the *Generic* HTP. Mendeleev created a framework, the periodic table of elements, and populated it with the

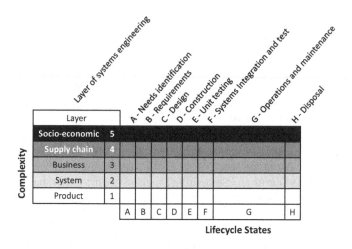

FIGURE 3.1   The HKMF.

then-known elements, leaving gaps which represented unknown elements. In a similar manner, the HKMF forms a framework for studying activities in the workplace[*] in the different layers and states of an SLC.

3.5   GENERIC

Perceptions from the *Generic* HTP include:

1. Risks can be generic to a class of system.[†] For example:
   - Generic defects in human males (e.g. prostate cancer) and females (e.g. breast cancer).
   - Ships that float on water are at risk of sinking.
   - Airships and aircraft are at risk of crashing into the surface.
   - Electric connects exposed to air and dampness have a risk of rusting or otherwise corroding.
   - Land vehicles are at risk of crashing into each other and into other objects.
   - These risks require domain knowledge of the class of system to identify, mitigate or prevent.
2. The risk management process is an instance of the traditional generic problem-solving process (Section 5.1.6.1).
3. The risk management process is the same in systems engineering and project management.
4. Similarities between risk management and failure management in systems engineering.

---

[*] Workplace analysis.
[†] A process is a system.

TABLE 3.1
Continuum of Risk Probability

Probability of Occurrence (%)	Condition
−100	Certain that it will not occur
Between −100 and 0	Might not occur
0	Not sure if it will or will not occur
Between 0 and 100	Might occur
100	Certain that it will occur

3.6 CONTINUUM

Perceptions from the *Continuum* HTP include:

1. The difference between risk management in project management and systems engineering is that project management performs risk management on the process, while systems engineering performs risk management on the product or system.
2. There are two types of risk management:
    1. *Proactive*: when risks are identified and mitigated/prevented ahead of time. This type of risk management is characterized by:
        1. Normal working hours.
        2. Projects meeting their budget and schedules.
    2. *Reactive*: when the events occur, are usually unforeseen and need to be dealt with urgently. This type of risk management is characterized by:
        1. Lots of frenzied activity and over time.
        2. Projects exceeding their budgets and schedules.
3. The common elements in both types of risk management are:
    1. Problems that need to be remedied.
    2. Decisions that need to be made.
4. The continuum of the probability of occurrence of a risk, ranging from 100% certainty that it will never to happen, through decreasing degrees of uncertainty, and then increasing degrees of certainty up to 100% certainty that it will happen as shown in Table 3.1.
5. The continuum of severity of impact, ranging from hardly noticeable to the end of the world as we know it. There is no single uniform standard metric for the probability of the occurrence of the event or the severity of the impact of that event. However, the levels can be defined as:
    - *Probability*: almost certain (5), likely (4), possible (3), unlikely (2) and rare (1).
    - *Severity*: extreme (5), major (4), moderate (3), minor (3) and insignificant (1).
6. If risks have a negative impact on the outcome of an action, then something that has a positive impact on the outcome of the same action is an opportunity. Opportunities can also be considered as having a probability of occurrence continuum and a range of benefits continuum.

7. The difference between traditional risk management and the systems approach to risk management where the:
   - Traditional approach teaches risk management as a separate parallel process to project management and systems engineering.
   - Systems approach integrates risk management into project management and systems engineering.
8. Generic and specific risks. Types of generic* and specific risks include:
   - *Business risks*: include customer demand.
   - *Known vs. unknown*: known with domain knowledge, unknown with insufficient domain knowledge.
   - *Organization risks*: include:
     - Insufficient management support.
     - Insufficient resources.
   - *Predictable vs. unpredictable*: predictable with domain knowledge, unpredictable with insufficient domain knowledge.
   - *Project risk*: generic and project-specific. Generic risks apply to all projects and include lack of resources (people, funding, equipment). A project-specific risk is an instance of a generic risk. For example, the project manager is injured in a traffic accident and will not be able to return to work for two weeks.
   - *Technical risks*: discussed in Section 8.4.
9. Strategic and operational risks where:
   1. Strategic risks:
      1. Typically affect the whole of an organization; consequently, strategic risks can potentially cause very high losses.
      2. Must be managed at the highest level in the organization.
   2. Operational risks:
      1. Refer to potential losses arising from the normal operations of the organization.
      2. Are managed in the appropriate section of the organization performing the operation.
10. Some high severity of impact risks are absolved because we don't have the capability to mitigate them. For example, the risk that the sun goes nova and destroys life on earth. It has a low probability at this time with severe impact on society, but the technology to manage the risk does not exist.
11. Some risks are mitigated irrespective of probability, e.g. the risk of assassination of a political leader even when the succession process is clearly defined. The risk is to the person, not the office.
12. Every domain has its own generic set of risks. Accordingly, it is possible to acquire a large list of risks from the literature such as one list of 130 project risks (Mar 2016). Given domain knowledge, risks may be identified by

---

* Inherited from the class of system.

asking the seven questions of reliability-centered maintenance (Moubray 1997) (Section 3.8.2.1) for any part of the process or product systems.

3.7 TEMPORAL

Perceptions from the *Temporal* HTP include:

1. Actions tend to make changes in the current situation discussed in Chapter 4.
2. Risks, when they become events, change the state of the system discussed in Section 6.2.1.

3.8 QUANTITATIVE

Perceptions from the *Quantitative* HTP include:

1. Perceptions of probability discussed in Section 3.8.1.
2. Perceptions of reliability discussed in Section 3.8.2.
3. Estimates associated with risks discussed in Section 3.8.3.
4. The risk rectangle discussed in Section 3.8.4.
5. Risk profiles discussed in Section 3.8.5.
6. Risk trees discussed in Section 3.8.6.

3.8.1 PERCEPTIONS OF PROBABILITY

Risks are probabilistic, not deterministic, namely they do not occur every time an action is taken unless the probability is 100%.

3.8.2 PERCEPTIONS OF RELIABILITY

Reliability can be defined as the probability that a system will perform its functions as required when used understated conditions for a given interval of time without failure.

OPAH (2019)

Risk management of a system or product ensures the system meets the operational requirement for availability when needed. From the perspectives of reliability, risk management can be defined as ensuring that any risk that affects the operation of a system is identified and mitigated or prevented. Perspectives from the *Generic* HTP note that the product and the process producing the product as well as the team are systems so the generic factors that make up reliability of a system apply to the product, process and the team (but perhaps in different ways). Accordingly, the reliability of a:

1. Product is discussed in Section 3.8.2.1.
2. Process is discussed in Section 3.8.2.2.
3. Team is discussed in Section 3.8.2.3.

3.8.2.1   Reliability of a Product

Perceptions from the *Quantitative* HTP note that the reliability of a product can be measured in terms of:

- *The mean* time between failures (MTBF) in operation*: such as breakdowns in a production line.
- *The mean time to repair (MTTR) a failure of the product in operation*: corrective maintenance.
- *Availability*: the probability that a system will be ready to perform its mission or function under stated conditions when called upon to do so at a random time. As such, availability is a function of how often the system fails and how long it takes to restore the system to an operational condition after a failure occurs. For systems for which no maintenance is possible or practical, availability is equal to the system reliability (OPAH 2019). In quantitative terms:

$$\text{Availability} = \frac{\text{MTBF}}{(\text{MTBF} + \text{MTTR})}$$

- *Downtime*: the time the system was not available for use, including the waiting time to repair and the time in which the system is not operating due to the lack of a resource, e.g. a car running out of fuel.
- *Maintainability*: the relative amount of time and cost in which a system can be retained in, or restored to, a specified condition when maintenance is performed by personnel having specified skill levels, using prescribed procedures and resources.
- *Mean time between maintenance (MTBM)*: the mean time between any maintenance: corrective maintenance (to repair a failure) and preventative maintenance (to prevent failures).
- *Mean downtime (MDT)*: the time in which the system is not operational measured by the actual time to perform maintenance and the delays in getting the parts, personnel spares, etc. to the location of the failure to perform the maintenance.
- *Operational availability*[†]: a measurement of how long a system was available (uptime) for use compared with how long it was supposed to be available (uptime+downtime). In quantitative terms:

$$\text{Operational availability} = \frac{\text{MTBM}}{\left(\text{MTBM} + \text{MDT}\right)}.$$

---

[*] As measured by the mean of a number of incidents of occurrence.

[†] Perceptions from the *Continuum* HTP note the difference between:
- Availability and operational of availability. Availability is a theoretical value based on the probability of something happening in the future. Operational availability is based on measurements of things that have happened.
- Reliability and availability. Reliability is a term based on the actual physical components and probabilities. Reliability looks to the future. Being a probability, it is usually expressed as a value between 0 and 1. It is often expressed as MTBF.

Operational availability is achieved by preventing the risk of a failure by a combination of reliability and preventative maintenance. Perceptions from the *Continuum* HTP note that there is a continuum of reliability that ranges from one end where systems have a low degree of reliability needing a large amount of preventative maintenance to another end where systems have a high degree of reliability and do not need any preventative maintenance. The particular combination for any system depends on the situation. For example:

- A planetary explorer spacecraft has to be extremely reliable because it is extremely difficult to send maintenance personnel after it to repair it.
- Desktop computers and offices can be easily maintained, and therefore, the ratio of reliability and maintenance is based on the function it performs.* If a computer system has to be available 24 hours a day seven days a week, it needs to be extremely reliable. On the other hand, if it is allowed to go down for an hour or so then it can be less reliable since preventative maintenance can be scheduled at times when normally there is little if any usage.
- Aircraft need a reliability that is longer than the flight times and a time to repair that is less than the time between flights.
- Breakdown maintenance or fix it when it breaks is a (lack of) preventative maintenance approach for systems that can be allowed to go down at random times as long as the operational availability can be achieved and the system going down or being down will not result in loss of life or loss of excessive resources.
- Buses and trains may use a combination of preventative and breakdown maintenance as long as the amount of downtime does not exceed passenger tolerance. This means that the occasional breakdown of one or two buses a month could be tolerated if the bus company would save significant costs using that concept.

Preventative maintenance can be based on a number of different concepts including:

- *The statistical probability of failure*: calculating the statistical probability of a failure in a part in terms of operating hours and scheduling preventative maintenance well before the number of hours has been accumulated. For example, if an engine is rated for 1000 hours of operation before it might start to fail, preventative maintenance should be scheduled well before that time to reduce the risk of failure. The actual number of hours at which the preventative maintenance would be scheduled will depend on the estimated severity of the impact of the failure.
- *Breakdown maintenance*: no preventative maintenance, but spares should be available at the appropriate locations so as to reduce the MDT to a minimum.

---

* Known as a mission task (MT) in analysis of alternatives (AoA) (Section 7.4.1.2).

- *Reliability-centered maintenance*: a concept that uses the following seven questions (Moubray 1997) as a basis for risk managment:
  1. What is the item supposed to do and its associated performance standards?
  2. In what ways can it fail to provide the required functions?
  3. What are the events that cause each failure?
  4. What happens when each failure occurs?
  5. In what way does each failure matter?
  6. What systematic task can be performed proactively to prevent, or to diminish to a satisfactory degree, the consequences of the failure?
  7. What must be done if a suitable preventive task cannot be found?

The systems approach builds these questions into the needs identification state of the SDP (Section 6.1.5.4.3) documenting the conceptual maintenance concept in the CONOPS (Section 6.5.9.1), quantifying the maintenance desired functions into a set of maintenance requirements in the system requirements state (6.1.5.4.4) and elaborating the impact of the requirements on the designs in the system design state (6.1.5.4.5). These activities require domain knowledge to truly mitigate the risks of failure. The responses to these questions then determine the appropriate risk mitigation strategy

Failure rates are calculated based on various assumptions and experience. There is a difference in the failure rates of hardware and software:

- *Hardware failures*: calculated based on finding the early failures, often known as infant mortality, then a reasonable operating life without failures followed by a time in which the number of failures starts to increase due to wear and tear and other factors that cause breakdowns. This timeline is known as the bathtub curve since it represents a cross section through a bathtub (u shape).
- *Software failures*: calculated based on finding the early failures. Once the early failures have been eliminated, software doesn't wear out so there are no operational failures.
- Both hardware and software suffer from latent defects which are unanticipated failures that show up once the system is in operation. Risk mitigation of the effect of early failures and early latent defects consists of operating the system for a period of time longer than the time calculated to find the early failures. This is often known as 'burn in'.

3.8.2.2   Reliability of a Process
The reliability of a process can be measured in terms of:

- *The repeatability of producing the same product every time*: ensured by conformance to process standards and measured and improved by such techniques as Six Sigma – an approach developed in Motorola in the 1980s based on the best elements taken from then existing quality ideas in use in Japan at the time dating back to the 1950s. Six Sigma is a methodology used to improve quality by measuring how many defects there are in a process and systematically eliminating the opportunities to reduce the number of defects being produced to as close to zero as possible (Tennant 2001: p. 6).

- *The percentage number of defects produced by the process*: often known as Quality according to Crosby's definition of Quality being conformance to specifications (Crosby 1979).
- *The MTBF in the process*: such as breakdowns in a production line.
- *The MTTR*: a breakdown in the process.
- The rest of the measurements of availability and operational availability discussed in Section 3.8.2.1.

Preventative maintenance should also be applied to the process, and the same seven questions of reliability-centered maintenance (Moubray 1997) should be asked. For example, typical questions might be:

1. What happens if scheduled test equipment is not available when needed?
2. What happens if the requirements are misinterpreted?
3. What happens if the requirement is vague?
4. What happens if the requirement is not testable?
5. What happens if a document does not contain the necessary information?

Question 1 should be asked as part of risk management in planning so that the risk can be mitigated or contingency plans (Section 7.4.2.2) can be prepared if there is a high risk of the equipment not being available.

The remaining questions don't seem to be asked in the current paradigm. The systems approach asks those questions to prevent the perennial problem of poor requirements (Section 6.8.3). The effects of poor requirements are cost and schedule escalations, and in many situations the risk of producing poor requirements is high, hence the perennial problem of poor requirements. This risk could be mitigated by a requirements workshop held in a just-in-time (JIT) manner to sensitize the personnel writing the requirements to the difficulty of writing good requirements (Kasser 2019b: Section 10.2).

3.8.2.3   Reliability of a Team

The reliability of a team tends to be neglected in the non-systems approach, except sometimes in the military when high-performance teams are used to achieve specific mission tasks (MTs). The systems approach recognizes the reliability of people, one of the quadruple constraints (Section 7.3.2) because the right people can make a project and the wrong people can break a project (Augustine 1986). The team is a system so the rules of reliability of a system apply to the team. The reliability of a team is made up of the reliability of:

1. *Each individual*: has a different reliability, some make more mistakes than others.
2. *The number of interactions between individuals*: this has been recognized in terms of the number of communications channels between the team members; the more communications channels, the lower the reliability.

3. *The number of individuals in the team*: the reliability goes up as the number increases. However, it reaches a turning point and the reliability then decreases as the number of individuals continues to increase.

Team members need maintenance to prevent failures such as burnout, reduced motivation and other factors which reduce people's reliability. These activities are generally handled by the human resources department if performed.

The reliability of a team can be increased by:

- Selecting reliable people who can get the complete job done in a timely manner with a minimum of errors.
- Optimizing the team size for the task.
- Ensuring the team has a clear and concise vision of the goal or MT.
- Providing the team with adequate resources in a timely manner.
- Regular maintenance as required including downtime.
- Ensuring the team has the required expertise or access to the expertise when required such as through the use of consultants.
- Employing some redundancy by having team members who are able to back up other team members if needed. Each team member has their own primary expertise, but they are also able to take over from another team member should that team member not be available due to an accident or other action. For example, one team member is the assigned driver, but other team members may be able to drive in the event the driver is not available.
- Purchasing key person insurance to mitigate financial losses if a key person is lost to the team. However, the insurance will not get the project done.

3.8.3 ESTIMATES ASSOCIATED WITH RISKS

Numeric levels are assigned to various probabilities of occurrence and severity of impact should the risk materialize. For example, there is a 45% probability that it will rain tomorrow afternoon.

The accuracy of numbers associated with risk occurrence and severity depends on who is making the estimate and what information is available (Section 3.8.3). There are a number of models available, particularly for cost (Kasser 2019a: Section 8.3) and reliability estimates.

The value of the estimated numbers should be rounded up to the nearest whole number. For example, a computer might define a risk as being 66.6667%, whereas 67% or even 70% is good enough for human purposes.

3.8.3.1 Probability of Occurrence

Every risk can be positioned on the continuum of the probability of occurrence shown in Table 3.1. Ways of estimating the position (Section 3.8.3) include:

- *Experience.* For example, if it rained at about 4 o'clock every day for the last week, what is the probability that it will rain at 4 o'clock today? Or if the number of days in which an event has occurred over a period of time is known, then the probability of that event occurring during the same period of time in the future can be considered to be the same as it was in the past, based on the assumption that nothing has changed.

- *Statistics*: a coin has two sides, so when tossing a coin and letting it land, there is a 50% probability of it landing on a predetermined side. So, the risk of it landing on the wrong side is 50%. Perceptions from the *Generic* HTP note the relationship between the probability of an event and the odds offered when gambling.

When dealing with these numbers, it is well to remember that they are based on generic assumptions, and it is important to understand if the generic assumptions apply in a specific situation and if the accuracy of the number makes a difference.

Perceptions from the *Generic* HTP note the similarity between probabilities of risk and probabilities of failure. If the probability of failure in a component is 40%, that just means that someone has calculated that given 100 of those components, 40 of them will have failed by the specific time. It does not mean that the component being used is going to fail any time in the near future; it might or it might not. Similarly, with a risk, if there is a 40% chance of the risk turning into an event, that doesn't mean say it will or it will not. On the other hand, would you take the chance of that failure or of that risk turning into an event? The answer is that it depends on the situation and the severity of the outcome.

When working out the probability of a risk, it is advisable to get some help or use a book of random numbers depending on the situation.

3.8.3.2   Severity of Occurrence

The severity of occurrence lies along a continuum with the end of the world or equivalent at the high end and insignificant at the low end. The severity is usually determined by thinking about the risk and the impact should it occur. Again, the numbers are subjective and are based on assumptions. For example, the severity of the risk of running out of sugar for sweetening a cup of coffee may be different for different people depending on their tolerance of the lack of sugar in their coffee. However, the severity of a landslide will be less subjective and based on the damage that the landslide could cause.

3.8.4   The Risk Rectangle

Knowing that individual numerical estimates are not that accurate, the traditional approach of summing up the risk and the severity of an event is to aggregate each risk probability and severity into five groups of increasing probability or severity and then multiply the probability by the severity and show the results in a table known as a risk rectangle such as the one shown in Table 3.2 (Kasser 2018) where:

- *E (extremely high)*: 25 (5*5) is coloured red.*
- *H (high risk)*: greater than or equal to 20 (4*4) is coloured orange.

---

* Shown in shades of grey in this book.

TABLE 3.2
A Risk Rectangle

Probability

Severity	Certain (5)	Likely (4)	Possible (3)	Unlikely (2)	Rare (1)
Extreme (5)					
Major (4)					
Moderate (3)					
Minor (2)					
Insignificant (1)					

- *M (moderate risk)*: between 5 and 20 is coloured yellow.
- *L (low risk)*: less than or equal to 4 (2*2) is coloured green.

In order for the risk rectangle to be a practical tool, the risk probability and severity must be quantified into specific numbers, which is not an easy thing to do. For example, numbers need to be assigned to the terms significant and severe. These numbers ought to be specified in terms of percentages. For example, a significant loss might be 90% of functionality and a severe loss might be 75%. Once these levels have been specified, the loss for a specific event can then be estimated and placed in the appropriate category. Perceptions from the *Generic* HTP show that this is the same process that is used to determine the values of the ranges for each of the categories in a categorized requirements in process (CRIP) chart (Section 7.5). For an example of how to assign numbers to the probabilities and using the risk rectangle to decide whether to purchase options in the stock market, see the *Systems Thinker's Toolbox: Tools for Managing Complexity* (Kasser 2018: Section 5.42).

3.8.5 RISK PROFILES

Risk profiles are:

- Attribute profiles (Kasser 2018: Section 9.1).
- An alternative to the risk rectangle (Section 3.8.4) that separates out the probability of occurrence and severity of impact into ranges in two separate histograms (Kasser 2018: Section 2.9):
  1. *Probability of occurrence profile*: a histogram which shows the number of risks associated with each range of probability.
  2. *Severity of the impact profile*: a histogram which shows the number of risks associated with each range of severity of the impact.
- Can be considered as the 'A' column in a CRIP chart (Section 7.5) for the categories of risk probability or severity for a project plotted as a histogram.
- Shown in Figure 3.2 (Kasser 2018).

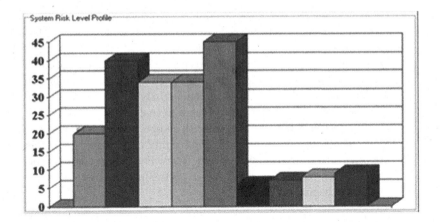

FIGURE 3.2   A typical project risk profile.

Risk profiles may be used in several ways including:

- As selection criteria for choosing between different alternatives. The selection criteria would be set up so that the alternative with the lowest risk profile is chosen.
- Selecting which risks must be mitigated and which risks must be prevented (the ones with the highest impact).
- Deciding if the project should be cancelled during the project planning state as being too risky.
- Comparing the risk profile of the project's system with a generic risk profile for the class of system to see if the project has a better-than-average chance of success.

3.8.6 RISK TREES

Risk trees are tools for risk analysis in certain circumstances. Perceptions of risk trees from the HTPs include:

- *Generic*: perceptions from this perspective note the similarity to:
  - *Decision trees*: (Kasser 2018: Section 4.6.1) which are:
    - A decision-making tool for making a choice between several options and combinations of options allowing weighted preferences at each decision point.
    - Concept maps (Kasser 2018: Section 6.1.2) showing all the pertinent factors to a decision in the form of a network of branches, hence the term decision tree.
  In a similar manner, risk trees are:
  - A decision-making tool for making a choice between several risk options and combinations of risk options, namely a specialized version of a decision tree.

- Concept maps (Kasser 2018: Section 6.1.2) showing all the pertinent factors to a decision between risks in the form of a network of branches, hence the term risk tree.
- *Quantitative*: both decision trees and risk trees enable a more accurate estimate to be made by separating out the components that contribute to the decision and estimating each individually.
- *Operational*: risk trees may also be used to analyse cascade risks.
- *Continuum*: the difference between risk and probability is often a matter of perspective. Each risk has a probability of occurrence, but they are often stated from opposite perspectives. For example, if the probability of completing a doctoral degree is 40% then the risk of not completing the degree is 60%.

Risk trees are created by examining timelines and annotating each work package (WP) or other element in each path with their risks. The total risk through a path can be then estimated in the same way as reliability of components in series and in parallel is computed.

3.9   SCIENTIFIC

Inferences from the *Scientific* HTP include:

1. Risks can be plotted in the areas of the risk framework (Section 1.2) and in the areas of the HKMF (Section 3.4.4) which can identify generic:
   - Categories of risk associated with the row and column.
   - Risk management techniques associated with the column. For example, a poorly worded instruction in a policy in layer 5 has the same risk of non-compliance as a poorly worded requirement in layers 1 and 2.
2. *The myths in risk management*: the traditional approach to risk management contains a number of myths (Kasser 2018: Section 12.7) including the selected myths summarized in Section 5.1.1.2.
3. *Risk management is a separate activity from system design*: in the systems approach, risk management is integrated into system design and operation. As such there is no separate risk management plan; risk management is a vital part of the project plan (Section 7.4.2.1).* See the exercise response in Section 7.13.1.1.4 for an example of an outline project plan incorporating a risk management section.
4. *Risk can be quantified as a single number*: the number is the product of the probability of occurrence of a risk and the severity of the potential outcome. This use is widespread encouraged by the US Department of Defense (DoD) in the form of the risk rectangle or traditional risk assessment matrix (Section 3.8.3). The single number quantification arose because project managers and decision-makers wanted simplicity when making high-risk decisions. The systems approach uses risk profiles (Kasser 2018: Section 9.1).

---

* If the project is large or complex, the risk management section of the project plan may be a sub-document.

5. *Maintain risk registers for all the risks*: This can lead to a very large and unmanageable risk register which was shown to be unworkable in behavioural psychology (Section 1.2). The systems approach maintains the top 6–10 risks at each level, or ~7±2 (Miller 1956).

6. *Risks are measured*: in reality they are estimated with an accuracy that may not be accurate.

7. *The flaw in the risk rectangle*: since the risk rectangle allocates a single number to a risk by multiplying probability by severity to assign the level of risk, risks with low probability of occurrence but high severity of consequences tend to be ignored in favour of risks with numbers. This is a flaw in the use of the risk rectangle because risks with a catastrophic severity of impact must be mitigated, avoided or prevented irrespective of their probability of occurrence if at all possible.

8. *Risk mitigation or risk prevention*: some risks must always be prevented, while others may be mitigated or even ignored if at all possible. For example:
   1. Truly catastrophic risks such as a nuclear power station meltdown must be prevented.
   2. Risks with high severity of impact and low probability of occurrence need to be mitigated or prevented.
   3. Risks with low probability of occurrence and low severity of impact are nuisances and may be ignored for a while but should be mitigated at some future date.

Traditional risk management seems to consider single risks. However, the systems approach cascades risks once a risk of failure is identified, by posing the question,[*] 'if this risk happens, will it stop the project?' If the answer to the question is:

1. *Yes*: ways of preventing or mitigating the risk need to be explored. For example, if an item of specialized equipment needs to be available for a certain WP, the consequences of not having that equipment when it is required need to be explored. The non-availability situation needs to be prevented or mitigated.

2. *No*: the risk analysis then has to determine the effect on the system to determine if the system can function with acceptable performance. If it can't, then ways of preventing or mitigating the risk need to be explored. If it can, then another round of risk analysis needs to be performed on this reduced performance process.

3.10   EXERCISES

This section provides a number of exercises as opportunities to practice risk identification, systems thinking and beyond (Section 2.5) and use the HTPs. A partial

---

[*] Perceptions from the *Generic* HTP note the similarity to questions in reliability-centered maintenance (Section 3.8.2.1).

acceptable solution is provided for the Engaporean drought relief policy exercise (Section 3.10.3). It is up to the reader to use the sample as a template when performing the remaining exercises.

3.10.1 THE ENU TRAFFIC LIGHT UPGRADE

In addition to meeting the generic requirements in Section 2.7:

1. Based on the information documented in the *Operational, Functional, Generic* and *Structural* HTPs identified in the previous exercise (Section 2.7.1), identify and document at least one generic and one specific risk (in each HTP) in as many areas of the risk framework shown in Table 1.2 as you can.
2. Use brainstorming (Section 2.3) or active brainstorming (Section 2.5.2) to pose questions from each of the three HTPs.

3.10.2 THE ENGAPOREAN MAID REDUCTION POLICY

In addition to meeting the generic requirements in Section 2.7:

1. Based on the information documented in the *Operational, Functional, Generic* and *Structural* HTPs identified in the previous exercise (Section 2.7.2), identify and document at least one generic and one specific risk (in each HTP) in as many areas of the risk framework shown in Table 1.2 as you can.
2. Use brainstorming (Section 2.3) or active brainstorming (Section 2.5.2) to pose questions from each of the three HTPs.

3.10.3 THE ENGAPOREAN DROUGHT RELIEF POLICY

In addition to meeting the generic requirements in Section 2.7:

1. Based on the information documented in the *Operational, Functional, Generic* and *Structural* HTPs identified in the previous exercise (Section 2.7.3), identify and document at least one generic and one specific risk (in each HTP) in as many areas of the risk framework shown in Table 1.2 as you can.
2. Use brainstorming (Section 2.3) or active brainstorming (Section 2.5.2) to pose questions from each from each of the three HTPs.

3.10.3.1 One Acceptable Solution

A completed exercise could take the form shown in this section. Note it provides one acceptable solution, not the single correct solution as there will be other acceptable solutions (Section 5.1.1.2.2).

TABLE 3.3

Compliance Matrix for the Exercise in Section 3.10.3

ID	Instruction	Deliverable	Section	Completed
1	Brainstorming or active brainstorming	No	All	Yes
2	Research	No	All	Yes
3	Use imagination	No	All	Yes
4	Use lists and charts	No	All	Yes
5	Spend less than 60 minutes	No	All	Yes
6	Assumptions	Yes	3.10.3.1.1	Yes
7	Preliminary compliance matrix	No	3.10.3.1.2	Yes
8	One generic risk in each HTP	Yes	3.10.3.1.3	Partial
9	One specific risk in each HTP	Yes	3.10.3.1.3	Partial
10	Lessons learned	Yes	3.10.3.1.4	Yes
11	Updated compliance matrix	Yes	3.10.3.1.2	Yes

3.10.3.1.1   The Assumptions

The assumptions include:

1. Tourists will be willing to donate water bottles since the Minister came up with the idea but did not commission a feasibility study.

3.10.3.1.2   The Compliance Matrix

The partially completed compliance matrix is shown in Table 3.3.*

3.10.3.1.3   Identify and Document At Least One Generic and One Specific Risk

Since the exercise requires one risk from four different HTPs in each area of the risk framework, the simplest way to tackle the exercise seems to be to modify Table 1.2 to include four columns in each state of the project lifecycle for the *Operational, Functional, Generic and Structural* HTPs and store pointers to the risks there as shown in Table 3.4.† The tabular method also provides an easy way to ensure compliance and no errors of omission.

The risks identified and listed in Table 3.4 are:

1. Not performing a feasibility study (Section 7.4.1.1) on the proposed change – a policy in layer 5.
2. Falling into the decision traps (Section 5.1.5.3) when deciding how to deal with the issues identified in the initialization state.
3. Implementing the wrong policy.

---

* Placing the section number in the completed column not only shows that the requirement has been met but shows where the response can be found, mitigating the risk of the instructor failing to find and grade the response.
† An alternative approach is to use four tables.

TABLE 3.4

Modified Risk Framework for Exercise

Lifecycle State

Layer of Complexity		Initialization				Planning				Performance				Closeout			
	HTP	O	F	G	S	O	F	G	S	O	F	G	S	O	F	G	S
5	Social	1	2	3	4					5	6						
4	Supply chain																
3	Business																
2	System																
1	Product																
0	Component																

4. Failing to create an effective team.
5. The water provided by tourists may contain chemical or biological contaminants.
6. Running out of storage space for the empty bottles.
7. Failing to provide the minister with a realistic assessment of the proposed policy for political and other reasons.

3.10.3.1.4 Lessons Learned

The lessons learned included:

- The compliance matrix is very handy for keeping track of what needs to be done and making sure that it is being done.
- Perceptions from the *Big Picture* HTP provided the context and the direct stakeholders (Kasser 2019b: Section 10.14).
- Perceptions from the *Generic* HTP provided ideas on the CONOPS (Section 6.5.9.1) for the policy in action.
- Perceptions from the *Operational, Functional* and *Structural* HTPs provided ideas on the policy in action once the basic concept was perceived from the *Generic* HTP.
- Perceptions from the *Continuum* HTP identified the risks associated with the perceptions from the other HTPs.

3.10.4 THE OFF-WORLD MINING POLICY

In addition to meeting the generic requirements in Section 2.7:

1. Based on the information documented in the *Operational, Functional, Generic* and *Structural* HTPs identified in the previous exercise (Section 2.7.4), identify and document at least one generic and one specific risk (in each HTP) in as many areas of the risk framework shown in Table 1.2 as you can.

2. Use brainstorming (Section 2.3) or active brainstorming (Section 2.5.2) to pose questions from each of the three HTPs.

3.11 SUMMARY

This chapter perceived risks and risk management from the nine HTPs and suggested generic risk mitigation and prevention activities. Systemic perceptions of risks using the HTPs included the risk management process, the risk rectangle, risk profiles and risk trees.

REFERENCES

Augustine, Norman R. 1986. *Augustine's Laws*. New York: Viking Penguin Inc.

Crosby, Philip B. 1979. *Quality is Free*. New York: McGraw-Hill.

DOD. 2001. Program manager's guide for managing software, Draft 0.4, 2001.

Hitchins, Derek K. 2000 *World class systems engineering – The five layer model* [Web site] 2000 [cited 3 November 2006]. Available from http://www.hitchins.net/5layer.html.

Hubbard, Douglas 2009. *The Failure of Risk Management: Why It's Broken and How to Fix It*. Hoboken, NJ: John Wiley & Sons.

Kasser, Joseph Eli. 2018. *Systems thinker's toolbox: Tools for managing complexity*. Boca Raton, FL: CRC Press.

Kasser, Joseph Eli. 2019a. *Systemic and Systematic Project Management*. Boca Raton, FL: CRC Press.

Kasser, Joseph Eli. 2019b. *Systems Engineering: A Systemic and Systematic Methodology for Solving Complex Problems*. Boca Raton, FL: CRC Press.

Mar, Ann. 2019. *130 Project risks (list). Simplicable 2016* [cited 30 August 2019]. Available from https://management.simplicable.com/management/new/130-project-risks.

Miller, George. 1956. The magical number seven, plus or minus two: Some limits on our capacity for processing information. *The Psychological Review* no. 63:81–97.

Moubray, John. 1997. *Reliability-Centered Maintenance*. 2nd edition. New York: Industrial Press Inc.

OPAH. 2019. *Operational availability handbook. Reliability analysis center (RAC)* [cited 26 September 2019]. Available from http://www.acqnotes.com/Attachments/Introduction%20to%20Operational%20Availability.pdf.

Tennant, Geoff. 2001. *Six Sigma: SPC and TQM in Manufacturing and Services*. Aldershot: Gower Publishing, Ltd.

4 Change

This chapter perceives change and change management (CM) from the holistic thinking perspectives (HTPs) because when risks turn into events change happens. In particular, it discusses some CM models, resistance to change and how to overcome it and some aspects of stakeholder management. Perceive change from the HTPs.

4.1 CONTINUUM

Perceptions from the *Continuum* HTPs include:

1. Desirable and undesirable changes discussed in Section 4.1.1.
2. Types of change discussed in Section 4.1.2.
3. Stakeholder grouping discussed in Section 4.1.3.
4. A difference between resistance to change and resistance to the approach to realize the change.
5. A difference between organizational change and product change.

4.1.1 DESIRABLE AND UNDESIRABLE CHANGES

Changes may be considered as lying along a continuum of desirability where the change is totally undesirable at one end and totally desirable at the other. There is a don't care point at the centre of the continuum. Some changes are:

- Desirable by some stakeholders, don't care and undesirable by others.
- Partially desirable and undesirable at the same time by a single stakeholder.

For example:

- Desirable changes include software upgrades to add needed features and fix defects (bugs).
- Undesirable changes include the user's need to upgrade* software packages when the package in use provides the user with all the needed functionality.

---

* At a price.

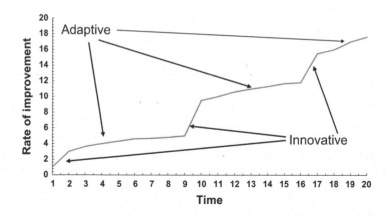

FIGURE 4.1   Adaptive and innovative changes.

4.1.2 TYPES OF CHANGE

Change can be adaptive, innovative (Kirton 1994) or a combination as shown in Figure 4.1 (Kasser 1995) where:

- *Adaptive changes*: improve the current paradigm.
- *Innovative changes*: tend to introduce a new paradigm or tend to be perceived as riskier, and consequently tend to be resisted more than adaptive changes.
- *Combination*: rather than either adaptive or innovative. Perceptions from the *Continuum* HTP note that there is actually a continuum of change with innovative changes at one end and adaptive changes at the other.

4.1.3 STAKEHOLDER GROUPING

When change takes place, the stakeholders to be affected can be grouped into the following three categories:

1. *For the change*: stakeholders who see the need for the change and some who are even eager both to implement it and to convert others to the cause.
2. *Wait and see*: stakeholders who have seen it all before and will wait to see which way the wind blows.
3. *Against the change*: stakeholders who tend to resist the change efforts and make them fail because they:
   - Feel they have nothing to gain and may even have something to lose by the change.
   - Agree with the need for the change but would like to see an alternative realization.
   - Have a different agenda.

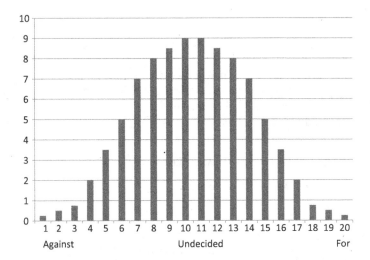

FIGURE 4.2 Degree of resistance to change.

The *Guide to the PMBOK* (PMI 2013) maps the degree of stakeholder interest in a proposed change as:

1. *Unaware*: stakeholders who are unaware of the project and the potential impacts.
2. *Resistant*: stakeholders who are aware of the project and the potential impacts and are resistant to the change.
3. *Neutral*: stakeholders who are aware of the project and are neither supportive nor resistant.
4. *Supportive*: stakeholders who are aware of the project and the potential impacts and are supportive of the change.
5. *Leading*: stakeholders who are aware of the project and the potential impacts and are actively engaged in ensuring the success of the project.

Perceptions from the *Continuum* HTP point out that except for the unaware degree of interest, the remaining degrees of interest may be considered as being on a continuum of interest ranging from resistant at the negative end to leading at the positive end. Figure 4.2 represents the magnitude of these stakeholders on the continuum as a normal distribution curve. Think of the figure as showing a fence. Most of the stakeholders are sitting on the fence (neutral) and need to be given an incentive to come down from the fence on the side of the change. Those on the resistant side of the fence need to be persuaded to climb the fence and acquiesce to the change. The stakeholders on the supportive side of the fence for the change may be harnessed to take actions to facilitate the change. The subsequent rewards for those actions will provide the incentives to convert the other stakeholders to favour or acquiesce to the change. There will generally always be a small resistive minority who will never agree to the change. How they are dealt with depends on the context, their position and influence.

4.2 TEMPORAL

Perceptions from the *Temporal* HTP note that:

1. There are four states in the change lifecycle discussed in Section 4.2.1.
2. Changes are risky discussed in Section 4.2.2
3. Changes take time discussed in Section 4.2.3.
4. Changes tend to fail discussed in Section 4.2.4.

4.2.1 THE FOUR STATES IN THE CHANGE LIFECYCLE

The four states in the change lifecycle are:

1. The change initialization state.
2. The change planning state.
3. The change performance or implementation state.
4. The change closeout state.

4.2.2 CHANGES ARE RISKY

For at least 500 years making a change has been perceived as being risky. For example:

> And it ought to be remembered that there is nothing more difficult to take in hand, more perilous to conduct, or more uncertain in its success, than to take the lead in the introduction of a new order of things. Because the innovator has for enemies all those who have done well under the old conditions, and lukewarm defenders in those who may do well under the new.

(Machiavelli 1515)

4.2.3 CHANGES TAKE TIME

While a system may appear as static at a particular point in time, in reality it is moving from the past through the present to the future. Someone with the authority deems a change in direction is needed and takes an action which will cause the system to change direction as shown in Figure 4.3. Once the action is taken, the following outcomes can appear:

1. The system moves in the desired or right direction.
2. The system moves in an undesired or wrong direction.
3. Nothing seems to happen.
4. Nothing happens for a while and then the system moves in the right direction.
5. Nothing happens for a while and then the system moves in the wrong direction.

Nothing may happen for a while so the outcomes of actions may not be seen immediately. There are often time delays between taking an action and seeing the effect of all the possible outcomes of that action. These delays have been grouped as (Kasser 2002):

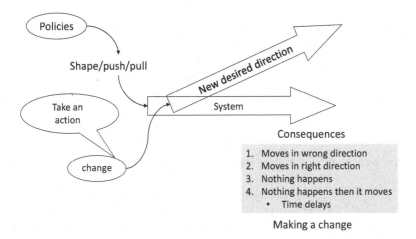

FIGURE 4.3   The effect of change on systems.

- *First order*: noticeable effect within a second or less.
- *Second order*: noticeable effect within a minute or less.
- *Third order*: noticeable effect within an hour or less.
- *Fourth order*: noticeable effect within a day or less.
- *Fifth order*: noticeable effect within a week or less.
- *Sixth order*: noticeable effect within a month or less.
- *Seventh order*: noticeable effect within a year or less.
- *Eighth order*: noticeable effect within a decade or less.
- *Ninth order*: noticeable effect within a century or less.
- *Tenth order*: noticeable effect after a century or more.
- And so on into the millennia.

Sometimes, due to the delays, the corrective action might cause the system to move past the intended direction, and further corrective action is needed to reverse the direction. Perceptions from the *Generic* HTP note that this is goal-seeking behaviour (Section 6.1.5.2) where the goal is set by the purpose of the corrective action.

4.2.4   CHANGES TEND TO FAIL

Mergers and acquisitions are said to have '75% failure rate' (Burke 2011) page 11). Beer and Nohria reported '70% of all change initiatives fail' (Beer and Nohria 2000) page 133). Kotter stated, 'few corporate change efforts have been very successful' (Kotter 2007) page 96) and he was talking about big companies such as Ford, General Motors, and British Airways with the resources to attract top talent like himself and Dr. Warner Burke. If elite change agents of the past four decades are reporting that change is near impossible then where's the hope for the 'everyday' change agent with limited resources (time, capital, talent, etc.)? What makes us think we can implement change more successfully than the elite? Theodore Roosevelt stated our challenge well over 100 years ago and provides inspiration and hope that even if we fail there is something to learn but without trying, there is only failure.

Stone (2015)

Machiavelli pointed out the perils of change making (Section 4.2) about 500 years ago.

4.3 BIG PICTURE

Perceptions from the *Big Picture* HTP note that change:

1. *Is not always an improvement*: some changes make things worse.
2. *Is usually resisted*: see quotations above (Machiavelli 1515, Stone 2015).
3. *Is continuous*: change is always taking place; however, the rate of change may be so slow as to be undetectable. For example, the Earth is undergoing changes as the tectonic plates move. This movement causes earthquakes. The ground only appears to be stable because we perceive it over a geological short time frame. Effects due to an action can appear within microseconds or can appear years later (Section 4.2.3).
4. *Can involve a paradigm shift*: namely a new way of looking at things (Kuhn 1970).

4.4 GENERIC

Perceptions from the *Generic* HTP note that:

1. The change lifecycle (Section 4.2.1) is identical to the policy lifecycle (Section 9.5.1).
2. Some changes constitute goal-seeking behaviour (Section 6.1.5.2) where the goal is set by the purpose of the corrective action which evolves the change.
3. CM is very similar if not identical to risk management.
4. The consequences of changes are generally the same as the consequences of decisions (Section 5.1.4.3) because decisions when implemented generally cause changes. The consequences can be grouped as:
    1. *Desired*: which also has two outcomes, namely:
        1. *Predictable*: that's the whole point of making the change.
        2. *Unpredictable* (serendipitous): wow, it's so much better than … and I want it.
    2. *Undesired*: which also has two outcomes, namely:
        1. *Predictable*: risks which must be prevented and/or mitigated.
        2. *Unpredictable*: according to the law of unintended consequences (Merton 1936) which become events when they occur and have to be dealt with.
    3. *Don't care*: a consequence that is of no interest irrespective of being predicted or not being predicted.

Alternatively, the consequences of change can also be grouped as:

1. *Predictable*: which also has three outcomes, namely:
   1. *Desired*: that's the whole point of making the change
   2. *Undesired*: risks which must be prevented and/or mitigated.
   3. *Don't care*: a consequence that is of no interest irrespective of being predicted or not being predicted.
2. *Unpredictable*: which also has three outcomes, namely:
   1. *Desired*: (serendipitous): wow, it's so much better than ... and I want it.
   2. *Undesired*: according to the law of unintended consequences which become events when they occur and have to be dealt with.
   3. *Don't care*: a consequence that is of no interest irrespective of being predicted or not being predicted.

4.5 FUNCTIONAL

Perceptions from the *Functional* HTP include:

1. Resistance to change discussed in Section 4.5.1.
2. Force field analysis discussed in Section 4.5.2.

4.5.1 RESISTANCE TO CHANGE

The literature abounds with various causes of resistance to change, which if not understood, doom change efforts to fail. Risk management is dealing with these causes and mitigating them or preventing them from happening, accordingly this section discusses several causes. One set of causes includes (Lippitt 1983):

1. Purpose of the change is not made clear.
2. Persons affected by the change are not involved in making the change.
3. Appeal for change is based on personal reasons.
4. Habit patterns of the individual are ignored.
5. Poor communication regarding the change.
6. Fear of failure.
7. Excessive pressure is involved.
8. 'Cost' is too high, or the reward inadequate.
9. Anxiety over personal security is not allowed.
10. Satisfaction with the status quo.

Other causes include:

1. *Fear of the effects of change* (Deevy 1995). The types of changes we need to make to transform today's organizations are cultural. When we fail to communicate the benefits of the change, and fail to reinforce desired behaviour according to the new rules, change fails.

TABLE 4.1
Table Template for Force Field Analysis

Proposed Change					
Forces for/against the Change					
#	Force	I	M	W	U
1					
2					
3					
4					
...					
	Total				

2. *Changes which change the rules of the game* and require people to shift their perspective (Campbell 1960). This is an emotional issue and cannot be settled by logic.
3. *Changes which require people to unlearn what they already know is correct* are resisted very strongly (Kuhn 1970).

4.5.2 FORCE FIELD ANALYSIS

Force field analysis is a tool which uses perceptions from the *Big Picture* HTP to identify the forces* and the magnitude of those forces that are both for and against the change at any point in time (Lewin 1943). Forces:

- For the change are called driving forces.
- Against the change are called restraining forces and are considered as risks.
- Can be stakeholders as well as those in the list below.
- Exist in different contexts and can be identified from the scenarios in the *Operational* HTP and other perspectives of the change.
- Exist at different times in the project lifecycle. The nine-system model (Section 6.7) helps identify forces in the different states of the project lifecycle.

The forces are listed in spreadsheets such as the one shown in Table 4.1 as a template. The columns are number, force, Importance (I), Magnitude (M), Weight (W), and Urgency (U). The situation is assumed to be in equilibrium balanced by the forces driving and restraining forces. If the driving forces can be increased, the situation will move in the desired direction, but if the restraining forces increase, the situation will move in the opposite direction. For example, during a rocket launch, the driving force from the exhaust pushes the rocket up into the air, while the restraining force of gravity tries to hold it back. The rocket rises as long as the force driving the rocket upwards overcomes the pull of gravity

---

* Many of those forces are in fact stakeholders. Force field analysis was published before the term 'stakeholder' came into vogue.

One way to perform force field analysis is to follow the following process:

1. Identify the stakeholders (Section 7.10).
2. Use brainstorming and active brainstorming (Sections 2.3 and 2.5.2) to generate ideas from the various HTPs about the driving and restraining forces. Forces to consider include the following list (ISU 2016) and HTP from which they are best perceived[*]:
   1. Available resources (*Big Picture*).
   2. Traditions (*Temporal*).
   3. Vested interests (*Big Picture, Operational*).
   4. Organizational structures (*Structural*).
   5. Relationships (*Operational, Functional*).
   6. Social or organizational trends (*Temporal*).
   7. Attitudes of people (*Functional*).
   8. Regulations (*Big Picture*).
   9. Personal or group needs (*Functional*).
   10. Present or past practices (*Temporal*).
   11. Institutional policies or norms (*Structural*).
   12. Agencies (*Big Picture*).
   13. Values (*Functional*).
   14. Desires (*Functional*).
   15. Costs (*Quantitative*).
   16. People (*Functional, Structural*).
   17. Events (*Operational*).

Typical active brainstorming questions for force field analysis include the starter questions in Section 2.5.2.2 which focus on risks as well as:

1. What is the current situation?
2. What will be the situation after the change?
3. What is the benefit of the change?
4. Who will benefit from the change?
5. Who can be counted on to support the change?
6. Who can be counted on to resist the change?
7. Why would someone support the change?
8. Why would someone resist the change?
9. How easy will it be to make the change?
10. How disruptive will the change process be?
11. How long will the change take to implement?
12. What would happen if the change was not made?
13. How soon will the benefits be felt?
14. What else will be affected by the change?
15. What else will not be affected by the change?

---

[*] Although they can show up from other perspectives as well.

3. Create two blank spreadsheets* using Table 4.1 as a template: one for the driving forces and the other for the restraining forces.
4. Number all the rows as shown in the tables.
5. Summarize the change in the 'Proposed change' title row.
6. Consider each idea generated in the brainstorming and active brainstorming session.
   1. List each driving force in a separate row in the 'Driving Forces' spreadsheet.
   2. List each restraining force in a separate row in the 'Restraining Forces' spreadsheet.
7. For each force in each row, estimate (Section 3.8.3) the magnitude of the force on a scale of 1–10 where 1 is 'weak' and 10 is 'extremely strong' and insert the number in the corresponding 'M' column.
8. Add up (sum) the numbers in the 'M' column for the driving and restraining forces.
9. For each force in each row, estimate the importance of the force on a scale of 1–10 where 1 is 'not important' and 10 is 'very important' and insert the number in the corresponding 'I' column.
10. For each force in each row, multiply the number in the 'I' column by the number in the 'M' column and insert the number in the corresponding 'W' column.
11. Add up (sum) the numbers in the 'W' column for the driving and restraining forces.
12. For each force in each row, locate the force in the risk framework (Section 1.2). Determine the urgency of dealing with that force on a scale of 1–10 where 1 is 'never' and 10 is 'yesterday' and insert the number in the corresponding 'U' column.

Perceptions from the *Continuum* HTP indicate that importance and urgency may be different (Covey 1989).

Perceptions from the *Generic* HTP note the similarity to multi-attribute variable analysis (MVA) in decision-making (Kasser 2018b: Section 4.6.2). Note that, in general, the earlier a restraining force can be dealt with, the higher the probability of preventing or mitigating it. However, this has to be balanced with the available resources for dealing with the restraining forces.

The information the spreadsheets forms the basis for the analysis. Any restraining force with a high 'W' (weighted value of resistance) needs to be dealt with; the urgency may be seen in the 'U' column. After performing the initial force field analysis, the ratio of weighted restraining forces to weighted driving forces provides an indication of whether the action should be:

1. *Taken*: when there are so many more driving forces than restraining forces, the risk of overcoming the restraining forces is minimal so the action will probably succeed, namely the restraining forces will have a negligible effect.

---

* Or use two tabs in a single spreadsheet.

2. *Taken with attention to the driving and restraining forces*: the driving forces should be supported and encouraged, whereas the restraining forces should be discouraged and converted to supporting forces. The importance and urgency of dealing with the forces can be seen in the spreadsheets. This is often the situation where stakeholder management (Section 7.10) is employed.
3. *Not taken*: when there are so many more restraining forces than driving forces, the risk of not being able to overcome the restraining forces is high so there is no point in undertaking the action.

The choice of action will depend a number of factors including:

- The resources available.
- The time available to deal with the stakeholders.
- Alternative courses of action that could deal with the situation.
- Actions with a better return on investment (ROI).
- The emotional attachment to the action in the initiator.

Perceptions from the *Generic* HTP also note the similarity of the table to a risk rectangle matrix (Section 3.8.4). Accordingly, it is tempting to multiply the value in the 'W' column by the value in the 'U' column to create a single number. However, the two numbers are in different dimensions, and while the product of the 'W' and 'U' columns would provide a single number, it would be meaningless.

4.6 STRUCTURAL

CM refers to various approaches to transitioning individuals, teams and organizations. Perceptions from the *Structural* HTP note the following typical definitions:

- CM is a systematic approach to dealing with change from the perspectives of both an organization and the individual (Rouse 2014).
- Transformational CM is the integrated management oversight of multiple ongoing organizational CM efforts (Craddock 2015).

Each definition seems to be based on a single different perspective.

4.7 OPERATIONAL

Perceptions from the *Operational* HTP note that CM is a risky process, since there are so many steps that could go wrong. One way to minimize the number of steps that could go wrong is to understand the process and develop models for the specific change the policy is to provide.

4.7.1 CM MODELS

The intent of CM models is to enable a successful planned CM process (Section 4.9.2), namely to mitigate or prevent the risks inherent in the change process. 'Essentially, all models are wrong, but some are useful' (Box and Draper 1987: p. 424).

The reason every model is wrong is that models are a simplification of reality and simplification may introduce errors of omission. Nevertheless, models are useful because they help us understand and explain situations.

The benefits of using a CM model include the following eight reasons for using a CM model (Connelly 2015):

1. *Forecasting*: change models forecast the process the change will take and prepare people for it.
2. *Measured results*: change models provide a good way to measure how individuals are managing change and what interventions may be most useful.
3. *Accountability*: change models create an intention for change that:
   1. Allow people to consider their role in the transition process.
   2. Hold people accountable for their own transition.
4. *Increased confidence*: change models give people confidence to talk to others about change.
5. *Reduce resistance*: change models help to:
   1. Identify potential areas of resistance and the pertinent stakeholders.
   2. Implement strategies designed to reduce or eliminate resistance before the change process starts.
6. *ROI*: Following a structured change model ensures that investments into the project are not lost and that budgets are managed.
7. *Role clarification*: change models provide clarity regarding the role of the stakeholders involved in or benefiting from the change.
8. *Shared approach*: change models provide a focus for all CM activities and help to align resources within the organization.

There are a number of models for organizational change in the literature. This section summarizes a sample of CM models in chronological order of their initial publication (*Temporal* HTP) and then compares them (*Scientific* HTP):

1. The Kurt Lewin CM model discussed in Section 4.7.1.1.
2. The Kübler-Ross model discussed in Section 4.7.1.2.
3. The 7S framework for organization change discussed in Section 4.7.1.3.
4. Checkland's soft systems methodology discussed in Section 4.7.1.4.
5. The Satir change model discussed in Section 4.7.1.5.
6. The William Bridges transition model discussed in Section 4.7.1.6.
7. The Burke-Litwin change model discussed in Section 4.7.1.7.
8. Kotter's 8-step change model discussed in Section 4.7.1.8.
9. The Nudge theory discussed in Section 4.7.1.9.
10. The ADKAR® CM model discussed in Section 4.7.1.10.

Further information about the models may be found in the cited texts and other sources.

4.7.1.1   The Kurt Lewin CM Model

The Kurt Lewin CM model is a simple three-state theory of change commonly referred to as unfreeze, change and freeze (or refreeze) (Lewin 1947), where:

1. *Unfreeze*: the state of getting ready for the change. This state includes the planning sub-states of deciding:
   * If the change should be made.
   * When the change should be made.
   * How the change should be made.
   * Overcoming the initial resistance to change by motivating people to desire or at least acquiesce to the change.
2. *Change*: the transition process from the undesirable situation to the feasible conceptual future desirable situation (FCFDS) (Section 5.1.7.2).
3. *Refreeze*: the state of establishing stability in the new changed situation. This is the stage in which the changed situation becomes the new normal situation.

4.7.1.2 The Kübler-Ross Model

The Kübler-Ross model is a notional five-stage model for the emotional reaction to changes due to an event such as death and serious trauma (Kübler-Ross 1969). The five stages are:

1. *Denial*: the initial stage of numbness and shock that gives the person time to absorb news of the event.
2. *Anger*: the stage in which the person gets angry and looks to blame something for making the event happen.
3. *Bargaining*: an attempt to postpone what is inevitable.
4. *Depression*: the person realizes that bargaining is not going to work the reality of the change sets in.
5. *Acceptance*: a resigned attitude towards the change, and a sense that the person must get on with it and start considering options.

The stages are not always sequential, and a person may be in more than one stage at the same time, move back and forth between the stages or even skip a stage. Although written as a model for a specific situation, the emotional reactions can be generalized cover reactions to other major changes in which a person has invested significant emotional resources such as not being accepted for a university degree and having a job application rejected.

4.7.1.3 The 7S Framework for Organization Change

The 7S framework (Pascale and Athos 1981, Waterman Jr., Peters, and Phillips 1980) is a management model based on the theory that effective organizational change is really the relationship between the following seven 'S' elements:

1. *Structure*: the structure of the organization before and after the change.
2. *Strategy*: the actions that a company plans in response to, or anticipation of, changes in its external environment – its customers, its competitors.
3. *Systems*: the procedures, formal and informal, that make the organization go, day by day and year by year: capital budgeting systems, training systems, cost accounting procedures, budgeting systems.

4. *Superordinate goals*: the guiding concepts – a set of values and aspirations, often unwritten, that go beyond the conventional formal statement of corporate objectives. Superordinate goals are the fundamental ideas around which a business is built. They are the broad notions of future direction that upper management wants to infuse throughout the organization. They are the way in which the team wants to express itself, to leave its own mark.
5. *Skills*: The dominating attributes, or capabilities of the company before and after the change.
6. *Style*: the corporate culture and the perception of the leadership within the organization.
7. *Staff*: the treatment of staff by managing what might be called the socialization process in their companies. This applies especially to the way they introduce young recruits into the mainstream of their organizations and to the way they manage their careers as the recruits develop into tomorrow's leaders.

The basic concept is that organization effectiveness stems from the interaction of several factors, namely a 'systems approach':

1. Change in one area will impact other areas.
2. It isn't obvious which of the seven factors will be the driving force (Section 4.5.2) in implementing a change in a particular organization at a particular point in time.

4.7.1.4   Checkland's Soft Systems Methodology

Checkland's soft systems methodology (SSM) (Checkland and Scholes 1990):

- Covers the change process with a focus on the front end of the process.
- Is a tool for gaining an understanding of a problematic situation in social situations – situations in which people are involved, in a systemic and systematic manner.
- Was developed for the express purpose of dealing with soft problems in layer 5 of the Hitchins-Kasser-Massie Framework (HKMF) (Section 3.4.4).
- Can be used to convert an ill-structured problem into a number of well-structured problems (Section 5.1.4.5).
- Is a seven-stage customized version of part of the traditional generic problem-solving process (Section 5.1.6.1). 'The function of Stages 1 and 2 is to display the situation so that the range of possible and, hopefully, relevant choices can be revealed, and that is the only function of those stages (Checkland 1991: p. 166).
- Seems to be the most widely known version of SSM; there are others such as Avison and Fitzgerald (2003).

This section describes this author's interpretation of Checkland's SSM. The interpretation is based on perceiving SSM from the HTPs (Section 2.5.1).

The seven stages of SSM are:

1. Recognizing the existence of a problematical or undesirable situation.
2. Expressing the real-world problematic situation.
3. Formulating root definitions (Section 4.7.1.4.3) of relevant systems of purposeful activity from different HTPs.
4. Building conceptual models of the systems named in the root definitions.
5. Comparing the conceptual models developed in Stage 4 with the real-world situations expressed in Stage 2.
6. Identifying feasible and desirable changes.
7. Taking actions to improve the problematic situation.

4.7.1.4.1 Stage 1: Recognizing the Existence of a Problematic or Undesirable Situation

This stage begins when a problem owner determines that an existing situation is undesirable and needs to be changed. The problem owner is someone who has or is given the authority and appropriate resources to initiate a project to study the situation to determine the cause or causes of undesirability and recommend appropriate actions to make a transition from the existing undesirable or problematic situation to an FCFDS (Section 5.1.7.2). The initial parts of the study are performed by the problem solver – often an analyst.

4.7.1.4.2 Stage 2: Expressing the Real-World Problematic Situation

This is the stage that examines and documents the real-world problematic situation. The emphasis is on gaining an understanding of the real-world situation and documenting it. The original SSM documented the situation in hand-drawn rich pictures (Kasser 2018a: Section 6.1.5) in accordance with the adage, 'a picture is worth 1000 words'. SSM suggested the following three types of analysis be carried out on organizations in which people are involved in this stage:

1. *Analysis one*: the role analysis identifies the roles of three of the actors in the transformation. These three roles are:
   1. *The* 'client': the person or persons who caused the study to take place.
   2. *The 'would be problem-solver'*: whoever wishes to do something about the situation in question and the intervention had been defined in terms of their perceptions, knowledge and readiness to make resources available, in other words in the client's language.
   3. *The 'problem owner'*.
2. *Analysis two*: is based on a model that assumes a social system to be a continually changing interaction between three elements: roles, norms and values. Each continually defines, redefines and is itself defined by the other two.
3. *Analysis three*: analyses the politics – 'the processes by which differing interests reach accommodation'. The analysis begins with the study of how power is expressed in the situation including authority (formal and informal), memberships of groups, committees, etc.

The analyst* perceives the problematic situation from a number of perspectives such as the HTPs (Section 2.5.1) collecting, sorting and documenting the information. For example, such perceptions from the various HTPs include:

1. *Big Picture*:
   - The environment.
   - The boundary of the problematic situation.
   - Adjacent systems to the problematic situation.
   - The issues that the people involved in the situation think are problematical.
2. *Operational*:
   - The processes or transformations taking place in the situation. Each transformation process may be considered as a black box system, the input being an entity and the output being a transformed entity.
   - The politics in the organization (Section 7.3.1.10).
3. *Functional*:
   - The internal functions inside the processes or transformations (black box) perceived from the *Operational* HTP.
4. *Structural*
   - Elements slow to change including:
     - The organization chart.
     - Physical elements such as buildings and locations.
     - Information used in the functions.
     - How power is expressed in the situation including authority (formal and informal), memberships of groups, committees, etc.
5. *Quantitative*:
   - The following are three minimal criteria for evaluating a process:
     1. *Efficacy*: how well the process performs its intended task. This may not be the lowest cost process.
     2. *Efficiency*: as measured by the amount of output divided by the amount of resources used.
     3. *Effectiveness*: how well the process meets the longer-term aim.
   - Other criteria such as ethics and elegance may be added depending on the situation. A minimum value for each criterion should be set at the time the project begins.

4.7.1.4.3   Stage 3: Formulating Root Definitions of Relevant Systems

The perceptions from the *Operational* and *Functional* HTPs are summarized in root definitions.

> A root definition expresses the core purpose of a purposeful activity system. That core purpose is always expressed as a transformation process in which some entity. The 'input' is changed, or transformed, into some new form of that same entity, the 'output'.

> **(Checkland and Scholes 1990: p. 33)**

---

* Depending on the nature and scope of the problematic situation, the problem-solver may be a single analyst or a team with different disciplinary skills.

Any system generally can be conceived as being different things and performing different functions by the stakeholders, many of which have no idea that a different perception exits. Different people have different ways of seeing things and belief systems, known as Weltanschauung (Checkland and Scholes 1990) worldviews or paradigms (Kuhn 1970). They may perceive problems or want different (and perhaps contradictory) remedies (solutions) to an undesirable situation or have different concepts of what the situation is all about.

In the development of SSM, the names of relevant systems had to be written in such a way that they made it possible to build a model of the system named. The names became known as 'root definitions' since they express the core or essence of the perception to be modelled.

The root definition is a perception from the *Operational* HTP (Section 2.5.1). The simplest version of a root definition is 'a system to do something' from the perspective of a stakeholder. However, perceptions from the *Continuum* HTP indicate that different stakeholders may have different root definitions, e.g. a hotel, pub or bar could be (Kasser 2015: p. 131):

1. *A profit-making system* from the perspective of the owners.
2. *An employment system* from the perspective of the (potential) employees.
3. *A recreational system* from the perspective of the customers.
4. *A social system* from the perspective of the local residents.
5. *A revenue generating system* from the perspective of the taxation authority.

Each of these systems performs a different transformation process and can be considered as a subsystem of the whole.

Checkland's root definitions are formulated as sentences that elaborate a transformation using six elements which are summed up in the template mnemonic CATWOE as follows:

- *Customers*: the victims or beneficiaries of a transformation process. These entities should be known as stakeholders since they have a stake in the process while the customers are the specific stakeholders who fund the transformation process.
- *Actors*: the stakeholders who perform, or are otherwise involved in, the transformation process.
- *Transformation process*: the conversion of the input to an output, namely the purpose of a system.
- *Weltanschauung*: this worldview or paradigm (Kuhn 1970) which makes the transformation process meaningful in context of the stakeholders. Perceptions from the *Continuum* HTP indicate that each stakeholder may have a different Weltanschauung.
- *Owner(s)*: the stakeholder(s) who has/have the power to start up and shut down the transformation process.
- *Environmental constraints*: elements that exist outside the system which it takes as given.

TABLE 4.2

Apparent Relationship between SSM's CATWOE and the HTPs

CATWOE	HTP
Client/customer	*Big Picture*
Actor	*Operational*
Transformation	*Functional and Quantitative*
Weltanschauung	*Big Picture*
Owner	*Big Picture*
Environment	*Big Picture*

Checkland's version of a root definition is 'a system to do something by means of something in order to achieve a goal or purpose'. Checkland provides the following example of a root definition, 'A householder-owned and manned system to paint a garden fence, by conventional-hand painting, in keeping with the overall decoration scheme of the property, in order to enhance the visual appearance of the property' (Checkland and Scholes 1990: p. 36). The CATWOE template points to the following:

C: householder.
A: actor or person performing the transformation.
T: activity which transforms an unpainted garden fence into a painted fence.
W: amateur painting can enhance the appearance of the house.
O: householder.
E: hand painting.

The core of CATWOE is the pairing of transformation process T and the W, the Weltanschauung or worldview which makes it meaningful. For any relevant purposeful activity there will always be a number of different transformations by means of which it can be expressed, these deriving from different interpretations of its purpose.

(Checkland and Scholes 1990: p. 35)

Perceptions from the *Generic* HTP indicate that this sentence is also an example of a mission statement (Kasser 2018a: Section 8.9), a statement which is used as a way of communicating the purpose of an organization. As inferred from the *Scientific* HTP, it seems that root definitions may be considered as mission statements and CATWOE can provide a template for mission statements. The CATWOE template seems to align with some of the HTPs (Section 2.5.1) as shown in Table 4.2* (Kasser 2018a). The *Quantitative, Structural, Generic* and *Continuum* HTPs are not implicitly invoked in SSM, and the *Scientific* HTP is implied by the findings of the study using SSM.

---

* The boundaries do not align directly because the decomposition of systems thinking is different.

4.7.1.4.4   Stage 4: Building Conceptual Models of the Relevant Systems

In this stage, the analyst builds a set of models of each transformation process which is considered as a system. Once the set of models is complete, if there are more than nine models in accordance with Miller's rule (Kasser 2018a: Section 3.2.5, Miller 1956), some of them should be combined into one or more complex models by appropriate grouping of functions. A useful tool for doing the grouping or aggregation task is the N^2 chart (Kasser 2018a: Section 2.10, Lano 1977).

The model description should be in a format or language understood by all the stakeholders in the situation (Checkland and Scholes 1990). Suitable tools to use include rich pictures (Kasser 2018a: Section 6.1.5), flowcharts (Kasser 2018a: Section 2.7), program evaluation review technique (PERT) charts (Kasser 2018a: Section 8.10), causal loops (Kasser 2018a: Section 6.1.1) or any process modelling language. Perceptions from the *Generic* HTP indicate that this model is also known as a 'to-be' model in business process reengineering (BPR).

4.7.1.4.5   Stage 5: Comparing Conceptual Models with Reality

The purpose of this stage is to generate debate about possible changes which might be made within the perceived problem situation. This is the stage in which the conceptual models built in Stage 4 are compared with the real-world expression from Stage 2. The work at this stage may lead to the reiteration of Stages 3 and 4. Perceptions from the *Generic* HTP indicate that this stage could also be known as a 'gap analysis' which identifies the difference between the real world and the conceptual model. Checkland identifies the following four approaches for doing the comparison: ordered questioning, comparing history with model prediction, general overall comparison and model overlay.

1. *Ordered questioning*: an approach which can be used when the real-world situation is very different from the conceptual model. The system models are used to open up debate about change. The model is used as a source of questions to ask about the existing situation. The questions are written down and answered systematically. The answers to the questions can provide illumination of the perceived problem. Perceptions from the *Generic* HTP indicate that active brainstorming (Section 2.5.2) which can be used irrespective of the relationship between the real world and the conceptual model encompasses ordered questioning.
2. *Comparing history with model prediction*: a comparison method which reconstructs a sequence of events in the past and compares what had happened with what should have happened if the relevant conceptual model has actually been implemented. In this way, the meaning of the models can be exhibited and a satisfactory comparison can be reached. Checkland also warned that this method of comparison should be used carefully because it could reveal the inadequacies of the actual procedure and people might interpret the results as offensive recrimination* concerning their past performance.

---

* Pointing the finger of blame.

3. *General overall comparison*: identifies the features of the conceptual models that are especially different from the real world and the reasons for the differences. This comparison is also generally discussed in terms of the 'What's' (*Operational* HTP) and 'How's' (*Functional* HTP). Stages 3 and 4 produced systems models which themselves derive from the careful naming, in root definitions, of human activity systems which are relevant to the problem situation and to its improvement.
4. *Model overlay*: overlays the conceptual model based on the chosen root definition with the second model of the real world. The second model should have as near as possible the same form as the conceptual model; i.e. it is based on the same template. The direct overlay of one model on the other reveals the gap.

4.7.1.4.6   Stages 6 and 7: Action to Improve the Existing Situation
Stage 6 is the stage in which feasible and desirable changes are identified and discussed. Once consensus is reached, Stage 7 is the realization stage in which the changes are implemented. Checkland recognized three kinds of changes:

1. *Changes in structure*: changes made to those parts of reality which do not change during normal operations.
2. *Changes in procedure*: changes to the dynamic elements.
3. *Changes in attitude*: changes to the behaviour appropriate to various roles, as well as changes in the readiness to rate certain kinds of behaviour as 'good' or 'bad' relative to others.

Changes in structure and procedure are easy to specify and relatively easy to implement. At least, these can be done by the people who have authority or influence. Attitude, on the other hand is relatively difficult to change because one must first understand why people have that attitude and what must be done to motivate them to change their attitude.

4.7.1.5   The Satir Change Model
The Satir change model is a five-stage model that describes the observed effects of each stage on a person's feelings, thinking, performance and physiology (Satir et al. 1991). The five stages are:

1. *Late status quo*: the situation before the change. People know and understand what is expected of them.
2. *Resistance*: people resist a *foreign element* that threatens the stability of the situation.
3. *Chaos*: the change begins but without a clear picture of where it is heading, people feel vulnerable and concerned about the future.
4. *Integration*: people discover a *transforming idea* that shows how the *foreign element* can benefit them. Resistance to change is minimized; the change takes place and the new situation comes into existence.
5. *New status quo*: the situation after the change.

4.7.1.6 The William Bridges Transition Model

The William Bridges transition model focuses on transition, not change (Bridges 1991). The model highlights three stages of transition that people go through when they experience change. These are:

1. *Ending, losing and letting go*: a stage is often marked with resistance and emotional upheaval, because people are being forced out of their comfort zone.
2. *The neutral zone*: the bridge between the old situation and the new situation; people will still be attached to the old, while they are also trying to adapt to the new. The people affected by the change are often confused and uncertain.
3. *The new beginning*: people have begun to embrace the change initiative, and starting to see early wins in the new situation.

Bridges states that people go through each stage at their own pace. Accordingly, those who are comfortable with the change will tend to move to Stage 3 quickly, while others will linger at Stage 1 or 2.

4.7.1.7 The Burke-Litwin Change Model

The Burke-Litwin change model is based on their consulting efforts over a period of about five years with British Airways which taught them a lot about what changes seemed to have worked and what did not (Burke and Litwin 1992). The model attempts to portray the 12 primary variables and linkages that need to be considered in any attempt to predict and explain the behaviour of an organization. The model organizes the variables and linkages into two major feedback loops: transformational and transactional. The 12 components of the model are:

1. Transformational elements:
   1. *External environment*: any outside conditions or situation that influence the performance of the organization (e.g. marketplaces, the economy or political considerations). The model considers that for the most part organization change is initiated by forces from the organization's external environment, a factor not considered in the 7S model (Section 4.7.1.3).
   2. *Mission and strategy*: (a) what the organization's upper management believes is and has declared is the organization's mission and strategy and (b) what employees believe is the central purpose of the organization.
   3. *Leadership*: the executives providing overall organizational direction and serving as behavioural role models for all employees. The model makes a distinction between leadership and management.
   4. *Organizational culture*: 'the way we do things around here' (Deal and Kennedy 1982) cited by Burke and Litwin (1992); the collection of overt and covert rules, values and principles that are enduring and guide organizational behaviour.

2. Transactional elements:
   5. *Structure*: the arrangement of functions and people into specific areas and levels of responsibility, decision-making authority, communication and relationships to assure effective implementation of the organization's mission and strategy.
   6. *Systems*: the standardized policies and mechanisms that facilitate work, primarily manifested in the organization's reward systems, management information systems (MIS), and in such control systems as performance appraisal, goal and budget development and human resource allocation.
   7. *Management practices*: what managers do in the normal course of events to use the human and material resources at their disposal to carry out the organization's strategy.
   8. *Work unit climate*: the collective current impressions, expectations and feelings that members of local work units have that, in turn, affect their relations with their boss, with one another and with other units.
   9. *Tasks and skills*: the required behaviour for task effectiveness, including specific skills and knowledge required of people to accomplish the work for which they have been assigned and for which they feel directly responsible. Essentially, this component concerns what is often referred to as the job–person match.
   10. *Individual values and needs*: the specific psychological factors that provide desire and worth for individual actions or thoughts.
   11. *Individual motivation level*: the aroused behaviour tendencies to move towards goals, take needed action and persist until satisfaction is attained.
3. Both transformational and transactional
   12. *Individual and overall performance*: the outcome or result as well as the indicator of effort and achievement (e.g. productivity, customer satisfaction, profit and quality).

The components are connected as shown in the N^2 chart in Table 4.3 which clearly shows the tight coupling between the groups of components.

4.7.1.8 Kotter's Eight-Step Change Model

Kotter examined the efforts of more than 100 companies to remake themselves into better competitors and identified the most common mistakes leaders and managers made in attempting to create change. They created the following eight-step process designed to overcome the obstacles (risks) to making policy changes in the corporate environment (Kotter 1996).

1. Establish a sense of urgency.
2. Create the guiding coalition.
3. Develop a vision and strategy.
4. Communicate the change vision.
5. Empower employees for broad-based actions.
6. Generate short-term wins.

TABLE 4.3

N² Chart for the Burke-Litwin Change Model

1	o	o	o								o
o	**2**	o		o			o	o			
o	o	**3**	o	o	o	o					
o		o	**4**		o		o		o		
	o	o		**5**	o	o	o	o			
	o	o	o		**6**	o	o		o		
	o		o	o		**7**	o				
o		o	o	o	o		**8**			o	
o				o				**9**		o	
		o			o				**10**	o	
							o	o	o	**11**	o
o										o	**12**

7. Consolidate gains and produce more change.
8. Anchor the changes in the corporate culture.

4.7.1.9 The Nudge Theory

The Nudge theory (Thaler and Sunstein 2008):

- Accepts that people have certain attitudes, knowledge, capabilities, etc., and allows for these factors.
- Advocates change in groups through indirect methods, rather than by direct enforcement or instruction.
- Seeks to improve understanding and management of the 'heuristic' influences on human behaviour which is central to 'changing' people.
- Is mainly concerned with the design of choices, which influences the decisions we make.

4.7.1.10 The ADKAR® CM Model

ADKAR® is a personal goal-oriented CM model that allows CM teams to focus their activities on specific human factor elements because problems with people dimension of change are the most commonly cited reason for project failures (Prosci 2016). The five parts of acronym ADKAR – awareness, desire, knowledge, ability and reinforcement – represent five milestones an individual must achieve for change to be successful. These milestones are:

1. *Awareness*: of the reasons or need to change.
2. *Desire*: to participate and support the change.
3. *Knowledge*: on how to change.
4. *Ability*: to implement the required skills and behaviours.
5. *Reinforcement*: to sustain the change

The ADKAR model is based on the premise that change happens in two dimensions, the business dimension and the people dimension, and successful change happens when both dimensions of change occur simultaneously.

TABLE 4.4
Comparison of CM Models

CM Model	Section	Best Used in Change Lifecycle State(s)
Kurt Lewin	3.7.1.1	All
Kübler-Ross	3.7.1.2	Initialization, planning
7S Framework	3.7.1.3	Initialization, planning
Checkland's SSM	3.7.1.4	Initialization, planning
Satir	3.7.1.5	All
William Bridges	3.7.1.6	Performance
Burke-Litwin	3.7.1.7	Initialization, planning
Kotter's eight-step	3.7.1.8	All
Nudge theory	3.7.1.9	Initialization, planning
ADKAR®	3.7.1.10	Performance

The business dimension contains the following phases:

1. Identify the business need or opportunity.
2. Define the scope and objectives of the project.
3. Design the business solution.
4. Develop the new processes and systems.
5. Implement the solution.

Effective management of the people dimension of change requires managing five key goals that form the basis of the ADKAR model:

1. Awareness of the need to change.
2. Desire to participate and support the change.
3. Knowledge of how to change (and what the change looks like).
4. Ability to implement the change on a day-to-day basis.
5. Reinforcement to keep the change in place.

4.8 SCIENTIFIC

Inferences from the *Scientific* HTP include:

1. Each model is based on a different partial perspective of the change lifecycle to facilitate understanding the type of problems the people who develop the models were facing.
2. When the CM models are compared, e.g. as shown in Table 4.4, perceptions from the HTPs indicate:

- *Generic*: many similarities.
- *Continuum*: many differences, e.g. some seem to be best used in the entire lifecycle of change: some in all states, some in the initialization and planning states and some best used in the performance change transition states.
- *Structural*: the descriptions are at a high high-level of abstraction so aspects are not necessarily missing; they are just not the focus in the summary in the table.

3. Each of the models are tools and were developed in specific situations. Since perceptions from the *Continuum* HTP note that many situations will not match the specific situations in the models, the models contain useful building blocks for generating a hybrid custom model for a specific situation, e.g. as in the maid reduction policy exercise (Section 4.10.2).
4. A specific instance of change can be assessed against a number of the models to increase the understanding of the situation and develop the lowest risk CM strategy.

4.9 CHANGE AND CM IN RISK FRAMEWORK LAYERS 3 AND BELOW

From the moment the requirements are accepted at the system requirement review (Section 6.1.5.4.2), changes can happen. They can be:

1. *Predicted*: as a result of an upgrade plan or as a result of a change request (Section 4.9.2).
2. *Unpredicted*: as a result of an unforeseen or unmitigated risk turning into an event.

4.9.1 IMPACT OF CHANGE

The impact of a change affects requirements, documents, work packages (WPs) (Section 7.3.1.1.4), builds and deliveries, as well as cost and schedule, depending on the point in the SDP (Section 6.1.5.4) in which the change occurred. For example:

- Changes in high-level requirements will affect lower level requirements and may affect implementation requirements.
- Documents affected include management plans, operations concepts, manuals, test plans and procedures.
- WP elements affected may include all SDP activities.
- For builds and deliveries, the implementation sequence may be changed to the point where a build does not add any value to the system, so the cost of testing, releasing and delivering the build may no longer be economical.
- The effects of the changes will show up as a variance in the cost and schedule and categorized requirements in process (CRIP) chart (Section 7.5).

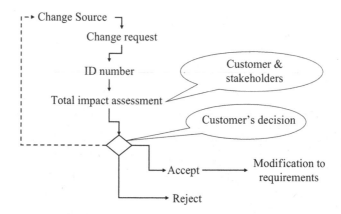

FIGURE 4.4   A functional view of the generic CM process.

4.9.2 THE CM PROCESS

In the CM process, requests for changes are made because something is undesirable due to the system:

1. Not doing what it should be doing, because:
   - Something is broken.
   - Something does not have capability any more (it is overloaded).
2. Not doing something it could be doing.
3. Doing something, but not as well as it could or should be doing it.
4. Doing something it should not be doing.

Perceived from the *Functional* HTP, the CM process shown in Figure 4.4 (Kasser 2019) consists of the following activities:

1. Convert the stakeholder area of concern into one or more requirement(s)/change request(s).
2. Assign a unique identification (ID) number to the requirement(s)/change request(s).
3. Prioritize the requirement request(s) with respect to other requirements/change requests.
4. Determine if a contradiction exists between the requirement(s)/change request(s) and existing accepted requirements/changes.
5. Perform an impact assessment which must:
   - Estimate the cost/schedule to implement the requirement(s)/change request(s).[*]

---

[*] In this pre-system requirement review (SRR) situation, there is no need to determine the cost and schedule for every requirement. Perceptions from the *Quantitative* HTP in the form of the Pareto principle perceive that the cost and schedule impact only needs to be determined for a few of the most expensive and longest time to realize requests (Hari, Shoval, and Kasser 2008).

- Determine the cost/schedule drivers: the factors that are responsible for the greatest part of the cost/schedule implementing the requirement(s)/change requests(s).
- Perform a sensitivity analysis on the cost/schedule drivers.
- Determine if the high cost/schedule drivers are really necessary and how much negotiating the requirement(s)/change request(s) with stakeholders can make modifications to the high cost/schedule drivers based on the results of the sensitivity analysis.

6. Make the customer's decision to accept, accept with modifications or reject the request.
7. Notify the stakeholder of the decision.
8. Document the decision(s) in the requirement/change repository to provide a history in case the same requirement(s)/change request(s) are received at some future time.
9. If the requirement(s)/change request(s) is/are accepted, allocate the implementation to a specific future version of the system and SDP, modifying the documentation appropriately.

Not having an efficient CM process constitutes a significant risk of incurring cost and schedule overruns as well as delivering products that do not comply with their changed specifications.

4.10   EXERCISES

This section provides a number of exercises as opportunities to think about the risks in change and practise aspects of CM. A partial acceptable solution is provided for the maid reduction policy exercise (Section 4.10.2). It is up to the reader to use the sample as a template when performing the remaining exercises.

4.10.1   ENU Traffic Light Upgrade

Using the information developed in the previous exercises and in addition to meeting the generic requirements in Section 2.7:

1. Identify the specific change that will take place. For example, this change will remove the uncertainty in the waiting time[*] by providing a countdown timer showing the number of seconds remaining until the light changes.
2. Identify at least three direct and indirect stakeholders.
3. Use force field analysis (Section 4.5.2) to determine at least five forces for and at least five forces against the change and estimate their strengths.
4. Discuss which CM model or combination of models (Section 4.7.1) would be most suitable in this situation and why.

---

[*] For pedestrians, or pedestrians and drivers? If the change will provide countdown timers for both stakeholders, then it will be more expensive to implement due to the additional countdown timer displays that will be needed.

4.10.2   THE ENGAPOREAN MAID REDUCTION POLICY

Using the information developed in the previous exercises and in addition to meeting the generic requirements in Section 2.7:

1. Identify the specific change that will take place. For example, this change will decrease the number of maids by p% within y years.
2. Identify at least three direct and indirect stakeholders.
3. Use force field analysis (Section 4.5.2) to determine at least five forces for and at least five forces against the change and estimate their strengths.
4. Discuss which CM model or combination of models (Section 4.7.1) would be most suitable in this situation and why.

4.10.2.1   One Acceptable Solution

A completed exercise could take the form shown in this section. Note it provides one acceptable solution, not the single correct solution as there will be other acceptable solutions (Section 5.1.1.2.2).

The response to this exercise contains assumptions in the operational functional and quantitative information due to the anticipated lack of domain knowledge in the students performing the exercise. The experience of thinking through the exercise is more important than the accuracy of the information. Although it might help to do enough of a sensitivity analysis on the assumptions to determine if they are reasonable. The analysis performed as part of the exercise shows the complexity and the interaction between the variables.

4.10.2.1.1   The Assumptions

The assumptions include:
1. The stakeholders may grumble about the policy but will not take any serious action (vote against the government in the next election which is two years in the future, organize street demonstrations or worse).

4.10.2.1.2   The Compliance Matrix

The compliance matrix is shown in Table 4.5.[*] Note that items 10 and 11 are listed separately to confirm that both have been addressed. Had both forces been combined on the same line, it would not have been clear that both types of forces had been addressed.

4.10.2.1.3   The Specific Change

This change will reduce the number of maids by 40% within two years.

4.10.2.1.4 The Stakeholders Affected by the Change

The direct and indirect stakeholders affected by the change include[†]:

---

[*] Placing the section number in the completed column not only shows that the requirement has been met but shows where the response can be found, mitigating the risk of the instructor failing to find and grade the response.

[†] For the purpose of this exercise, the number of stakeholders has been limited to facilitate explaining the issues. In a real situation, more time would be spent on identifying direct and indirect stakeholders.

TABLE 4.5

Compliance Matrix for the Exercise in Section 4.10.2

ID	Instruction	Deliverable	Section	Completed
1	Brainstorming or active brainstorming	No	All	Yes
2	Research	No	All	Yes
3	Use imagination	No	All	Yes
4	Use lists and charts	No	All	Yes
5	Spend less than 60 minutes	No	All	Yes
6	Assumptions	Yes	4.10.2.1.1	Yes
7	Preliminary compliance matrix	No	4.10.2.1.2	Yes
8	Identify the specific change	Yes	4.10.2.1.3	Yes
9	The stakeholders affected by the change	Yes	4.10.2.1.4	Yes
10	The forces for the change	Yes	4.10.2.1.5	Partially
11	The forces against the change	Yes	4.10.2.1.5	Partially
12	CM models	Yes	4.10.2.1.6	Yes
13	Lessons learned	Yes	4.10.2.1.7	Yes
14	Updated compliance matrix	Yes	4.10.2.1.2	Yes

4.10.2.1.4.1 The Direct Stakeholders In the traditional approach, the direct stakeholders (Kasser 2019: Section 10.14) would be identified as the stakeholders affected by the change including:

1. The maids.
2. The employers.
3. The agencies who supply the maids.

These are the stakeholders in the problem and solution domains (Section 5.1.4.6). In the systems approach, there are also stakeholders in the implementation domain who can also have an effect, namely be a directing or an opposing force. These stakeholders include the government bureaucrats who will develop the policy and the other stakeholders involved in implementing the change.

In general, any stakeholder inside the system or who interfaces directly with the system is a direct stakeholder.

4.10.2.1.4.2 The Indirect Stakeholders The indirect stakeholders (Kasser 2019: Section 10.14) are identified from use cases or scenarios in both the undesirable situation ('as is' in BPR) and the FCFDS (Section 5.1.7.2) ('to be' in BPR). Typical scenarios include what the maids do on the job and what they do on their day off, e.g. babysitting, meeting with their friends and other maids in cafés, going shopping and going to the cinema or theatre. Each current and FCFDS scenario needs to be analysed from the *Functional* and *Operational* HTPs and understood to identify the stakeholders or the forces for and against the change and assess the appropriate *Quantitative* HTP value of the strength or magnitude of the forces and the importance and urgency of dealing with the stakeholder. For example, in:

- *The going shopping scenario*: the types of shops visited and the amount spent in those shops need to be determined to sufficient accuracy to assess the impact of the loss of sales revenue to the retail establishments.
- *The entertainment scenario*: cinema, theatre and café scenarios need to be analysed in the same way to determine the same information. The indirect stakeholders faced with a loss of revenue can be expected to resist the change, namely the new policy; the strength of the resistance might be estimated based on the effect of the loss of revenue. It will be necessary to work out what percentage of the revenue is contributed by the maids.

Other scenarios need to be investigated as well, e.g. the effect on families that no longer have maids. These scenarios include what the maid does as her daily job. Examples include babysitting, shopping, childcare, cooking, cleaning and washing. How will these scenarios be affected? Some families might decide to use day-care facilities for their younger children, while other families might decide to have a stay home parent. The impact on this scenario is a reduced family income due to either the fees paid to the day-care facility or the lack of the second income. Thinking about this scenario identified more indirect stakeholders associated with the employers of the maids, including:

1. *Owners of day-care facilities*: who can be expected to support the change, namely be a driving force for the change. The whole set of similar scenarios needs to be identified and analysed.
2. *Owners of retail establishments*: shops, cinemas, theatres, cafés, travel agents and restaurants where the employers of the maids currently spend some if not all of the discretionary part of the second income. These stakeholders faced with the loss of revenue can be expected to oppose the change.
3. *Travel agents*. One additional scenario assumes that the maids send part of their salary to their families in the home country. There would be an effect due to the lack of income and also the lack of employment for potential maids. However, this scenario would generally be outside the boundary of the system being analysed, namely outside the scope of the analysis.

4.10.2.1.4.2.1 The Forces for and against the Change   The forces for and against the change (stakeholders) and the estimates associated with those stakeholders in force field analysis are summarized in Tables 4.6 and 4.7. The tables also serve as a compliance matrix (Kasser 2018b: Section 9.5.2) to check compliance with the instructions for the exercise (the requirements). The tables contain a subset of one driving force and five restraining forces (stakeholders). Given the disparity, the systems approach suggests that there are other reasons not immediately obvious for the proposed change. These reasons might include:

TABLE 4.6

Driving Forces for the Maid Reduction Policy (Partial)

Proposed Change: Maid Reduction Policy per the Exercise Instructions					
Driving Forces for the Change					
#	Force	I	M	W	U
1	Day-care facilities	1	3	3	6
2					
3					
4					
...					
	Total		3	3	

- *Political*: the government would like to cut down on the number of alien residents for some reason.
- *Fiscal*: the government would like to cut down on the transfer of money out of the country.
- *Personal*: the minister or someone close to the minister has some reason for creating the policy.

The magnitude of the forces is estimated as follows:

- *Driving force*: is estimated as follows:
  - *Day-care facilities*: a three because there are not too many of them.
- *Restraining forces*: are estimated as follows:
  - *Maids*: a one because they constitute a very small percentage of the population.

TABLE 4.7

Restraining Forces for the Maid Reduction Policy (Partial)

Proposed Change: Maid Reduction Policy per the Exercise Instructions					
Restraining Forces against the Change					
#	Force	I	M	W	U
1	Maids	1	1	1	1
2	Maid employment agencies	3	4	12	8
3	Employers	8	6	48	8
4	Café owners	5	5	25	2
5	Bureaucrats who will have to enforce the policy	9	5	45	9
	Total		21	130	

- *Maid employment agencies*: a four because there are more of them than day-care facilities.
- *Employers*: a six because while there are a significant number of employers, they only constitute less than 10% of the population.
- *Café and restaurant owners*: a five because while they are quite a few of them, the maids constitute a small percentage of their revenue.
- *Bureaucrats*: a five because this will be perceived as additional work without any additional reward.

The importance of the forces is estimated as follows:

- *Driving force*: is estimated as follows:
    - *Day-care facilities*: a one because it is expected they will not bother to take any action to prevent the change from being implemented.
- *Restraining forces*: are estimated as follows:
    - *Maids*: a one because they have no political clout.
    - *Maid employment agencies*: a three because they can be expected to protest and gain some publicity for their protest when they hear about the proposed change, but again they probably will not take any action about it.
    - *Employers*: an eight because they are expected to protest initially, but the protest will decline as time goes by and the situation will become accepted.
    - *Café and restaurant owners*: a five for the same reason as the number allocated to the employers; they only get a five because there are a lot fewer.
    - *Bureaucrats*: a nine because they won't protest, but they will probably find other ways to delay or fail to implement the change.

The urgency of dealing with the forces is estimated as follows:

- *Driving force*: is estimated as follows:
    - *Day-care facilities*: a six because they may have to prepare for an expansion, but they don't have to be dealt with right away.
- *Restraining forces*: are estimated as follows:
    - *Maids*: a one because the maid employment agencies will take care of dealing with the maids. That can be part of the implementation approach to minimize the workload of the bureaucrats.
    - *Maid employment agencies*: an eight because they will have to handle the administrative details, but they don't need to be informed until the change is about to go into effect.
    - *Employers*: an eight because they will have to make changes to their lives to compensate for the loss of the maids.
    - *Café and restaurant owners*: a two because they will find out about the change from the maids as they discuss it during their days off.
    - *Bureaucrats*: a nine because they will have to handle the work.

The values are inserted into Tables 4.6 and 4.7 according to the procedure in Section 4.5.2. When the information in Tables 4.6 and 4.7 is studied, the balance between driving and restraining forces is weighted very much in favour of the restraining forces. The most influential and urgent restraining force constitutes the bureaucrats who will be tasked with implementing the policy. One way of reducing their resistance would be to minimize their workload. This puts a constraint on the way the policies are implemented. For example, the change could be implemented in three steps:

1. Not providing visas for any new maids to enter the country to replace maids leaving the country or take up new positions.
2. Not extending any existing expiring visas for the maids currently in the country.
3. Significantly increasing the employment taxes on maids.

This implementation approach would put a minimal workload on the bureaucrats, because there would be:

- Nothing for the Ministry of Interior to do other than refuse visas.
- Little for the taxation department in the Ministry of Finance to do other than change the amount of the employment tax.

However, before adopting this approach, the Ministry should perform a feasibility study focusing on perceptions from the:

- *Quantitative HTP* to determine if the approach would meet the reduction target.
- *Temporal and Generic HTPs* to determine if the proposed increase in the employment tax would significantly increase the number of undocumented alien workers as the employers seek to avoid the tax increases. And if so, what steps would have to be taken to ensure the increase does not happen.

Any implementation approach needs to minimize the opposition to the change from the employers of the maids.

4.10.2.1.4.2.2 THE APPROPRIATE CM MODEL   After some study of the models, it was determined that*:

1. The Kurt Lewin CM model (Section 4.7.1.1) is not suitable in this situation because the decision to make the change has already been made.
2. The Kübler-Ross model (Section 4.7.1.2) is not suitable in this situation because it really seems to deal with individuals. It could be used as part of the analysis of the reaction of the stakeholders expected to oppose the policy during any negotiation and discussion.

---

* The choice of CM model depends on the situation and the change team. This section discusses the reasons for the choice as made by one specific team. A different team might choose a different model which is an alternative acceptable response as long as it is relevant and pertinent.

3. The 7S framework for organization change (Section 4.7.1.3) is not suitable in this situation because it's not an organizational change.
4. Checkland's SSM (Section 4.7.1.4) might be suitable in this situation during the change initialization and planning states to gain an understanding of the various stakeholders' point of view (Weltanschauung).
5. The Satir change model (Section 4.7.1.5) is not suitable in this situation for the same reasons that the Kübler-Ross model (Section 4.7.1.2) is not suitable.
6. The William Bridges transition model (Section 4.7.1.6) is not suitable in this situation for the same reasons that the Kübler-Ross model (Section 4.7.1.2) is not suitable.
7. The Burke-Litwin change model (Section 4.7.1.7) is not suitable in this situation for the same reasons that the Kübler-Ross model (Section 4.7.1.2) is not suitable.
8. Kotter's eight-step change model (Section 4.7.1.8) is suitable because most of it is applicable.
9. The Nudge theory (Section 4.7.1.9) is suitable for the change implementation state.
10. The ADKAR® CM model (Section 4.7.1.10) is not suitable in this situation for the same reasons that the Kurt Lewin CM model (Section 4.7.1.1) is not suitable.

Conclusion: there isn't really one CM model that is suitable for this specific change. Accordingly, a hybrid CM model based on a combination of parts of the Nudge theory (Section 4.7.1.9) and the ADKAR® CM model (Section 4.7.1.10) was created to study the changes during the policy lifecycle.

4.10.2.1.4.2.3 LESSONS LEARNED   Lessons learned from the exercise include:

• The compliance matrix is very handy for keeping track of what needs to be done and making sure that it is being done.
• Changes that affect direct stakeholders can also affect indirect stakeholders. This phenomenon is sometimes known as the domino effect or ripples of change.
• The stakeholders tasked with implementing the change can have a major impact on the change. If they are in favour of the change, the changes will happen speedily and smoothly. However, if they are against the change, then the change will be delayed or may never happen.
• When existing tools don't meet the need, create a new tool based on appropriate parts of existing tools, with some original additions if necessary.

4.10.3 THE ENGAPOREAN DROUGHT RELIEF POLICY

Using the information developed in the previous exercises and in addition to meeting the generic requirements in Section 2.7:

1. Identify the specific change that will take place. For example, this change will increase the amount of water capacity w% by providing a ...
2. Identify at least three direct and indirect stakeholders.
3. Use force field analysis (Section 4.5.2) to determine at least five forces for and at least five forces against the change and estimate their strengths.
4. Discuss which CM model or combination of models (Section 4.7.1) would be most suitable in this situation and why.

4.10.4 THE OFF-WORLD MINING POLICY

Using the information developed in the previous exercises and in addition to meeting the generic requirements in Section 2.7:

1. Identify at least five specific changes that will take place. For example, this change will decrease pollution due to oil spills by 100% by providing an alternate economical way of gathering hydrocarbons from the atmosphere of Jupiter.
2. Identify at least three direct and indirect stakeholders.
3. Use force field analysis (Section 4.5.2) to determine at least five forces for and at least five forces against the change and estimate their strengths.
4. Discuss which CM model or combination of models (Section 4.7.1) would be most suitable in this situation and why.

4.11 SUMMARY

This chapter perceived change and CM from the HTPs because processes in which risks occur generally cause change. In particular, the chapter discussed some CM models, resistance to change and how to overcome it and some aspects of stakeholder management.

REFERENCES

Avison, David, and Guy Fitzgerald. 2003. *Information Systems Development: Methodologies, Techniques and Tools*. Maidenhead: McGraw-Hill Education (UK).
Beer, M., and Nitin Nohria. 2000. Cracking the code of change. *Harvard Business Review* no. 78 (3):131–141.
Box, George Edward Pelham, and Norman R. Draper. 1987. *Empirical Model Building and Response Surfaces*. New York: John Wiley & Sons.
Bridges, William. 1991. *Managing Transitions*. Reading, MA: Addison Wesley.
Burke, W. Warner. 2011. *Organization Change: Theory and Practice*. 3rd edition. Thousand Oaks, CA: Sage.
Burke, W. Warner, and George H. Litwin. 1992. A causal model of organizational performance and change. *Journal of Management* no. 18 (3):523–545.
Campbell, John W. 1960. Editorial: the word and the truth. Astounding Science Fact and Fiction no. LXIV (6):4–7, 176. Street and Smith Publications.
Checkland, Peter. 1991. *Systems Thinking, Systems Practice*. Chichester: John Wiley & Sons.

Checkland, Peter, and Jim Scholes. 1990. *Soft Systems Methodology in Action*. Chichester: John Wiley & Sons.

Connelly, Mark. 2016. *Eight reasons to use a change management model 2015* [cited 29 February 2016]. Available from http://www.change-management-coach.com/change-management-model.html.

Covey, Steven R. 1989. *The Seven Habits of Highly Effective People*. New York: Simon & Schuster.

Craddock, William T. 2015. *Change Management in the Strategic Alignment of Project Portfolios*. Newtown Square, PA: Project Management Institute, Inc.

Deal, Terrence E., and Allan A. Kennedy. 1982. *Corporate Cultures: The Rites and Rituals of Corporate Life*. Reading, MA: Addison-Wesley.

Deevy, Edward. 1995. *Creating the Resilient Organization. A Rapid Response Management Program*. Englewood Cliffs, NJ: Prentice Hall Inc.

Hari, Amihud, Shraga Shoval, and Joseph Eli Kasser. 2008. *Conceptual design to cost: A new systems engineering tool*. In the 18th International Symposium of the INCOSE. Utrecht, Holland.

ISU. 2016. *Force field analysis. Iowa State University extension and outreach 2016* [cited 17 February 2016]. Available from http://www.extension.iastate.edu/communities/force-field-analysis.

Kasser, Joseph Eli. 2002. Configuration management: The silver bullet for cost and schedule control. In the *IEEE International Engineering Management Conference (IEMC 2002)*. Cambridge, UK.

Kasser, Joseph Eli. 2015. *Holistic Thinking: Creating Innovative Solutions to Complex Problems*. Vol. 1, Solution Engineering. 2nd edition. Charleston, SC: Createspace Ltd.

Kasser, Joseph Eli. 2018a. *Systems Thinker's Toolbox: Tools for Managing Complexity*. Boca Raton, FL: CRC Press.

Kasser, Joseph Eli. 2018b. Using the systems thinker's toolbox to tackle complexity (complex problems). In *SSSE Presentation at Roche*. Zurich: Swiss Society of Systems Engineering.

Kasser, Joseph Eli. 2019. *Systems Engineering: A Systemic and Systematic Methodology for Solving Complex Problems*. Boca Raton, FL: CRC Press.

Kirton, Michael J. 1994. *Adaptors and Innovators: Styles of Creativity and Problem Solving*. London: Routledge.

Kotter, John P. 2007. Leading change: Why transformation efforts fail. *Harvard Business Review* no. 73 (1):55–67.

Kotter, John P. 1996. *Leading Change*. Boston, MA: Harvard Business School Press.

Kübler-Ross, Elisabeth. 1969. *On Death and Dying*. New York: MacMillan Company.

Kuhn, Thomas S. 1970. *The Structure of Scientific Revolutions*. Second Edition, Enlarged Edition. Chicago, IL: The University of Chicago Press.

Lano, Robert. 1977. *The N2 Chart*. TRW Software Series. Redondo Beach, CA: TRW Inc.

Lewin, Kurt. 1943. Defining the field at a given time. *Psychological Review* no. 50 (3):292–310.

Lewin, Kurt. 1947. Frontiers of group dynamics: Concept, method and reality in social science, social equilibria, and social change. *Human Relations* no. 1:5–41. doi: 10.1177/001872674700100103.

Lippitt, Gordon 1983. *A Handbook for Visual Problem Solving: A Resource Guide for Creating Change Models*. Bethesda, MD: Development Publications.

Machiavelli, Nicollo. 1515. *The Prince, Translated by W. K. Marriott, Electronically Enhanced Text*. Washington, DC: World Library, Inc. (1991).

Merton, Robert King. 1936. The unanticipated consequences of social action. *American Sociological Review* no. 1 (6):894–904.

Miller, George. 1956. The magical number seven, plus or minus two: Some limits on our capacity for processing information. *The Psychological Review* no. 63:81–97.

Pascale, Richard Tanner, and Anthony G. Athos. 1981. *The Art of Japanese Management: Applications for American Business.* New York: Simon & Schuster.

PMI. 2013. *A Guide to the Project Management Body of Knowledge.* 5th Edition. Newtown Square, PA: Project Management Institute, Inc.

Prosci, Josef. 2016. *ADKAR change management model overview.* Prosci Inc. 2016 [cited 29 February 2016]. Available from https://www.prosci.com/adkar/adkar-model.

Rouse, Margaret 2019. *Change management.* TechTarget 2014 [cited 2 September 2019]. Available from http://searchcio.techtarget.com/definition/change-management.

Satir, Virginia, John Banmen, Jane Gerber, and Maria Gomori. 1991. *The Satir Model: Family Therapy and Beyond.* Palo Alto, CA: Science and Behavior Books.

Stone, Kyle B. 2015. Burke-Litwin organizational assessment survey: Reliability and validity. *Organization Development Journal* no. 33 (2):33–50.

Thaler, Richard H., and Cass R. Sunstein. 2008. *Nudge – Improving Decisions about Health, Wealth and Happiness.* New Haven, CT: Yale University Press.

Waterman Jr., Robert H., Thomas J. Peters, and Julien R. Phillips. 1980. Structure is not organization. *Business Horizons* no. 23 (3):14–26.

5 Problems and Problem-Solving

The risk management process is riddled with changes, problems and solutions because managing risks poses problems which need to be solved or remedied and change is an outcome, mostly desired but sometimes undesired. A risk mitigation strategy is a solution to the problem of managing risk but poses a problem to the people who will have to implement the strategy. Accordingly, this chapter* discusses problems and problem-solving to explain how problem-solving deals with risks. The chapter:

1. Discusses perceptions of problem-solving from the HTPs in Section 5.1.
2. Discusses complexity and how to use the systems approach to manage complexity in Section 5.2 specifying the minimum number of elements in a system for it to be defined as, and managed as, complex.
3. Shows how to remedy well-structured problems in Section 5.3.
4. Shows how to remedy ill-structured and wicked problems in Section 5.4.
5. Shows how to remedy complex problems in Section 5.5.
6. Discusses remedying multiple problems simultaneously in Section 5.6.
7. Discusses generic risks in the problem-solving process in Section 5.7.

5.1 PERCEPTIONS OF PROBLEM-SOLVING

This section explains the problem-solving aspects of making changes by discussing perceptions of problem-solving from the following HTPs:

1. *Big Picture* in Section 5.1.1.
2. *Quantitative* in Section 5.1.2.
3. *Structural* in Section 5.1.3.
4. *Continuum* in Section 5.1.4.
5. *Functional* in Section 5.1.5.
6. *Operational* in Section 5.1.6.
7. *Scientific* in Section 5.1.7.

5.1.1 BIG PICTURE

Perceptions of problem-solving from the *Big Picture* HTP include:

1. Some assumptions underlying formal problem-solving in Section 5.1.1.1.
2. Selected myths about problem-solving in Section 5.1.1.2.

---

* This chapter is a modified version of Chapter 3 in *Systems Engineering* (Kasser 2019b).

5.1.1.1   Assumptions Underlying Formal Problem-Solving

Problem-solving like most other things is based on a set of assumptions. Waring provided the following four assumptions underlying formal problem-solving (Waring 1996):

1. The existence of the problem may be taken for granted.
2. The structure of the problem can be simplified or reduced so as to make its definition, description and solution manageable.
3. Reduction of the problem does not reduce the effectiveness of the solution.
4. Selection of the optimal solution* is a rational process of comparison.

Waring seems to be discussing well-structured problems (Section 5.1.4.5.1). With respect to Waring's first point, while the existence of the problem may be taken for granted, it may take a while for the stakeholders to agree on the nature of the problem because different stakeholders may have a different Weltanschauung and may view the same set of facts through different cognitive filters and perceive different undesirable elements. The literature on decision-making, one of the key elements in problem-solving, has two schools of thought on Waring's fourth point:

1. *Agree*: decision-making is logical.
2. *Disagree*: decision-making is emotional.

The systems approach eschews the either–or view, rather perceiving decision-making as being spread along a continuum with 100% logical at one end and 100% emotional at the other end. The middle of the continuum is 50% emotional and 50% logic. Each decision is made using a mixture of emotion and logic. Different types of decisions are made at different places on the continuum. Moreover, a decision that is made mostly logically one day may be made mostly emotionally under a different set of circumstances.

5.1.1.2   Selected Myths about Problem-Solving

There are a number of myths about problem-solving that hinder problem-solving and need to be exposed (Kasser and Zhao 2016a). These myths include:

1. The word 'problem' has an unambiguous meaning discussed in Section 5.1.1.2.1.
2. The fixation on a single correct solution discussed in Section 5.1.1.2.2.
3. All problems can be solved discussed in Section 5.1.1.2.3.
4. The traditional generic problem-solving process is a linear time-ordered sequence discussed in Section 5.1.1.2.4.
5. One problem-solving approach can solve all problems discussed in Section 5.1.1.2.5.

---

* Decision-making.

FIGURE 5.1 The single correct solution.

5.1.1.2.1 The Word 'Problem' Has an Unambiguous Meaning

The myth is that the word 'problem' has an unambiguous meaning. This is an incorrect assumption. The reality is that the many definitions in the literature of the word 'problem' can be grouped into the following three different meanings:

1. A problem is a question or matter involving doubt, uncertainty or difficulty. For example:
   1. 'Making this presentation is going to be a problem'.
   2. 'Getting my wife to listen to me is a problem'.
   3. The need to determine the necessary sequence of activities to transform an initial undesirable situation into a desirable situation.*
   4. Mitigating this risk is a problem.
2. A problem is a question proposed for solution or discussion. For example,
   1. 'What are 1+1?'
   2. 'What time shall we have dinner?'
   3. 'Is my suitcase overweight?'
   4. How do we mitigate this risk?
3. A problem is the underlying cause of an undesirable situation. For example:
   1. Someone ends a sentence with '... and that's the problem' when they mean '... and that's the undesirable situation' or 'and that's the cause'.
   2. 'Policy problems are unrealized needs, values, or opportunities for improvement' (Dery 1984) cited by Dunn (2012: p. 67).

5.1.1.2.2 The Fixation on a Single Correct Solution

The myth is that there is always a single correct solution. This is an incorrect assumption. The reality is that perceptions from the *Continuum* HTP show that most of the time there is more than one acceptable solution.

In school, generally, we are taught to solve problems by being given a problem and then asked to find the solution as shown in Figure 5.1. The assumption is that there is a well-structured problem (Section 5.1.4.5.1) with a single well-defined correct solution. This is a myth that does not apply in the real world.

The reality is that risk management deals with undesirable situations that generally have more than one equally acceptable solution. For example, Fred is hungry which is generally an undesirable situation. The problem is to figure out a way to remedy that undesirable situation by consuming some food to satisfy the hunger. There are a number

---

* Once the necessary sequence of activities is determined, the subsequent problem is to plan the process to perform the necessary sequence of activities. Once the plan is created, the subsequent problem is to realize the desirable situation by carrying out the plan.

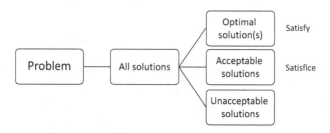

FIGURE 5.2   The full range of solutions.

of solutions to this problem including cooking something, going to a restaurant, collecting some takeaway food and telephoning for home delivery. Then there is the choice of what type of food: Italian, French, Chinese, pizza, lamb, chicken, beef, fish, vegetarian, etc. Now consider the vegetables, sauces and drinks. There are many solutions because there are many combinations of types of food, meat, vegetables and method of getting the food to the table. Which solution is the correct one? The answer is that the correct solution is the one that satisfies his hunger in a timely and affordable manner.[*] If several of the solution options can perform this function and he has no preference between them, then each of them is just as correct or acceptable as any of the other options that satisfy his hunger. The words 'right solution' or 'correct solution' should be thought of as meaning 'one or more acceptable solutions' as shown in Figure 5.2.

Conventional wisdom based on the mythical need for a single correct solution suggests that when a decision cannot be made because two choices score almost the same in the decision-making process, the decision-maker should perform a sensitivity analysis by varying the parameters and/or the weighting to see if the decision changes. By recognizing the reality that there may be more than one acceptable solution, the 'don't care which solution option is chosen' may eliminate the need for the sensitivity analysis.

Figure 5.2 can also be used to explain the dictionary definitions of 'satisfy' and 'satisfice' where:

- *Satisfy* means provide solutions that are optimal.
- *Satisfice* means provide solutions that are acceptable.

However, this still leaves open the question of who defines the problem.

5.1.1.2.3   All Problems Can Be Solved

One of the myths associated with the problem-solving process is that all problems can be solved. The reality is that:

1. *Problems are solved, resolved, dissolved or absolved*: where only the first three actually remedy the problem (Ackoff 1978). The word 'solve' is often misused in the literature to mean solved, resolved or dissolved, when a better word is 'remedy'. The four ways of dealing with a problem are:

---

[*] And does not cause any gastric problems.

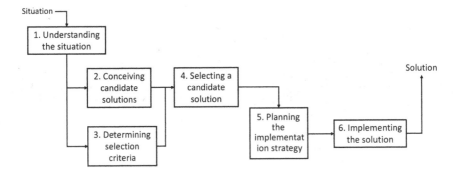

FIGURE 5.3   The traditional generic problem-solving process.

1. *Solving*: when the decision-maker selects those values of the control variables (Section 5.1.2.1) which maximize the value of the outcome (satisfies the need, an optimal solution).

2. *Resolving*: when the decision-maker selects values of the control variables which do not maximize the value of the outcome but produce an outcome that is good enough or acceptable (satisfices the need, one acceptable solution).

3. *Dissolving*: when the decision-maker reformulates the problem to produce an outcome in which the original problem no longer has any actual meaning. Dissolving the problem generally leads to innovative solutions.

4. *Absolving*: when the decision-maker ignores the problem or imagines that it will eventually disappear on its own. Problems may be intentionally ignored for reasons that include:

   1. They are too expensive to remedy.
   2. The technical or social capability needed to provide a remedy is unavailable; it may not be known, affordable or available.

2. Only well-structured problems (Section 5.1.4.5.1) can be remedied.

5.1.1.2.4   The Traditional Generic Problem-Solving Process Is a Linear Time-Ordered Sequence

The traditional generic problem-solving process (Section 5.1.6.1) is taught as a linear time-ordered sequence as shown in Figure 5.3 (Kasser 2019b). This is simple to teach but incorrect or a myth in the real world. For example, consider what happens if:

- None of the solutions options conceived in Step 2 meet the selection criteria conceived in Step 3 when performing Step 4 of the generic problem-solving process shown in Figure 5.3.
- None of the solution options conceived in Step 2 remedies the undesirable situation.
- All the solutions options conceived in Step 2 are too expensive or will take too long to realize or unacceptable for any other reason.

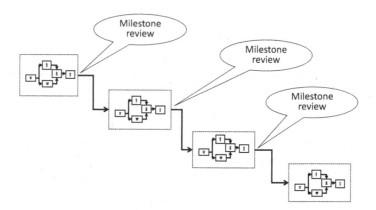

FIGURE 5.4   Iteration of the problem-solving process across the SDP in an ideal project.

Then the choices to be made include:

- Absolve the problem (Section 5.1.1.2.3) for a while until something changes.
- Absolve parts of the problem. Decide to remedy parts of the undesirable situation, sometimes known as reducing the requirements, until the remedy is feasible. Ways of doing this include:
  - Removing the lower priority aspects of the undesirable situation and determining the new cost/schedule information until the solution option becomes affordable or can be realized in a timely manner. This is a holistic approach to the concept of designing to cost and used in conceptual design to cost (Hari, Shoval, and Kasser 2008) (Kasser 2019a: Section 8.3.1) and in cost as an independent variable (CAIV) (Rush 1977).
  - Remedy the causes of undesirability with the highest priorities and absolve the problem posed by the remaining causes.
  - Continue to look for an acceptable feasible solution that will remedy the undesirable situation in a timely manner.

The reality is that the problem-solving process is iterative in several different ways including:

1. *Iteration of the problem-solving process across the SDP*: each state of the SDP (Section 6.1.5.4) contains the problem-solving process shown in Figure 5.3. The traditional waterfall chart does not show the inside of the blocks in the waterfall; however, if the blocks inside the waterfall chart were visible, a partial waterfall would be drawn as shown in Figure 5.4* (Kasser 2019b).
2. *Iteration of the problem-solving process within a state in the SDP*: each state in the SDP (Section 6.1.5.4) contains a problem-solving process with two exit conditions:

---

* Note since the figure contains two levels of the hierarchy, it should only be used to show the repetition of the problem-solving process in each state of an ideal SDP – one in which no changes occur.

1. The normal planned exit at the end of state.
2. An anticipated abnormal exit anywhere in the state that can happen at any time in any state and necessitates either a return to an earlier state or a move to a later state and skipping intermediate states.

In addition, the entire SDP may map into a single iteration of the problem-solving process.

5.1.1.2.5 One Problem-Solving Approach Can Solve All Problems

The reality is that there are different types of problems that need different versions of the problem-solving process, for example:

- Research and intervention problems discussed in Section 5.1.4.4.
- Well-structured problems discussed in Section 5.1.4.5.
- Ill-structured problems discussed in Section 5.1.4.5.2.
- Wicked problems discussed in Section 5.1.4.5.3.
- Complex problems discussed in Section 5.5.

5.1.2 QUANTITATIVE

Perceptions of problem-solving from the *Quantitative* HTP include the five components of a problem.

5.1.2.1 Components of Problems

The five components of a problem (Ackoff 1978: pp. 11–12) are:

1. *The decision-maker*: the person faced with the problem.
2. *The control variables*: aspects of the problem situation the decision-maker can control.
3. *The uncontrolled variables*: aspects of the problem situation the decision-maker cannot control which constitute the problem environment.[*] The uncontrolled variables may give rise to unanticipated negative emergent properties of the solution often called undesirable outcomes.
4. *Constraints*: imposed from within or without on the possible values of the controlled and uncontrolled variables.
5. *The possible outcomes*: desired, undesired and don't care produced jointly by the decision-maker's choice and the uncontrolled variables[†] (Section 5.1.4.3). The desired outcome may be represented in several ways including:
   - A specified relationship between the controlled variables and the uncontrolled variables.
   - A design or architecture.
   - An FCFDS (Section 5.1.7.2).

---

[*] There may be unknown uncontrolled variables; see Simpson's paradox (Savage 2009).
[†] Desired and undesired.

5.1.3   STRUCTURAL

Perceptions of problem-solving from the *Structural* HTP include:

1. The level of difficulty of a problem discussed in Section 5.1.3.1.
2. Various definitions of complexity discussed in Section 5.2.1.1.

5.1.3.1   The Level of Difficulty of the Problem

Ford introduced four categories of increasing order of difficulty (subjective complexity) (Section 5.2.2.2) for mathematics and science problems: easy, medium, ugly and hard (Ford 2010). These categories may be generalized and defined as follows (Kasser 2015b):

1. *Easy*: problems that can be solved in a short time with very little thought.
2. *Medium*: problems that:
   • Can be solved after some thought.
   • May take a few more steps to solve than an easy problem.
   • Can probably be solved without too much difficulty, perhaps after some practice.
3. *Ugly*: problems are ones that will take a while to solve. Solving them:
   • Involves a lot of thought.
   • Involves many steps.
   • May require the use of several different concepts.
4. *Hard*: problems usually involve dealing with one or more unknowns. Solving them:
   • Involves a lot of thought.
   • Requires some research.
   • May also require iteration through the problem-solving process as learning takes place (knowledge that was previously unknown becomes known (Section 5.2.4.2)).

Classifying problems by level of difficulty is difficult in itself because difficulty is subjective since one person's easy problem may be another person's medium, ugly or hard problem. For example, consider an undesirable situation faced by Fred who arrives in a foreign country for a visit and lodges in an apartment where he has to do his own cooking. As Fred cannot speak the local language, he is in a number of undesirable situations. Consider the one in which the kitchen has a gas cooker, but he has no way to ignite the gas. The corresponding desirable situation is that Fred has something to ignite the gas.[*] Assuming Fred has local currency or an acceptable credit card, is the difficulty of the problem of purchasing something that will ignite the gas easy, hard or something in between? The answer is 'it depends'. Classifying the difficulty of the problem depends on a number of issues including:

---

[*] Note the use of functional language 'ignite' instead of solution language 'matches' in solution.

- If Fred has faced this problem before in the same country. If so, what did he do then?
- If Fred knows where to purchase matches or a gas lighter.
- If Fred even knows how to say 'matches' or 'gas lighter' in the local language. If he does not know the words, he may not be able to ask anyone to provide the items.

Thus, as far as Fred is concerned, the problem is:

- *None existent*: if Fred already has matches, a gas lighter, a cigarette lighter or another instrument with which to light the gas.
- *Easy*: if Fred knows where to purchase matches or a gas lighter and knows the local words.
- *Medium*: if Fred knows where to purchase matches or a gas lighter and does not know the local words. After all, he can go to the store or relevant location and look around until he sees matches or lighters on a shelf and then purchase them.
- *Hard*: if Fred does not know where to purchase matches or a gas lighter and does not know the local words. The problem is hard because two unknowns have to become known for a solution to be realized.

5.1.4 CONTINUUM

Perceptions of problem-solving from the *Continuum* HTP included:

1. The difference between problems and symptoms discussed in Section 5.1.4.1.
2. The difference between the quality of the decision and the quality of the outcome discussed in Section 5.1.4.2.
3. The different decision outcomes discussed in Section 5.1.4.3.
4. The difference between research and intervention problems discussed in Section 5.1.4.4.
5. The different categories of problems discussed in Section 5.1.4.5.
6. The different domains of problems discussed in Section 5.1.4.6.
7. The system implementation continuum discussed in Section 5.1.4.7.
8. The four levels of difficulty of a problem discussed in Section 5.1.3.1.
9. Some decisions are made emotionally and others are made logically discussed in Section 5.1.7.1.[*]

5.1.4.1 Problems and Symptoms

Perceiving the difference between problems, symptoms and causes from the *Continuum* HTP, the undesirable situation manifests itself as symptoms which are used to diagnose the underlying problem. Having diagnosed the problem, action is then taken to remedy the problem. The traditional problem-solving feedback

---

[*] And the same person can make the same decision emotionally at one time and logically at another time.

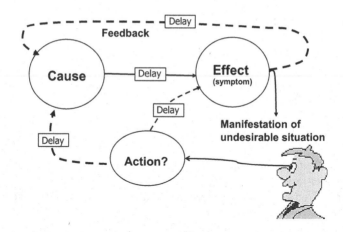

FIGURE 5.5 Problems, causes and effects (symptoms).

approach is represented by the causal loop shown in Figure 5.5 (Kasser 2002a). An action can tackle the problem or a symptom. If the root cause of the problem is not found, a solution may not work or may only work for a short period of time. In addition, even if the implemented solution works, it may introduce further problems (causes of undesirability) that only show up after some period of time (Section 4.2.3).

The analysis of the requested change has to consider all of the above as applicable. While the higher order time delays may not be applicable in a computer-based system, they are applicable in long-lived systems such as large engineering projects (LEPs) (Chapter 8). Sometimes, a second action is taken before the effect of the first one is observed leading to the need for a further action to remedy the effect of the second one. Sometimes, the action partially remedies the problem; sometimes, the action only mitigates the symptoms and produces a new undesirable situation.

5.1.4.2 The Difference between the Quality of the Decision and the Quality of the Outcome

Perceptions of decision-making from the *Continuum* HTP note, 'We need to differentiate between the quality of the decision and the quality of the outcome' (Howard 1973: p. 55). A good decision can lead to a bad outcome and conversely a bad decision can lead to a good outcome. The quality of the decision is based on doing the best you can to increase the chances of a good outcome, hence the development and use of decision-making tools. Decisions can be made using quantitative and qualitative methods (Kasser 2018a: Chapter 4).

5.1.4.3 The Different Outcomes and Consequences of Decisions

All decision outcomes have consequences. Perceptions from the *Generic* HTP include:

1. The consequences of decisions are generally the same as the consequences of changes (Section 4.4) because decisions when implemented generally cause changes.

TABLE 5.1
Decision Table for Known Outcomes of Actions

	Certain	Uncertain	Certain
Probability of occurrence	0% (will never happen)	0% < 100% (might happen)	100% (will always happen)
Desired	Need to conceptualize an alternative action	Opportunity that should be planned for, depending on the probability of occurrence	Preferred outcome
Don't care	Ignore	Opportunity that might be considered depending on the probability of occurrence	Opportunity that could be taken advantage of
Undesired	Can be ignored	Risk that should be mitigated depending on the probability of occurrence and the severity of consequences	Outcome that must be prevented or mitigated depending on the severity of consequences

Perceptions from the *Continuum* HTP include:

1. Outcomes and consequences lie on a probability of possibilities continuum as shown in Table 5.1 ranging from 0% to 100% where an outcome with a probability of occurrence of 100% is a certain outcome and an outcome of 0% is one that is never going to happen (negative certainty). Anything in between is an uncertain outcome. The difference between certain and uncertain outcomes is:
   - *Certain*: is deterministic since you can determine what the outcome will be before it happens. For example, if you toss a coin into the air, you are certain that it will come down* and come to rest with one side showing if it lands on a hard surface.
   - *Uncertain*: is non-deterministic since while you know there may be more than one possible outcome from an action, you can't determine which one it will be. For example, if you toss a coin, the outcome is non-deterministic or uncertain because while you know that the coin will show one of two sides when it comes to rest, you cannot be sure which side will be showing. However, you could predict one side with a 50% probability of being correct.†
2. Outcomes and consequences can be anticipated and unanticipated where:
   - *Anticipated*: can be:
     - *Desired*: where the result is something that you want. For example, you want the coin to land showing 'heads' and it does.
     - *Undesired*: where the result is something that you don't want. For example, you don't want the coin to land showing 'tails' and it lands

---

* Unless you toss it so fast that it escapes from the Earth's gravity.

† So how can tossing a coin be certain and uncertain at the same time? It depends on the type of outcome you are looking for. Which side it will land on is uncertain; but that it will land on a side is certain. It is just a matter of framing the issue from the proper perspective.

showing 'tails'. This type of outcome and its consequences are known as risks before they occur and events once they have occurred.

- – *Don't care*: where you have no preference for the result. For example, if you have no preference as to which side is showing when the coin lands, you have a 'don't care' situation (Kasser 2018a: Section 3.2.2).
- • *Unanticipated*: can also be desired, undesired and don't care once discovered.

There can also be more than one outcome and consequence from an action; for example, each of the outcomes may be:

- • Dependent on, or independent from, the other outcomes.
- • Acceptable or not acceptable.
- • Desired, undesired or 'don't care' (Kasser 2018a: Section 3.2.2).
- • Unanticipated the first time that the action is taken.
- • A combination of the above.

Table 5.1 shows the links between the known outcomes of decisions and uncertainty. Many of the decision-making tools in the literature deal with the decisions being made in the desired certain area. Don't care outcomes should not be neglected but should be looked at as opportunities. For example, if you are considering purchasing a commercial-off-the-shelf (COTS) item and initially don't care about the colour, then from the holistic perspective you might want to think about what additional benefits you might get from a specific colour.

Typical active brainstorming questions (Kasser 2018a: Section 7.1) when considering risks include:

- • What if it is late?
- • What if it performs below specification?
- • What if it fails before the specified time?

When considering opportunities, the questions would be the opposite of those asked from the perspective of risk, namely:

- • What if it is early?
- • What if it performs above specification?
- • What if it lasts longer than specified?

The answers and the resulting actions taken would depend on the situation.

5.1.4.3.1   Sources of Unanticipated Consequences or Outcomes of Decisions
Unanticipated consequences or outcomes of decisions need to be avoided or minimized.[*] In the systems approach, if we can identify the causes of unanticipated consequences, we should be able to prevent them from happening. A literature search found Merton's analysis which discussed the following five sources of unanticipated consequences in social interventions (Merton 1936):

---

[*] This also applies to unanticipated emergent properties (*Generic* HTP).

1. Ignorance.
2. Error.
3. Imperious immediacy of interest.
4. Basic values.
5. Self-defeating predictions.

These sources may be generalized as discussed below.

1. *Ignorance*: deals with unanticipated consequences or outcomes due to the type of knowledge that is missing or ignored in making the decision. Ignorance in the:
   - *Problem domain*: may result in the identification of the wrong problem.
   - *Solution domain*: may produce a solution system that will not provide the desired remedy.
   - *Implementation domain*: may produce a conceptual solution that cannot be realized.
2. *Error*: there are two types of errors: errors of commission and errors of omission (Ackoff and Addison 2006) where:
   1. *Errors of commission*: do something that should not have been done. There are also two types of errors of commission: design errors and implementation errors.
      1. *A design error*: an error which produces an undesired outcome. For example:
         - *A logic error in a computer program*: which can be found by comparing the software source code to the specification.
         - *A hardware error*: which can be found by conceptualizing what the part as designed would do in operation and comparing it to the specification for operation. A difference represents an error.
      2. *An implementation error*: a mistake was made in creating the design. For example, a syntax error in a computer program, a failure to test something under realistic operating conditions, the wrong part was installed, or a part was installed backwards. This type of error is usually found by:
         - Compiling the software source code and allowing the compiler to point out the syntax errors.
         - Visual inspection of the parts.
         - Testing the design against specifications in the design state of the SDP.
         - Subsystem testing.
         - Using proven test procedures.
   2. *Errors of omission*:
      1. Fail to do something that should have been done such as in instances where only one or some of the pertinent aspects of the situation which influences the solution are considered. This can range from the case of simple neglect (lack of systematic thoroughness in examining the situation) to 'pathological obsession where there is a

determined refusal or inability to consider certain elements of the problem' (Merton 1936).

2. Are more serious than errors of commission because, among other reasons, they are often impossible or very difficult to correct. 'They are lost opportunities that can never be retrieved' (Ackoff and Addison 2006: p. 20). Merton adds that a common fallacy is the too-ready assumption that actions, which have in the past led to a desired outcome, will continue to do so. This assumption often, even usually, meets with success. However, the habit tends to become automatic with continued repetition so that there is a failure to recognize that procedures, which have been successful in certain circumstances, need not be successful under any and all conditions.*

3. *Imperious immediacy of interest*: the paramount concern with the foreseen immediate consequences excludes the consideration of further or other consequences of the same act, which does in fact produce errors.†

4. *Basic values*: there is no consideration of further consequences because of the felt necessity of certain action enjoined by certain fundamental values. For example, the Protestant ethic of hard work and asceticism paradoxically leads to its own decline in subsequent years through the accumulation of wealth and possessions.

5. *Self-defeating predictions*: the public prediction of a social development proves false precisely because the prediction changes the course of history. Merton later conceptualized the 'the self-fulfilling prophecy' (Merton 1948) as the opposite of this concept.

5.1.4.4 Research and Intervention Problems

Perceptions from the *Continuum* HTP indicate a difference between research and intervention problems (Kasser and Zhao 2016a). Consider both of them.

5.1.4.4.1 Research Problems

This type of problem manifests when the undesirable situation is the inability to explain observations of phenomena or the need for some particular knowledge. In this situation, using the problem formulation template (Section 5.1.7.2):

1. *The undesirable situation*: the inability to explain observations of phenomena or the need for some particular knowledge.

2. *The assumptions*: the research is funded and the researcher is an expert in the field.

3. *The FCFDS*: the outcome is the ability to explain observations of phenomena or the particular knowledge.

4. *The problem*: how to gain the needed knowledge.

---

* This assumption also applies to component reuse.
† Perceptions from the *Generic* HTP perceive the similarity to the decision traps (Section 5.1.5.3).

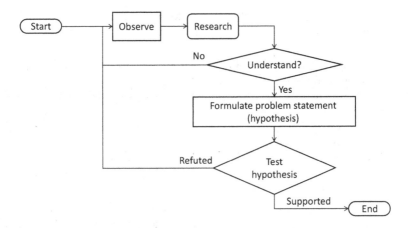

FIGURE 5.6   The scientific method.

5. *The solution*: use the scientific method which works forwards from the current situation in a journey of discovery towards a future situation in which
the knowledge has been acquired.

The scientific method:

- Is a variation of the generic problem-solving process.
- Is summarized in Figure 5.6 (Kasser 2019b).
- Is a systemic and systematic way of dealing with open-ended research
problems.
- Has been stated as different variations of the following sequence of activities:
  1. Observe an undesirable situation.
  2. Perform research to gather preliminary data about the undesirable
  situation.
  3. Formulate the hypothesis to explain the undesirable situation using
  inductive reasoning (Kasser 2018a: Section 5.1.3.1.2).
  4. Plan to gather data to test the hypothesis. The data gathering may take
  the form of performing an experiment, using a survey, reviewing literature or some other approach depending on the nature of the undesirable
  situation and the domain.
  5. Perform the experiment or otherwise gather the data.
  6. Analyse the data (experimental or survey results) using deductive reasoning (Kasser 2018a: Section 5.1.3.1.1) to test the hypothesis.
  7. If the hypothesis is supported, then the researcher often publishes the
  research. If the hypothesis is not supported, then the process reverts
  back to Step 1.

In the real world, the hypothesis is often created from some insight or a 'hunch'
in which the previous steps are performed subconsciously. The researcher then

designs the data collection method, collects and examines the data to determine if the hypothesis is supported. In this situation:

- The publication is generally written as if the steps in the scientific method have been performed as described above.
- Half the data may be used in defining the hypothesis and half the data may be used in testing the hypothesis.
- There is also an unfortunate tendency to ignore or explain away data which do not support the hypothesis for reasons that include:
  - The researcher may only be looking for data to support the hypothesis. See the decision traps (Section 5.1.5.3) – factors that lead to bad decisions in (Kasser 2018a: Section 8.3).
  - The data sample may be defective. It is important to verify such data, because if the data are valid, they may indicate an instance of Simpson's paradox (Savage 2009) (Kasser 2018a: Section 4.3.2.6.12) and provide the opportunity for further research which could lead to the identification of one or more previously unknown and accordingly unconsidered variables in the situation which would then provide a better understanding and perhaps a Nobel prize or equivalent in the specific research domain.

5.1.4.4.2   Intervention Problems

This type of problem manifests when a current real-world situation is deemed to be undesirable and needs to be changed over a period of time into an FCFDS. In this situation, using the problem formulation template (Section 5.1.7.2):

1. *The undesirable situation*: may be a lack of some functionality that has to be created or some undesirable functionality that has to be eliminated.
2. *The assumptions*: about the situation, constraints, resources, etc.
3. *The FCFDS*: one in which the undesirable situation no longer exists*.
4. *The problem*: how to realize a smooth and timely transition from the current situation to the FCFDS minimizing resistance to the change.
5. *The solution*: create and implement the lowest risk transition process to move from the undesirable situation to the FCFDS together with the solution system operating in the situational context.

The decision-maker or problem-solver is faced with an undesirable situation. Once given the authority to proceed:

1. The decision-maker should use the relevant starter questions from the *Generic* HTP for active brainstorming (Section 2.5.2) to determine if anyone else has faced the same or a similar problem, what they did about it, what results they achieved and the similarities and differences between

---

* And may have improvements to make it even more desirable.

their situation and the current situation. This is the concept behind TRIZ[*] (Barry, Domb, and Slocum 2007) and the copycat systems thinking tool.

2. The decision-maker uses the research problem-solving process (Section 5.1.4.4.1) to conceptualize a vision of the solution system operating in the FCFDS, which becomes the target or goal to achieve.
3. Then the problem the decision-maker faces is to create the transition process and the solution system that will be operational in the FCFDS.
4. The decision-maker uses imagination and visualization (Kasser 2018a: Section 7.11) to work backwards from the FCFDS to the present undesirable situation creating the transition process.
5. The decision-maker then documents the process in the project plan (Section 7.4.2.1) as a sequential process working forwards from the present undesirable situation to the FCFDS. Risk management is incorporated into the project plan. This version of the problem-solving process is a variation of the SDP (Section 6.1.5.4).

5.1.4.5   Different Structures of Problems

Perceived from the *Continuum* HTP, problems lie on a continuum of structures which range from 'well-structured' through 'ill-structured' to 'wicked'. Consider each of them.

5.1.4.5.1   Well-Structured Problems

Well-structured problems are problems where the existing undesirable situation and the FCFDS are clearly identified. These problems may have a single solution or sometimes more than one acceptable solution (Section 5.1.1.2.2). Examples of well-structured problems with single correct solutions are:

- Mathematics and other problems posed by teachers to students in the classroom. For example, in mathematics, $1+1=2$ every time.
- Making a choice between two options. For example, choosing between drinking a cup of coffee and drinking a cup of tea. However, the answer may be different each time.
- Finding the cheapest airfare between Singapore and Jacksonville, Florida, if there is only one cheapest fare. However, the answer may be different depending on the time of the year.

Examples of well-structured problems with several acceptable but different solutions are:

- What brand of coffee to purchase? Although the solution may depend on price, taste and other selection criteria, there may be more than one brand (solution) that meets all the criteria.
- Which brand of automated coffee maker to purchase?

---

[*] TRIZ is the Russian acronym for 'teoriya resheniya izobretatelskikh zadatch' which has been translated into English as 'theory of inventive problem-solving'.

- What type of transportation capability to acquire?
- Finding the cheapest airfare between Singapore and Jacksonville, Florida, if two airlines charge the same fare.

Well-structured problems:

- May be formulated using the problem formulation template (Section 5.1.7.2).
- With single solutions tend to be posed as closed questions.
- With multiple acceptable solutions tend to be posed as open questions.

The traditional problem-solving approach for well-structured complex problems elaborates a complex problem into a number of non-complex problems so that when the non-complex problems have been remedied,[*] the complex problem has also been remedied.

5.1.4.5.2   Ill-Structured Problems

Ill-structured problems:

- Sometimes called 'ill-defined' problems are problems where either or both the existing undesirable situation and the FCFDS are unclear (Jonassen 1997).
- Cannot be solved (Simon 1973); they have to be converted to one or more well-structured problems.

5.1.4.5.3   Wicked Problems

Wicked problems, also known as 'messy' problems,[†] are extremely ill-structured problems[‡] first stated in the context of social policy planning (Rittel and Webber 1973). Wicked problems:

- Cannot be easily defined so that all stakeholders cannot agree on the problem to solve.
- Require complex judgements about the level of abstraction at which to define the problem.
- Have no clear stopping rules (since there is no definitive 'problem', there is also no definitive 'solution', and the problem-solving process ends when the resources, such as time, money or energy, are consumed, not when some solution emerges).
- Have better or worse solutions, not right and wrong ones.
- Have no objective measure of success.
- Require iteration – every trial counts.
- Have no given alternative solutions – these must be discovered.
- Often have strong moral, political or professional dimensions.

---

[*] Which may require a number of iterations.
[†] When complex.
[‡] Technically, there is no problem since while the stakeholders may agree that the situation is undesirable, they cannot agree on the cause and the remedy.

5.1.4.6 Different Domains of a Problem

Remedying a problem requires competency in three different domains, namely (Kasser 2015a):

1. *Problem domain*: the situation in which the need for the activity has arisen.
2. *Solution domain*: the situation and solution system that will be created as a result of the activity.
3. *Implementation domains*: the environment in which the activity (research or transition process) is being performed.

It is tempting to assume that the three domains are the same, but they are not necessarily so. For example, the problem domain may be urban social congestion, while the implementation domain is tunnel boring and the solution domain may be a form of underground transportation system to relieve that congestion. Accordingly:

1. *Lack of problem domain competency*: may lead to the identification of the wrong problem.
2. *Lack of implementation domain competency*: may lead to schedule delays due to preventable problems.
3. *Lack of solution domain competency*: may lead to selection of a less than optimal, or even an unachievable, solution system.

5.1.4.7 The Technological System Implementation Continuum

When considering candidate designs for a technological system, each candidate will lie on a different point on the implementation continuum with a different mixture of people, technology, a change in the way something is done, etc. At one end of the implementation continuum is a completely manual solution; at the other end, is a completely automatic solution. The concept of designing a number of solutions and determining the optimal solution, which may be either one of the solutions or a combination of parts of several solutions, comes from the *Continuum* HTP and performs a risk management function since a benefit of producing several solutions is that one of the design teams conceptualizing the solutions may pick up on matters that other teams missed.

A benefit of recognizing the system implementation continuum is that the system can be delivered in builds wherein Build 1 is manual and the degree of automation increases with each subsequent build.

5.1.5 FUNCTIONAL

Perceptions of problem-solving from the *Functional* HTP included:

1. Decision-making discussed in Section 5.1.5.1.
2. Decision-making tools discussed in Section 5.1.5.2.
3. Decision traps discussed in Section 5.1.5.3.

5.1.5.1 Decision-Making

The most important use of systems thinking in problem-solving is in decision-making. When thinking about anything, we perceive it from different viewpoints, process the information and then infer a conclusion. That inference process is a decision-making process.

Since policy creators and implementers, project managers and systems engineers spend a lot of their time making decisions, understanding the decision-making process will help them to make better decisions or lower the risk of making bad decisions.

The decision-making process is the front end (Steps 2–4) of the traditional generic problem-solving process (Steps 2–6) shown in Figure 5.3* based on Hitchins (2007: p. 173). Perceived from the decision-making perspective, Figure 5.3 depicts the series of activities which are performed in series and parallel that transform the undesirable situation into the strategies and plans to realize the solution system operating in its context.

1. The milestone to start the problem-solving process.
2. The authorization to make the decision.
3. The process to define the problem.
4. The process to conceive several solution options.
5. The process to identify ideal solution selection criteria.
6. The process to perform trade-offs to find the optimum solution.
7. The process to select the preferred option.
8. The process to formulate strategies and plans to realize the selected option.

For a more detailed discussion on decision-making and decision-making tools including decision traps (Section 5.1.5.3), see *Systems Thinker's Toolbox: Tools for Managing Complexity* (Kasser 2018a: Chapter 4).

Note:

- *Risks, opportunities and benefits are selection criteria for decisions*: the degree of risks and benefits are solution selection criteria allowing the degree of susceptibility to a specific risk or the possibility of taking advantage of a specific opportunity to be evaluated in conjunction with other criteria.
- *Make decisions using advice from appropriate personnel*: making decisions about the probability and severity of risks requires knowledge in the problem, solution and implementation domains. This means that even though the project manager makes the decision or approves and is responsible for the decision, the decision-making team needs to include personnel with the appropriate domain knowledge. It also helps to avoid the decision traps (Section 5.1.5.3) and unanticipated negative consequences.

---

* Hitchins' version of the problem-solving process has been modified to add milestones at the beginning and end of the process.

5.1.5.2 Decision-Making Tools

The literature tends to discuss each tool and decision-making approach as being used in an either–or case, namely use one tool or the other to make a decision. However, in the real world we use perceptions from the *Continuum* HTP to develop a mixture of tools or parts of tools as appropriate. For example, determination of the selection criteria for the decision is often a subjective approach, even when those criteria are later used in a quantitative manner. For a more detailed discussion of decision-making, see *Systems Thinker's Toolbox: Tools for Managing Complexity* (Kasser 2018a: Chapter 4).

5.1.5.3 Decision Traps

Russo and Schoemaker provided ten decision traps or risk factors that lead to bad decisions (Russo and Schoemaker 1989). They are mainly due to poor critical thinking and need to be avoided. In summary, they are:

1. *Plunging in*: jumping to conclusions[*] without taking the time to think through the issue, gather information and reach an informed decision.
2. *Frame blindness*: defining the wrong problem.
3. *Lack of frame control*: failing to define the problem in more than one way or being unduly influenced by other people's frames.[†]
4. *Overconfidence in your judgement*: failing to collect key facts because you are sure of your assumptions and opinions.[‡]
5. *Short-sighted shortcuts*: relying inappropriately on 'rules of thumb' such as implicitly trusting the most readily available information or anchoring too much on convenient facts.
6. *Shooting from the hip*: believing that you can keep all the information in your head and therefore 'winging it' rather than following a systematic procedure when making the final choice.
7. *Group failure*: assuming that with many smart people involved, good choices will follow automatically, and therefore failing to manage the group decision-making process.[§]
8. *Fooling yourself about feedback*: failing to interpret the evidence from the past outcomes for what it really is,[¶] either because you are protecting your ego or because you are tricked by hindsight.
9. *Not keeping track*: assuming that experience will make its lessons available automatically, and therefore failing to keep systematic records to track the results of your decisions and failing to analyse these results in ways that reveal their key lessons learned.
10. *Failure to audit your decision-making process*: failing to create an organized approach to understanding your own decision-making, so you remain constantly exposed to repeating the previous nine mistakes.

---

[*] Also known as 'jumping the gun'.
[†] Also known as 'cognitive filters'.
[‡] Perceptions from the *Generic* HTP perceive the similarity to the 'imperious immediacy of interest' source of unanticipated consequences discussed in Section 5.1.4.3.1.
[§] The assumption is that the group has the appropriate domain knowledge to make a good decision.
[¶] It is never the decision-maker's fault; it is always someone else's fault.

5.1.6 OPERATIONAL

Perceptions of problem-solving from the *Operational* HTP included:

1. The traditional generic problem-solving process discussed in Section 5.1.6.1.
2. The extended problem-solving process discussed in Section 5.1.6.2.

5.1.6.1   The Traditional Generic Problem-Solving Process

The traditional generic problem-solving process:

- Considers the problem-solving process as a linear sequence of activities, starting with a problem and ending with a solution as in Figure 5.3. The traditional generic process contains the following six steps:
  1. Understand the problem and define the problem space to bound the problem.
  2. Conceive at least two candidate solutions.
  3. Determine selection criteria for choosing between the candidate solutions.
  4. Select a candidate solution (preferred acceptable solution option).
  5. Plan the implementation strategy.
  6. Implement the solution.
- Is recursive as each part contains an iteration of the problem-solving process.
- Is based on assumptions that include:
  - The problem is solved after one pass through the process, but often the outcome is 'oops' and the problem-solving process has to repeat. Accordingly, iteration is implied but not explicitly called out.
  - Someone has already defined the correct problem. This is often an incorrect assumption.
- Is accordingly incomplete. Further steps are required as shown below:
  7. Verify that the solution remedies the problem.
  8. If the solution has not remedied the problem, go back to Step 1.

However, the modified traditional generic problem-solving process is still incomplete.

5.1.6.2   The Extended Problem-Solving Process

Problems do not present themselves as givens; they must be constructed by someone from problematic* situations which are puzzling, troubling and uncertain.

Schön (1991)

Accordingly, the traditional generic problem-solving process shown in Figure 5.3 lacks a step or steps that define the problem to be remedied. Unlike the traditional

---

* Or undesirable.

generic problem-solving process which begins with a problem and ends with a solution (Section 5.1.6.1), the systems approach takes a wider perspective and begins with an undesirable situation (Schön 1991) that needs to be explored (Fischer, Greiff, and Funke 2012). From this perspective, the observer becomes aware of an undesirable situation that is made up of one or more undesirable factors which have to be converted into an FCFDS by remedying a series of problems, for example:

1. *Problem*: to do something about the undesirable situation.
   Solution: get the authorization for a project to do something about the undesirable situation together with adequate resources.
   Outcome: the authorization and the adequate resources.
2. *Problem*: to understand the undesirable situation, determine what makes the situation undesirable and define the problem space (bound the problem).
   Solution: follow the research problem-solving process (Section 5.1.4.4.1) to gain the understanding.
   Outcome: a statement of the correct problem.
3. *Problem*: to create a vision of an FCFDS that would remedy the undesirable situation.
   Solution: follow the problem-solving process to create the FCFDS.
   Outcome: the FCFDS.
4. *Problem*: how to transition from the undesirable situation to the FCFDS.
   Solution: create the specifications and project plan (Section 7.4.2.1) to realize the outcome.
   Outcome:
   - A set of requirements (specifications) for a solution system that when operating in the situation would remove the undesirability.
   - A plan for the remedial action that would realize the solution system specified by the requirements.
5. *Problem*: to perform the transition (intervention) process that creates the specified solution system according to the plan.
   Solution: follow the transition process according to the plan.
   Outcome: the undesirable situation has been transformed into what should be the desirable situation.
6. *Problem*: to create the test plan to test the solution system in operation in the actual situation existing at time t_1 to determine if it remedies the undesirable situation.
   Solution: follow the process that creates the test plan.
   Outcome: the test plan.
7. *Problem*: to create the test procedure to perform the planned tests.
   Solution: follow the process that creates the test process.
   Outcome: the test procedure.
8. *Problem*: to follow the test procedure and test for compliance to requirements.
   Solution: follow the test procedure, investigate and correct anomalies.
   Outcome: a system that is compliant to requirements and remedies the undesirable situation.

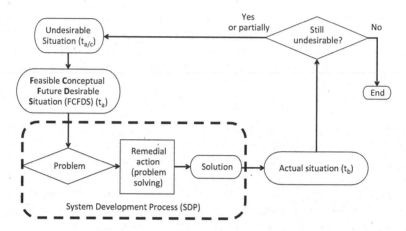

FIGURE 5.7 The extended problem-solving process. (©2016 IEEE. Reprinted, with permission, from Wicked problems: Wicked solutions, Joseph Kasser.)

However, should the remedial action take time, the undesirable situation may change during that time from what it was at t_0 to a new undesirable situation existing at t_2. If the undesirable situation has been remedied, then the process ends; if not, the process iterates from the undesirable situation at t_2 to a new undesirable situation as shown in Figure 5.7 (Kasser and Zhao 2016c).[*]

In summary, in general:

- There is an undesirable or problematic situation.
- An FCFDS is created.
- The problem is how to transition from the undesirable situation to the FCFDS.
- The solution is made up of two parts:
  1. The transition process.
  2. The solution system operating in the context of the FCFDS.
- It may require more than one iteration of Figure 5.7 to evolve a remedy.

If the solution requires a project to realize it, the FCFDS may be designed by system engineers, the transition process may be architected by system engineers and process architects (often project management) and the transition process is generally then managed by project management, while the solution system is developed by engineers.

5.1.7 SCIENTIFIC

Inferences from perceiving problem-solving from the *Scientific* HTP included:

1. The two schools of thought (Section 5.1.1) perceiving decision-making from different single perspectives on each end of a continuum.
2. The problem formulation template discussed in Section 5.1.7.2.

---

[*] Which is actually Figure 6.10 with a feedback loop.

5.1.7.1 The Two Schools of Thought

The two schools of thought (Section 5.1.1) perceive decision-making from different single perspectives on a continuum. Accordingly, each of the two schools of thought is perceiving the decision-making from a different single perspective. Moreover, a perception from the *Continuum* HTP indicates that decisions are not either logical or emotional, rather they are spread along a continuum with 100% logical at one end and 100% emotional at the other end. The middle of the continuum is 50% emotional and 50% logical. Each decision is made using a mixture of emotion and logic. Different types of decisions are made at different places on the continuum. Moreover, a decision that is made mostly logically one day may be made mostly emotionally under a different set of circumstances. For example, once upon a time when my children were very young, Susie said to me that she had noticed that some days they were told off for doing something that they were not told off for doing on other days. She asked me why that was the case. I thought about it for a moment and then I came up with the concept of a 'tolerance level'.[*] I explained what a tolerance level was and how it could go up and down depending on what was going on in my life that had nothing to do with the children. So, on days when my tolerance level was high, I decided that they could get away with doing stuff, and on days when my tolerance level was low, I decided that they needed to be told off for doing the same thing. She accepted that explanation. A few days later I was working in my home office when a little head stuck itself around the corner of the doorway and a little voice asked, 'Daddy, what's your tolerance level today?'.[†]

5.1.7.2 A Problem Formulation Template

A five-part problem formulation template (Kasser 2018a: Section 14.3) is:

- A tool to overcome the generic problem of 'poor problem formulation'.
- A tool to think through the whole problem-solving process.
- Based on the traditional generic problem-solving process (Section 5.1.6.1).
- A way to assist the problem-solving process by encouraging the planner to think through the problem and ways to realize a solution when formulating the problem.
- Made up of the following five parts:
  1. *The undesirable situation*: as perceived from each of the pertinent descriptive HTPs. This is the 'as-is' situation in business process reengineering (BPR).
  2. *Assumptions*: about the situation, problem, solution, constraints, etc. that will have an *impact* on developing the solution. One general assumption is that there is enough expertise in the group formulating

---

[*] If the tolerance level was low, I reacted emotionally; otherwise, I reacted logically or in this case didn't react.

[†] Is that or is that not risk management? I burst out laughing and my tolerance level was extremely high for the rest of the day.

the problem to understand the undesirable situation and specify the correct problem. If this assumption is not true, then that expertise needs to be obtained for the duration of the activity. Providers of the expertise include other people within the organization, consultants and members of the group looking it up on the Internet.

3. *The FCFDS* (*Scientific* HTP) or *desired outcome* as described by the appropriate descriptive HTPs. Something that remedies the undesirable situation and is to be interoperable with evolving adjacent systems over the operational life of the solution and adjacent systems (the outcome). This is generally:

    1. A conceptual 'as-it should be' situation. This is known as the 'to-be' situation in BPR.
    2. The undesirable situation without the undesirability and with improvements.

4. *The problem*: how to convert the FCFDS into reality.

5. *The solution*:

    1. Is inferred from the *Scientific* HTP.
    2. Remedies the undesirable situation.
    3. Has to be interoperable with evolving adjacent systems over the operational life of solution and adjacent systems (ideal).
    4. Is made of two interdependent parts:
        - *Process*: create and follow the SDP or transition process that converts the undesirable situation to a desirable situation[*] by first visualizing the FCFDS and then realizing the conceptual system that will operate in the context of the FCFDS. This is generally done using the research problem-solving methodology to visualize the FCFDS and create the SDP or transition process (intervention process) which is then performed.
        - *Product or system*: the solution system operating in the context of the FCFDS.

If the problem is objectively complex, the first version of the problem formulation template generally does not include a description of the solution. The problem-solving process creates the solution.

At the time the problem is being identified, the interdependent parts of the solution (at some time in the future) consist of two sets of functions;

1. *The product mission and support functions*: to be performed by the solution system once realized (F_s).
2. *The process functions*: which have to be performed to realize the solution system (F_w).

---

[*] Or a less undesirable situation if the situation is complex and requires iterations of the problem-solving process.

This concept can be represented by the following relationship:

$$\text{Total solution}(S) = F_s + F_w \tag{5.1}$$

where

$$F_s = F_d - F_c \tag{5.2}$$

which gives

$$S = (F_d - F_c) + F_w \tag{5.3}$$

In summary,

F_c = complete set of current functions; functionality provided in the existing situation which may range from zero (nothing exists) to some functionality in an existing system deemed as not providing a complete solution.

F_d = complete set of desired functions to be developed.

F_s = functions performed by the solution system.

F_w = functions (needed to be) performed to realize the solution system (create the desired functions that do not exist at the time the project begins).

Moreover, F_s and F_w can both consist of mission and support functions as discussed above.

5.2 COMPLEXITY

Consider perceptions of complexity from the following HTPs to gain an understanding of their nature and the problem posed by the need to remedy complex problems:

1. *Structural* discussed in Section 5.2.1.
2. *Continuum* discussed in Section 5.2.2.
3. *Temporal* discussed in Section 5.2.3.
4. *Scientific* discussed in Section 5.2.4.

5.2.1 STRUCTURAL

One perception of complexity from the *Structural* HTP is:

1. The various definitions of complexity discussed in Section 5.2.1.1.

5.2.1.1 The Various Definitions of Complexity

The scientific community cannot agree on a single definition of a complex problem (Quesada, Kintsch, and Gomez 2005) cited by Fischer, Greiff and Funke (2012). The literature contains many different definitions of complexity, e.g.:

- 'A complex system usually consists of a large number of members, elements or agents, which interact with one another and with the environment' (ElMaraghy et al. 2012). According to this definition, the only difference between a system and a complex system is in the interpretation of the meaning of the undefined word 'large'.

- 'The classification of a system as complex or simple will depend upon the observer of the system and upon the purpose he has for constructing the system' (Jackson and Keys 1984).
- 'A simple system will be perceived to consist of a small number of elements, and the interaction between these elements will be few, or at least regular. A complex system will, on the other hand, be seen as being composed of a large number of elements, and these will be highly interrelated' (Jackson and Keys 1984).
- 'A complex system is an assembly of interacting members that is difficult to understand as a whole' (Allison 2004: p. 2).

The attributes associated with the different definitions of complexity include:

- Number of issues, functions or variables involved in the problem.
- Degree of connectivity among those variables.
- Type of relationships among those variables.
- Stability of the properties of the variables over time.

However:

- There are no specific numbers that can be used to distinguish complex systems from non-complex systems; accordingly, it does seem that complexity is in the eye of the beholder (Jackson and Keys 1984).
- Inferences from the *Scientific* HTP based on a perception from the *Generic* HTP have suggested a number (Section 5.2.4.3).

5.2.2 CONTINUUM

Perceptions of complexity from the *Continuum* HTP include:

1. The complexity dichotomy discussed in Section 5.2.2.1.
2. Partitioning complexity discussed in Section 5.2.2.2.

5.2.2.1 The Complexity Dichotomy

Perceptions of complexity from the *Continuum* HTP (Kasser 2015a) indicate that there is a dichotomy on the subject of how to remedy the complex problems associated with complex systems. There is literature that states:

1. *We have a problem*: there is a need to develop new tools and techniques to remedy complex problems.
2. *What's the problem?* There is no need for new tools and techniques. Complex problems are being remedied successfully including:

1. The successful Atlas Intercontinental Ballistic Missile (ICBM) development of the 1950s where 'systems engineering was the methodology used to manage the problem of scheduling and coordinating hundreds of contractors developing hundreds – even thousands – of subsystems that eventually would be meshed into a total system' (Hughes 1998: p. 118).*

2. The problems posed by the development of the Public Housing System, industrial development and the Air Defence System (ADS) in Singapore were successfully remedied by systems engineering (Lui 2007).

3. Decision-makers in civilian organizations solving the complex problems posed in managing fleets of cruise ships, financial and banking networks, oil rigs, airlines, transportation systems and hospitals on a daily basis.

5.2.2.2 Partitioning Complexity

Complexity can be partitioned in various ways including:

- Thirty-two different complexity types in 12 different disciplines and domains such as projects, structural, technical, computational, functional and operational complexity (Colwell 2005) cited by ElMaraghy et al. (2012).
- Fifty technical, 21 organizational and 14 environmental which can be expanded to include additional elements in the future (Bosch-Rekveldt et al. 2010).

However, lessons learned from behavioural psychology have shown that long lists are not useful and the items in such lists need to be aggregated into a short list (Section 1.2). One such short list is subjective and objective complexity (Sillitto 2009) where:

1. *Subjective complexity*: or difficulty; people don't understand it and can't get their heads round it, e.g. (Allison 2004: p. 2) discussed in Section 5.2.2.2.1.
2. *Objective complexity*: the problem situation or the solution has an intrinsic and measurable degree of complexity, e.g. (ElMaraghy et al. 2012, Jackson and Keys 1984) discussed in Section 5.2.2.2.2.

There does not appear to be unique words that uniquely define the concepts of 'subjective complexity' and 'objective complexity' in the English language. Hence, the literature accordingly uses the words 'complicated' and 'complex' both as synonyms to mean both subjective complexity and objective complexity and to distinguish between subjective complexity and objective complexity. To further muddy the situation, some authors use the word 'complex' to mean subjective complexity, while other authors use the word 'complicated' to mean subjective complexity and vice versa.

---

* The development was systems engineered not "managed".

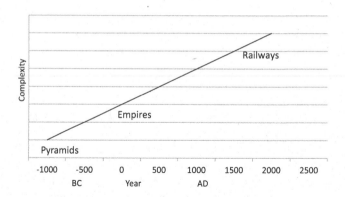

FIGURE 5.8 Perceptions of objective complexity from the *Temporal* HTP.

5.2.2.2.1 Subjective Complexity

Perceptions from the *Generic* HTP note that the subjective complexity of an issue noted by a person is the level of difficulty of the issue. Accordingly, the four categories of increasing order of difficulty (Section 5.1.3.1) can provide a measurement of subjective complexity.

5.2.2.2.2 Objective Complexity

Perceptions from the *Continuum* HTP differentiate the various definitions of objective complexity in the literature into two types as follows (Kasser and Palmer 2005):

1. *Real-world complexity*: elements of the real world are related in some fashion and made up of components which are aggregated according to the Hitchins layers (Section 1.2). This complexity is not reduced by appropriate abstraction, it is only hidden.
2. *Artificial complexity*: arising from either poor aggregation or failure to abstract out elements of the real world that, in most instances, should have been abstracted out when drawing the internal and external system boundaries, since they are not relevant to the purpose for which the system was created (Kasser 2018a: Section 13.3.1.1). For example, in today's paradigm, complex drawings are generated that contain lots of information,* and the observer is supposed to abstract information as necessary from the drawings. The natural complexity of the area of interest is included in the drawings; hence, the system is thought to be complex.

5.2.3 TEMPORAL

Perceptions of complexity from the *Temporal* HTP note that the objective complexity of systems that humanity can manage has grown over the centuries roughly as shown in Figure 5.8 (Kasser 2018b).

---

* The Department of Defense Architecture Framework (DoDAF) Operational View (OV) diagrams provide wonderful examples of artificial complexity (Section 5.2.2.2.2).

TABLE 5.2

Summary of Reasons for the Complexity Dichotomy from Various Perspectives

	Perspective	Unsuccessful	Successful
1	Solution paradigm	Looks for a single correct solution	Looks for acceptable solutions
2	HKMF column	B–F	A and G
3	HKMF layer	Layer 2 moving up to layer 3	In layer 3
4	Subjective complexity	Hard to understand	Easy to understand
5	Degree of confusion	Confusing ill-structured problems with complexity	No confusion
6	Structure of the problem	Ill-structured, wicked	Well-structured
7	Systems engineering camp	Process	Problem-solving, enabler
8	Number of entities	Up to and including 7 ± 2	More than 7 ± 2
9	Boundary of knowledge	Outside	Inside

5.2.4 SCIENTIFIC

Inferences of complexity from the *Scientific* HTP include:

1. Real and subjective complexity discussed in Section 5.2.4.1.
2. Resolving the complexity dichotomy discussed in Section 5.2.4.2.
3. The number of elements that makes a system (objectively) complex discussed in Section 5.2.4.3.
4. The problem classification framework discussed in Section 5.2.4.4.

5.2.4.1 Real and Subjective Complexity

Using the analogy to complex numbers in mathematics (a perception from the *Generic* HTP), inferences from the *Generic* HTP consider objective complexity as the real part of complexity and subjective complexity as the imaginary part which can be reduced by education and experience (Kasser and Zhao 2016a) This would allow problems to be plotted in a two-dimensional matrix with objective complexity along the vertical axis and subjective complexity along the horizontal axis (Section 5.2.4.4).

5.2.4.2 Resolving the Complexity Dichotomy

The complexity dichotomy (Section 5.2.2.1) may be resolved by perceiving from the *Continuum* HTP that each side is focused on one or more different non-contradictory aspects of the situation as summarized in Table 5.2. From a risk management perspective, for a specific problem-solver, the greater the number of aspects of the situation in the successful column, the lower the risk of the complex problem not being remedied. The aspects of the situation summarized in Table 5.2 are:

1. *The solution paradigm*: the unsuccessful side may be talking about the need to develop new tools and techniques to solve the problems associated with producing a single correct optimal solution that satisfies the problem, while the successful side consists of those who are willing to settle for an acceptable solution that satisfies the problem.

2. *The Hitchins-Kasser-Massie framework (HKMF) column*: the unsuccessful side may be talking about developing new complex systems in the HKMF columns A–F (Section 3.4.4), and the successful side may be taking about managing complex systems in operation in the HKMF column G.[*]

3. *The HKMF layer*: the unsuccessful side may be positioned in the HKMF layer 2, while the successful side is positioned in the HKMF layers 3–5. The theory of integrative levels (Needham 1937) cited by Wilson (2002) recognizes that system behaviour is different in the different levels or layers of the hierarchy so that tools and techniques that work in layer 2 may not work in layer 3. Moreover, people on the layer 2 side are used to dealing with their system in layer 2, the metasystem[†] in layer 3 and the subsystem in layer 1. When they move up into layer 3, they add layer 4 to their area of concern but do not drop layer 1 increasing the artificial complexity (Section 5.2.2.2.2). Those on the other side of the dichotomy already in layer 3 have dropped or never considered layer 1 simplifying their area of concern.

4. *Subjective complexity*: the unsuccessful side perceives the problem from a different level of subjective complexity (Section 5.2.2.2) than the successful side. Namely, one side of the dichotomy has experience[‡] and education working with the class of problems and so understand the problem, while the other side is facing the problem for the first time.

5. *Degree of confusion*: the unsuccessful side is confusing wicked problems (Section 5.1.4.5.3) with complexity, e.g. (APSC 2007), while the successful side is not.

6. *Structure of the problem*: the unsuccessful is trying and failing to manage ill-structured and or wicked problems, e.g. (APSC 2007), which cannot be solved (Simon 1973), while the successful side is successfully managing well-structured problems (Section 5.1.4.5.1).

7. *Systems engineering camp*: the process campers (Section 6.5.5) have trouble dealing with complexity, while the problem and enabler campers ask, 'what's the problem?'

8. *The number of entities being addressed*: as noted in Section 5.2.4.3, managing complexity is basically a thinking process. Miller's rule sets a limit on the number of items a human brain can keep track of (Miller 1956). The magic number is 7 ± 2. Accordingly, people who are successfully managing complexity are generally managing less than 7 ± 2 entities,[§] while those having problems with complexity are generally attempting to manage more than 7 ± 2 entities. The successful side deals with high levels of complexity by abstracting the system at as high a level as possible and then progressively reducing the level of abstraction (Maier and Rechtin 2000: p. 6).

---

[*] The purpose and function of operations research.

[†] Which they call a system of systems and then seek to find a definition that will differentiate between the system of system in layer 3 and a system in layer 2 ignoring the fifth principle for systems engineered systems (Kasser 2019b Section 4.12).

[‡] In operations research.

[§] Using the principle of hierarchies (Section 6.3.1) to abstract out lower level aspects of complexity.

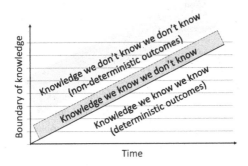

FIGURE 5.9 Humanity's learning curve: the boundary of knowledge.

9. *Boundary of knowledge* (Kasser 2018b): the successful side is:
   - Working inside the boundary of knowledge.
   The unsuccessful side is:
   - Working beyond the boundary of knowledge.

Donald Rumsfeld articulated the following three types of knowledge (Rumsfeld 2002):

1. Knowledge we know we know.
2. Knowledge we know we don't know.
3. Knowledge we don't know we don't know.

Mark Twain would have added a fourth type, as in, 'It ain't what you don't know that gets you into trouble. It's what you know for sure that just ain't so'.[*]

Let the line that represents the level of complexity in Figure 5.8 also represent the boundary of knowledge; Rumsfeld Type 1 knowledge lies below the line, while Type 2 and Type 3 knowledge lie above the line[†] as shown in Figure 5.9 (Kasser 2018b). This means that any complex situation:

1. *Below that line*: can be managed by posing well-structured problems (Section 5.1.4.5.1) using Rumsfeld Type 1 knowledge with deterministic results if all goes well. For example, the problems faced by cruise ship companies, oil rigs, airlines and railway systems.
2. *Just above the line*: cannot be managed irrespective of the structure of the problem because it will take Rumsfeld Type 2 knowledge to manage it. This situation leads to applied research to convert Rumsfeld Type 2 knowledge to Rumsfeld Type 1 knowledge before the complex system may be managed. For example, the problem posed by sending a man to the moon and returning him to earth alive and well, had elements just above and below the boundary of knowledge.

---

[*] Also, Russo and Shumaker's decision traps (Section 5.1.5.3).
[†] Referring to the types makes the argument clearer than repeating lots 'knowledge that you …'.

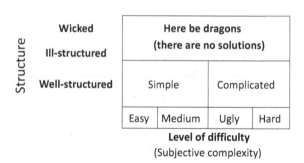

FIGURE 5.10   A problem classification framework.

3. *Well above the line*: cannot be managed because it will take Rumsfeld Type 3 knowledge to manage it. Accordingly, these problems must be absolved until research is carried out that converts that knowledge to Type 2 or better still Type 1.

Questions arise as we convert Rumsfeld Type 2 knowledge to Rumsfeld Type 1 knowledge.[*] The act of thinking up these questions has converted some Rumsfeld Type 3 knowledge to Rumsfeld Type 2 knowledge.[†] The systems that we can manage successfully become more complex as time passes and the amount of Rumsfeld Type 1 knowledge increases.

5.2.4.3   The Number of Elements That Make a System (Objectively) Complex

There are no specific numbers that can be used to distinguish complex systems from non-complex systems in all the studied definitions of a complex system in the literature (Section 5.2.1.1). Namely complexity is in the eye of the beholder (Jackson and Keys 1984). Since managing complexity is basically a thinking process, it seems that Miller's rule should apply.[‡] Miller's rule sets a limit on the number of items a human brain can keep track of (Miller 1956). The magic number is 7 ± 2. Managing complexity begins by abstracting the system at as high a level as possible to less than 7 ± 2 components at that level and then progressively reducing the level of abstraction in each level to 7 ± 2 components.

5.2.4.4   The Problem Classification Framework

The structure of the problem and the level of difficulty of the problems are combined in the two-dimensional problem classification framework shown in Figure 5.10 (Kasser 2018a: Section 5.3). The two dimensions are:

---

[*] The learning process.
[†] We now know there is some knowledge we know we didn't know that we didn't know we didn't know before.
[‡] An inference from the *Scientific* HTP.

1. *Structure of the problem*: ranges from non-complex through complex well-structured problems to complex ill-structured problems and wicked problems (Section 5.1.4.5.3).
2. *Level of difficulty (subjective complexity)*: four levels of subjective complexity ranging from easy to hard (Section 5.1.3.1); where problems with low subjective complexity can be easy and medium, while problems with higher levels of subjective complexity are those that are ugly and hard.

Different people may position the same problem in different places in the framework. This is because as knowledge is gained from research, education and experience, a person can reclassify the subjective difficulty of a problem down the continuum from 'hard' towards 'easy'.

5.3   REMEDYING WELL-STRUCTURED PROBLEMS

The traditional problem-solving approach manages large and complex well-structured problems by breaking them out into smaller and simpler problems. Each of these problems is remedied which in turn provides a remedy to the large and complex problem if all goes well. When dealing with small problems, the process used to find a remedy is called the problem-solving process or the decision-making process. When faced with large and often complex problems, the same generic process is known as the SDP (Section 6.1.5.4) as shown in Figure 6.11* (Kasser and Hitchins 2013). The words 'remedial action (problem-solving)' in Figure 5.7 have been replaced by the lower-level term 'Series of (sequential and parallel) activities' in Figure 6.11. The figure shows that the lower-level term 'Series of (sequential and parallel) activities' is known as the SDP for large or complex systems which breaks up a complex problem into smaller less-complex problems (analysis), then solves each of the smaller problems and hopes that the combination of solutions to the smaller problems (synthesis) will provide a solution to the large complex problem. The SDP:

- Contains sub-processes for developing and testing products and parts of products as discussed in the three streams of activities (Section 7.3.1.1.2). Each of these processes needs to be under configuration control so that the version of the process used at any time to develop or test any product is known and the processes are repeatable.
- May be considered as:
  - Two sequential problem-solving processes (Section 5.3.1).
  - One iteration of the multiple-iteration problem-solving process for remedying complex problems (Section 5.3.2).

5.3.1   THE TWO-PART SDP

The top-level SDP is a two-part sequential process as shown in Figure 5.11 (Kasser and Zhao 2016a): planning and doing or implementing.

---

* The forward reference is because the figure is discussed in Section 6.5.7.

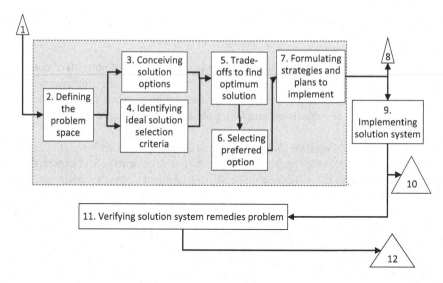

FIGURE 5.11   Modified Hitchins' view of the problem-solving decision-making process.

The first problem-solving process:

- Is a research problem-solving process (Section 6.5.5), much of which was documented by Hall (1962).
- Takes place in the needs identification state of the SDP.
- Shown in Figure 5.3 is the front end of the traditional generic problem-solving process. The process contains the following major milestones (identified in triangles) and activities or processes (shown in rectangles):
  1. The milestone to provide authorization to proceed.
  2. The process for defining the scope of the problem and gaining an understanding.
  3. The process for conceiving several solution options.
  4. The process for identifying ideal solution selection criteria.
  5. The process for performing trade-offs to find the optimum solution.
  6. The process for selecting the preferred option.
  7. The process for formulating strategies and plans to implement the preferred option.
  8. The milestone to confirm consensus to proceed with implementation.

The second problem-solving process:

- Is an intervention or a transition problem-solving process.
- Begins once the stakeholder consensus is confirmed at Milestone 8 in Figure 5.11.
- Covers the remaining states in the SDP; the project moves on to the implementation states shown from the *Functional* HTP in Block 9 of Figure 5.11 where the additional following major milestones and activities are:

9. The process for implementing the solution system often using the SDP.
10. The milestone review to document consensus that the solution system has been realized and is ready for validation.
11. The process for validating the solution system remedies the evolved need in its operational context, often known as operational test and evaluation (OT&E) for complex systems.
12. The milestone to document consensus that the solution system remedies the evolved need in its operational context.

5.3.2 THE MULTIPLE-ITERATION PROBLEM-SOLVING PROCESS

The various problem-solving processes in the literature are parts of a meta-problem-solving process that starts with ill-defined problems, converts them to well-defined problems and evolves a remedy to the set of well-defined problems recognizing that the problem may change while the remedy is being developed.

Kasser and Zhao (2016a)

From the perspective of the problem-solving paradigm of systems engineering, the standard approach to evolving a solution can be reworded to become 'an evolutionary approach to remedying the undesirability in a situation by turning an ill-structured problem into a number of well-structured problems, remedying the well-structured problems and then integrating the partial remedies into a whole remedy' (Kasser and Hitchins 2012). This definition is derived from the following definition of systems engineering: 'An iterative process of top down synthesis, development and operation of a real-world system that satisfies in a near optimal manner, the full range of requirements for the system' (Eisner 1988).

By observing the management of complex systems in industry from the HTPs, it was possible to infer a meta-process for solving the problems associated with complex systems based on a modified version of the extended problem-solving process (Section 5.1.6.2). Let this meta-process shown in Figure 5.12 (Kasser and Zhao 2016b) be called the multiple-iteration problem-solving process. The multiple-iteration problem-solving process consists of two sequential problem-solving processes (Section 5.3.1) embedded in an iterative loop.

1. *The first problem-solving process*: converts the ill-structured problem posed by the situation into one or more well-structured problems (Section 5.1.4.5.1).
2. *The second problem-solving process*: is tailored to remedy specific type of well-structured problems since one problem-solving approach does not fit all problems (Section 5.1.1.2.5).

The choice of which of the well-structured problems identified by the first problem-solving process to tackle in the second problem-solving process will depend on a number of factors, including urgency, impact on undesirable situation, the need to show early results and available resources.

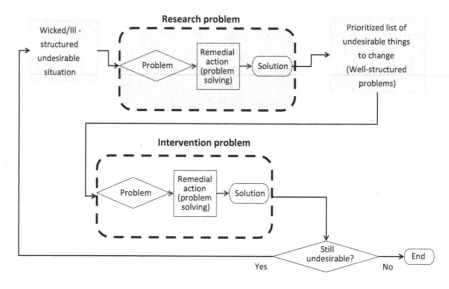

FIGURE 5.12   The two-part multiple-iteration problem-solving process.

This sequential evolutionary process is sometimes known as 'build a little test a little' (Section 7.4.3.1) and evolves the solution from a baseline or known state to the subsequent milestone, which then becomes the new baseline.

The causes of the undesirability may be remedied at different levels and locations in the situation hierarchy, simultaneously or sequentially.

5.3.2.1   The First Problem-Solving Processes

The first problem-solving process in the multiple-iteration problem-solving process remedies a research problem (Section 5.1.4.4.1) using an adaptation of the scientific method. This process:

- Is a research problem-solving process.
- Takes place in the needs identification state of the SDP (Section 6.1.5.4).
- Figures out the nature of the problematic situation and what needs to be done about it.
- Creates a prioritized list of symptoms to change. The problem-solvers create the prioritized list of symptoms to change by:
  - Gaining a thorough understanding of situation.
  - Identifying the undesirable aspects of the situation.
  - Performing research to gather preliminary data about the undesirable situation.
  - Formulating the hypothesis to explain the causes of the undesirable situation.
  - Estimating the approximate contribution of each cause to the undesirable situation.
  - Developing priorities for remedying the causes.

- Deciding which causes to remedy.
- Conceptualizing the FCFDS
- Performing feasibility studies.

Even when consensus on the causes of the undesirability cannot be achieved, it is often possible to achieve consensuses on what is undesirable and on the FCFDS.

Feasibility studies (Section 7.4.1.1) on the FCFDS are performed on the FCFDS because there is no point in creating an FCFDS if it is not feasible. Examples include:

- *Operational feasibility*: at least one solution or combination of partial solutions is achievable.
- *Quantitative feasibility*: Cost is affordable; risk and uncertainty are acceptable.
- *Structural (technical) feasibility*: suitable technologies exist at the appropriate technology readiness levels (Section 8.4).
- *Temporal feasibility*: Schedule (the solution system will be ready to operate in the FCFDS when needed).

As an example, the complex problem may be associated with undesirable traffic congestion in an urban area. The mayor, feeling under pressure to do something about the growing traffic congestion in her city, provides the authorization to initiate the problem-solving process which begins with the ill-structured problem of the need to remedy the undesirable effects of traffic congestion. An understanding of the situation might produce a number of causes, e.g. commuting to work and school, deliveries and tourists, etc. The analysis would also provide quantitative information such as an estimate of the degree of the contribution by the cause to the undesirable situation, e.g. commuting to work (40%) and school (30%), deliveries (20%) and tourists (10%). The need to remedy each cause would then be prioritized according to selection criteria that might include cost, schedule, political constraints, performance and robustness.

The often-forgotten domain knowledge needed to gain consensus on and prioritize the cause is 'human nature'. Each of the stakeholders needs to know 'what's in it for me?' in implementing the change (Kasser 1995). So, the policy creator and implementer, the systems engineer and the project manager need to identify and communicate that information.

Tools developed for gaining an understanding of the system (situation) and the nature of its undesirability include:

- The systems engineering mathematical and analytical tools of the 1960s (Alexander and Bailey 1962, Chestnut 1965, Wilson 1965) in their modern incarnations.
- Avison and Fitzgerald's interventionist methodology (Avison and Fitzgerald 2003).
- Checkland's SSM (Section 4.7.1.4).
- Active brainstorming (Section 2.5.2).
- The nine-system model (Section 6.7).
- The HTPs (Section 2.5.1).

5.3.2.2 The Second Problem-Solving Process

The second problem-solving process in the multiple-iteration problem-solving process is:

- The traditional SDP (Section 6.1.5.4).
- An intervention problem-solving process (Section 5.1.4.4.2) and remedies (some of) the aspect(s) of the undesirable situation identified by the first problem-solving process. Since one problem-solving approach does not fit all problems (Kasser and Zhao 2016a), the second problem-solving process is tailored to remedy the specific type of problems. Once the second problem-solving process is completed, the entire process may iterate back to the beginning for a new cycle as shown in Figure 5.12 because the second problem-solving process may:
  - Only partially remedy the original undesirable or problematic situation.
  - Contain unanticipated undesirable emergent properties from the *solution system* and its interactions with its adjacent systems.
  - Only partially remedy new undesirable aspects that have shown up in the situation during the time taken to develop the solution system.
  - Produce new unanticipated undesired emergent properties of the solution system and its interactions with its adjacent systems, which in turn produce new undesirable outcomes.

5.4 REMEDYING ILL-STRUCTURED PROBLEMS

Ill-structured problems cannot be remedied; they must first be converted to well-structured problems (Simon 1973). Perceptions from the *Generic* HTP show that is what we do in the research problem-solving process depicted in Figure 5.12. The undesirable or problematic situation poses an ill-structured problem. As we gain an understanding of the situation through research, we convert the ill-structured problem into one or more well-structured problems.

5.5 REMEDYING COMPLEX PROBLEMS

A complex problem may be defined as 'one of a set of problems posed to remedy the causes of undesirability in a situation in which the solution to one problem affects another aspect of the undesirable situation' (Kasser and Zhao 2016b). Consequently, remedying a complex problem will depend on the structure of the problem.

5.5.1 Remedying Well-Structured Complex Problems

Well-structured complex problems:

- Consist of a set of interconnected well-structured problems. The choice of which of the symptoms identified by the first problem-solving process (Section 5.3.2) to tackle in the second problem-solving process will depend on a number of factors (selection criteria), including urgency, impact on undesirable situation, the need to show early results and available resources.

- Are remedied using the multiple-iteration problem-solving process (Section 5.3.2). This is a time-ordered multi-phased evolutionary approach that provides remedies to one or more of the well-structured problems, integrates the remedies, re-evaluates the situation and then repeats the process for the subsequent set of problems (Kasser 2002b).

5.5.2   REMEDYING ILL-STRUCTURED COMPLEX PROBLEMS

The undesirable situation causing the ill-structured complex problem cannot be remedied until the ill-structured complex problem has been transformed into one or more well-structured problems. Consequently, finding a solution requires converting the ill-structured complex problem into a well-structured complex problem or a series of well-structured complex problems. Determining the real cause(s) of the undesirable situation and finding solutions sometimes means doing both functions in an iterative and interactive manner. In this situation, initially:

1. *The undesirable situation*: an ill-structured problem.
2. *The assumptions*: there may be than one cause of undesirability. Ill-structured problems may be a result of stakeholders perceiving the situation from single different HTPs in the manner of the fable of the blind men perceiving the elephant (Yen 2008); they identified different animals – in this situation, the stakeholders are identifying different causes. Even if the stakeholders cannot agree on the causes of the undesirability, they can usually agree on the nature of the undesirability.
3. *The FCFDS*: (desired outcome) one or more well-structured problems.
4. *The problem*: how to convert the ill-structured problem into one or more well-structured problems.
5. *The solution*: follow the multiple-iteration problem-solving process (Section 5.3.2).

The research problem-solving process (Section 5.1.4.4.1) is used to convert the ill-structured problem into one or more well-structured problems, which can then be remedied either singly or as a group (Section 5.5.1). However, take care when converting ill-structured problems into a series of well-structured problems because the process can produce different and sometimes contradictory well-structured problems which would generate different and sometimes contradictory solutions.

As an example, consider the ill-structured complex problem of how to win a war. This problem is broken out into two lower level ill-structured problems.

1. *Defence*: how to defend the nation.
2. *Attack*: how to destroy the other side and end the war.

Each of these problems is then further broken out into a set of well-structured problems, which if remedied successfully in a timely manner* will end the war successfully.

The first approach to creating the set of well-structured problems goes beyond systems thinking (2.5) and uses perceptions from the *Continuum, Temporal* and *Generic* HTPs often using lessons learned in command school. Questions to be researched include†:

- Who has faced this situation before?
- What did they do about it?
- Why would that solution work/not work in this case?
- How can we improve on the previous instance?

5.5.3 REMEDYING WICKED PROBLEMS

Wicked problems (Section 5.1.4.5.3) are considered impossible to solve using the current problem-solving paradigm. When faced with insolvable problems, the best way to approach them is to dissolve the problems (Section 5.1.1.2.3) or bypass them by finding an alternative paradigm (Kuhn 1970). Using inferences from the *Scientific* HTP, instead of trying to solve or resolve wicked problems, dissolve the problem by changing the paradigm from 'problem' to 'situation' (Kasser and Zhao 2016c). Instead of dealing with wicked problems, deal with wicked situations. Then non-deterministic behaviour:

1. Is not a characteristic of complex problems.
2. Is a characteristic of:
    1. Ill-structured and wicked situations.
    2. The initial state in the scientific method (Section 5.1.4.4.1) as perceived from the *Generic* HTP due to:
        1. A lack of understanding of the situation which precludes determining the behaviour.
        2. Being beyond the boundary of the current body of knowledge, the line in Figure 5.8.

One of the characteristics of wicked problems (Section 5.1.4.5.3) is, 'Cannot be easily defined so that all stakeholders cannot agree on the problem to solve'. Accordingly, assume multiple unknown causes of undesirability in the undesirable situation.‡ The assumption of multiple causes leads to perceiving that there may be multiple solutions (perhaps even at different levels in the hierarchy of systems); one or more for each cause.

1. Elaborate the undesirable situation into one or more undesirable situations.

---

* Before the enemy does the same.
† These are standard questions to be posed in similar problematic situations.
‡ The hypothesis in the scientific method.

2. Use the multiple-iteration problem-solving process (Section 5.3.2) to create wicked solutions (Kasser and Zhao 2016c) which have similar characteristics to wicked situations.

When creating wicked solutions, the initial solution may not be the needed solution, since wicked solutions:

- Evolve via the multiple-iteration problem-solving process (Section 5.3.2).
- Each instance of the wicked solution:
  1. May only remedy part of the undesirability in the whole wicked situation.
  2. May satisfice and not necessarily satisfy the problem in a single pass through the multiple-iteration problem-solving process.
  3. May apply simultaneously in the wicked situation hierarchy at more than one level and more than one location at a particular level.

5.6 SIMULTANEOUSLY REMEDYING MULTIPLE PROBLEMS

Complex undesirable situations generally in layers 4 and 5 of the risk framework (Section 1.2) generally contain a large number of controlled variables (Section 5.1.2.1) and a number of undesirable situations. The systems approach recognizes this complex situation and looks to see if the undesirability in one situation can be remedied by the undesirability in a different situation. Perceptions from the *Generic* HTP note the similarity between the hierarchy of systems and the hierarchy of problems. The complex situation is similar to the system, and the different undesirable situations are similar to the different subsystems of that system. A system engineer for a system responsible for developing the system should be monitoring the development of the subsystems to ensure that the subsystems do not exceed their power, weight and other budgets as well as meeting the functional specifications. In a similar manner, at the highest level in the complex undesirable situation, the person responsible for the complex undesirable situation needs to be aware of the individual problem-solving activities remedying parts of the undesirability and influence the choice of solutions so that the undesirability in more than one situation is remedied by the same choice of solution. For example:

- Federated Aerospace is finishing a project and will shortly have surplus personnel. The systems approach looks to see if there are other projects either existing or about to begin that will be able to use the surplus personnel to avoid having to lay them off.
- One part of a very large country has produced a food surplus, while there is a food shortage in another part of the country. The systems approach to remedying the undesirability of having a surplus looks to see if there are shortages elsewhere. Conversely, the systems approach to remedying the undesirability of having a shortage of food looks to see if there is a surplus elsewhere.
- The government needs to be aware of the policies in action to ensure that the goal of a proposed new policy does not contradict the goal of one or more existing policies.

5.7   GENERIC RISKS IN THE PROBLEM-SOLVING PROCESS

Generic risks in the traditional problem-solving process may be organized by the different steps in Figure 5.3. These generic risks include:

1. *Formulating the wrong problem*: due to not understanding the problem or falling into one or more of the decision traps (Section 5.1.5.3).
2. *Selecting unacceptable candidate solutions*: even though they are feasible.
3. *Selecting infeasible solutions*: because they were not analysed and to ensure they were feasible.
4. *Selecting impractical solutions*: politically incorrect or culturally unacceptable.
5. *Determining the wrong selection criteria*: for the situation.
6. *Planning risks*: all the risks involved in planning a project (Section 7.4.2.1).
7. *Implementing the solution risks*: all the generic risks in the SDP (Section 6.1.5.4) including the risk of creating undesired unanticipated outcomes (Section 5.1.4.3.1).

5.8   EXERCISES

This section provides a number of exercises as opportunities to think about problems, solutions, risks and complexity; apply risk management in problem-solving and practice using the problem formulation template (Section 5.1.7.2). A partial acceptable solution is provided for the Engaporian drought relief policy (Section 5.8.3). It is up to the reader to use the sample as a template when performing the remaining exercises.

5.8.1   THE ENU TRAFFIC LIGHT UPGRADE

In addition to meeting the generic requirements in Section 2.7:

1. Format the problem posed by the ENU traffic light upgrade in Section 2.7.1 in the problem formulation template (Section 5.1.7.2).
2. Format the problem posed by the ENU traffic light upgrade in Section 3.10.1 in the problem formulation template (Section 5.1.7.2).
3. Format the problem posed by the ENU traffic light upgrade in Section 4.10.1 in the problem formulation template (Section 5.1.7.2).
4. Format the problem posed by this exercise in the problem formulation template (Section 5.1.7.2).
5. For each of the four formatting problems:
   1. State the level of difficulty of the problem (Section 5.1.3.1) with the reasons for the choice.
   2. State the level of objective complexity of the problem (Section 5.2.2.2.2) with the reasons for the choice.
   3. State the type of problem (Section 5.1.4.4) with the reasons for the choice.
   4. State the structure of the problem (Section 5.1.4.5) with the reasons for the choice.

5.8.2   The Engaporean Maid Reduction Policy

In addition to meeting the generic requirements in Section 2.7:

1. Format the problem posed by the maid reduction policy in Section 2.7.2 in the problem formulation template (Section 5.1.7.2).
2. Format the problem posed by the maid reduction policy in Section 3.10.2 in the problem formulation template (Section 5.1.7.2).
3. Format the problem posed by the maid reduction policy in Section 4.10.2 in the problem formulation template (Section 5.1.7.2).
4. Format the problem posed by this exercise in the problem formulation template (Section 5.1.7.2).
5. For each of the four formatting problems:
   1. State the level of difficulty of the problem (Section 5.1.3.1) with the reasons for the choice.
   2. State the level of objective complexity of the problem (Section 5.2.2.2.2) with the reasons for the choice.
   3. State the type of problem (Section 5.1.4.4) with the reasons for the choice.
   4. State the structure of the problem (Section 5.1.4.5) with the reasons for the choice.

5.8.3   The Engaporian Drought Relief Policy

In addition to meeting the generic requirements in Section 2.7:

1. Format the problem posed by the Engaporian drought relief policy in Section 2.7.3 in the problem formulation template (Section 5.1.7.2).
2. Format the problem posed by the Engaporian drought relief policy in Section 3.10.3 in the problem formulation template (Section 5.1.7.2).
3. Format the problem posed by the Engaporian drought relief policy in Section 4.10.3 in the problem formulation template (Section 5.1.7.2).
4. Format the problem posed by this exercise in the problem formulation template (Section 5.1.7.2).
5. For each of the four formatting problems:
   1. State the level of difficulty of the problem (Section 5.1.3.1) with the reasons for the choice.
   2. State the level of objective complexity of the problem (Section 5.2.2.2.2) with the reasons for the choice.
   3. State the type of problem (Section 5.1.4.4) with the reasons for the choice.
   4. State the structure of the problem (Section 5.1.4.5) with the reasons for the choice.

5.8.3.1   One Acceptable Solution

A completed exercise could take the form shown in this section. Note it provides one acceptable solution, not the single correct solution as there will be other acceptable solutions (Section 5.1.1.2.2).

*5.8.3.1.1   The Problem Posed by the Engaporian Drought
             Relief Policy Exercise in Section 2.7.3*

This section provides one acceptable solution, not the single correct solution as there will be other acceptable solutions (Section 5.1.1.2.2).

5.8.3.1.1.1   The Assumptions   The assumptions include:

1. Information already identified in previous exercises should be referenced instead of being repeated.

5.8.3.1.1.2   The Formatted Problem   The problem formatted according to the problem formulation template (Section 5.1.7.2) is:

1. *The undesirable situation*: is the need to:
   1. Comply with the generic requirements in Section 2.7.
   2. Draw a context diagram of a bottle of water and the entities associated with the bottle.
   3. Document the conceptions of variations of the policy in action (tourists bringing and not bringing bottles) from the descriptive HTPs.
2. *The assumptions*: are:
   1. The exercise can be completed within the 60 minutes.
   2. The assumption listed in Section 5.8.3.1.1.1.
3. *The FCFDS*: the exercise has been completed and the desired grade has been achieved.
4. *The problem*: how to do the exercise, namely:
   1. Meet the generic requirements in Section 2.7.
   2. Draw a context diagram of the water bottle and the entities associated with it.
   3. Document the scenarios of the policy in action (tourists bringing and not bringing bottles) for the concept of operations (CONOPS) (Section 6.5.9.1) from the descriptive HTPs.
   4. Verify that all the requirements for the exercise have been completed and update the compliance matrix accordingly.
5. *The solution*: consists of working through each of the problems. A partial solution is presented below.

5.8.3.1.1.3   The Compliance Matrix   The partially completed compliance matrix is shown in Table 5.3.[*]

5.8.3.1.1.4   The Context Diagram   Perspectives from the *Temporal* HTP note that the water bottle timeline contains:

- The plastic raw material.
- The manufacturer who made the bottle.
- The water.

---

[*] Placing the section number in the completed column not only shows that the requirement has been met but shows where the response can be found, mitigating the risk of the instructor failing to find and grade the response.

TABLE 5.3

Compliance Matrix for the Exercise in Section 5.8.3.1.1

ID	Instruction	Deliverable	Section	Completed
1	Brainstorming or active brainstorming	No	All	Yes
2	Research	No	All	Yes
3	Use imagination	No	All	Yes
4	Use lists and charts	No	All	Yes
5	Spend less than 60 minutes	No	All	Yes
5	Format problem	Yes	5.8.3.1.1.2	Yes
6	Assumptions	Yes	5.8.3.1.1.1	Yes
7	Preliminary compliance matrix	No	5.8.3.1.1.3	Yes
8	Context diagram	Yes	5.8.3.1.1.4	Partial
9	CONOPS scenarios	Yes	5.8.3.1.1.4	Partial
10	Level of difficulty (Section 5.1.3.1)	Yes	5.8.3.1.1.5	Yes
11	Level of objective complexity (Section 5.2.2.2.2)	Yes	5.8.3.1.1.6	Yes
12	Type of problem (Section 5.1.4.4)	Yes	5.8.3.1.1.7	Yes
13	Structure of the problem (Section 5.1.4.5)	Yes	5.8.3.1.1.8	Yes
14	Lessons learned	Yes	5.8.3.1.1.9	Yes
15	Updated compliance matrix	Yes	5.8.3.1.1.3	Partial

- The entity that filled the bottle with water.
- The lorry that transported the bottle to the shop.
- The shopkeeper that displayed the bottle.
- The shop that sold the bottle to the tourist.
- The tourist who brought it to Engaporia.
- The person who accepted the bottle from the tourist.
- The person who transported the bottle to the disposal facility.
- The entities involved in disposing of the water and the bottle.

The CONOPS scenarios include:

1. Presenting the bottle upon arrival at the airport. Options include:
   - Collection bins (attended and unattended) at the arrival gates.
   - Collection bins (attended and unattended) in the baggage collection area.
   - Collection bins (attended and unattended) in the immigration area.
2. Transporting the bottle to the disposal facility. Options include:
   - Collections at various times of the day.
   - Transportation by road or rail.
3. Disposing of the water. Options include:
   - Directly into the reservoir.
   - Into the water recycling plant for purification.
4. Disposing of the bottle. Options include:
   - Recycling.
   - Dumping in the landfill.
   - Burning to generate electrical power.

5.8.3.1.1.5   The Level of Difficulty of the Problem   The level of difficulty of the problem is easy because the activities performed by the policy in action already exist, e.g.:

- Notifying potential travellers of the policy via travel agents, websites and writing it on the immigration forms passed out to travellers at the port of departure by the airlines.
- Collecting and disposing of the water and bottles.

5.8.3.1.1.6   The Level of Objective Complexity   The problem is not very complex or medium because there are only few variables (water bottle, tourists, etc.)

5.8.3.1.1.7   The Type of Problem   The problem is an intervention problem because it will change an existing system using existing personnel and technology, so no research is needed.

5.8.3.1.1.8   The Structure of the Problem   The problem is well structured because both the problem and the FCFDS are known.

5.8.3.1.1.9   Lessons Learned   Lessons learned include:

- Perceptions from the *Temporal* HTP identified the entities involved as well as the scenarios.
- Perceptions from the *Operational* HTP provided more detail about the scenarios.
- Perceptions from the *Structural* HTP identified the type of equipment needed in the scenarios.
- Perceptions from the *Functional* HTP identified the functions being performed in the scenarios.
- Perceptions from the *Quantitative* HTP provided estimates of the number of water bottles to be processed based on a set of assumptions.
- Inferences from the *Scientific* HTP provided the conceptual solution system.
- The compliance matrix is very handy for keeping track of what needs to be done and making sure that it is being done.
- The lessons learned from the perceptions of the HTPs should have been learned when performing the exercise in Section 2.7.3.

5.8.3.1.2   The Problem Posed by the Engaporian Drought
*   Relief Policy in Section 3.10.3*

This section provides one acceptable solution, not the single correct solution as there will be other acceptable solutions (Section 5.1.1.2.2).

5.8.3.1.2.1   The Formatted Problem   The problem formatted according to the problem formulation template (Section 5.1.7.2) is:

TABLE 5.4

Compliance Matrix for the Exercise in Section 5.8.3.1.2

ID	Instruction	Deliverable	Section	Completed
1	Brainstorming or active brainstorming	No	All	Yes
2	Research	No	All	Yes
3	Use imagination	No	All	Yes
4	Use lists and charts	No	All	Yes
5	Spend less than 60 minutes	No	All	Yes
6	The preliminary compliance matrix	No	5.8.3.1.3.2	Yes
7	One generic risk in area[a]	Yes	3.10.3.1.3[b]	Partial
8	One specific risk in area[a]	Yes	3.10.3.1.3[b]	Partial
9	Lessons learned	Yes	5.8.3.1.2.3	Yes
10	Updated compliance matrix	Yes	5.8.3.1.3.2	Yes

[a] This should be one line per area in the risk framework to verify completion of all the requirements.
[b] There is no need to repeat the information here, so it has been referenced.

1. *The undesirable situation*: is the need to:
   1. Comply with the generic requirements in Section 2.7.
   2. Identify and document at least one generic and one specific risk (in each HTP) in as many areas of the risk framework shown in Table 1.2 based on the information documented in the *Operational*, *Functional*, *Generic* and *Structural* HTPs identified in the previous exercise (Section 2.7.2).
2. *The assumption*: the exercise can be completed within the 60 minutes.
3. *The FCFDS*: the exercise has been completed and the desired grade has been achieved.
4. *The problem*: how to do the exercise, namely:
   1. Meet the generic requirements in Section 2.7.
   2. Identify and document at least one generic and one specific risk
   3. Verify that all the requirements for the exercise have been completed and update the compliance matrix accordingly.
5. *The solution*: consists of working through each of the problems. A partial solution is presented below.

5.8.3.1.2.2 The Compliance Matrix The partially completed compliance matrix is shown in Table 5.4.

5.8.3.1.2.3 Lessons Learned Lessons learned include:

- Those learned in the exercise in Section 3.10.2.
- The compliance matrix is very handy for keeping track of what needs to be done and making sure that it is being done.

5.8.3.1.3 The Problem Posed by the Engaporian Drought Relief Policy in Section 4.10.3

This section provides one acceptable solution, not the single correct solution as there will be other acceptable solutions (Section 5.1.1.2.2).

5.8.3.1.3.1 The Formatted Problem   The problem formatted according to the problem formulation template (Section 5.1.7.2) is:

1. *The undesirable situation*: is the need to:
    1. Comply with the generic requirements in Section 2.7.
    2. Identify the specific change that will take place. For example, this change will increase the amount of water capacity w% by providing a ...
    3. Use force field analysis (Section 4.5.2) to determine at least five forces for and at least five forces against the change and estimate their strengths.
    4. Discuss which CM model or combination of models (Section 4.7.1) would be most suitable in this situation and why.
2. *The assumption*: the exercise can be completed within the 60 minutes.
3. *The FCFDS*: the exercise has been completed and the desired grade has been achieved.
4. *The problem*: how to do the exercise, namely:
    1. Meet the generic requirements in Section 2.7.
    2. Identify the specific change that will take place
    3. Use force field analysis (Section 4.5.2) to determine at least five forces for and at least five forces against the change and estimate their strengths.
    4. Discuss which CM model or combination of models (Section 4.7.1) would be most suitable in this situation and why.
    5. Verify that all the requirements for the exercise have been completed and update the compliance matrix accordingly.
5. *The solution*: consists of working through each of the problems. A partial solution is presented below.

5.8.3.1.3.2 The Compliance Matrix   The partially completed compliance matrix is shown in Table 5.5.

5.8.3.1.4 The Problem Posed the Exercise in Section 5.8.3 (This Exercise)

This section provides one acceptable solution, not the single correct solution as there will be other acceptable solutions (Section 5.1.1.2.2).

5.8.3.1.4.1 The Formatted Problem   The problem formatted according to the problem formulation template (Section 5.1.7.2) is:

1. *The undesirable situation*: is the need to:
    1. Comply with the generic requirements in Section 2.7.
    2. Remedy the problem posed by this exercise.
2. *The assumption*: the exercise can be completed within the 60 minutes.
3. *The FCFDS*: the exercise has been completed and the desired grade has been achieved.

TABLE 5.5

Compliance Matrix for the Exercise in Section 5.8.3.1.3

ID	Instruction	Deliverable	Section	Completed
1	Brainstorming or active brainstorming	No	All	Yes
2	Research	No	All	Yes
3	Use imagination	No	All	Yes
4	Use lists and charts	No	All	Yes
5	Spend less than 60 minutes	No	All	Yes
5	The preliminary compliance matrix	No	5.8.3.1.3.2	Yes
6	Identify the specific change	Yes	4.10.2.1.3[a]	Yes
7	The stakeholders affected by the change	Yes	4.10.2.1.4[a]	Yes
8	The forces for the change	Yes	4.10.2.1.5[a]	Yes
9	The forces against the change	Yes	4.10.2.1.5[a]	Yes
10	CM models	Yes	4.10.2.1.6[a]	Yes
11	Lessons learned	Yes	4.10.2.1.7[a]	Yes
12	Updated compliance matrix	Yes	5.8.3.1.3.2	Yes

[a] There is no need to repeat the information here.

4. *The problem*: how to do the exercise, namely:
   1. Meet the generic requirements in Section 2.7.
   2. Remedy the problem posed by this exercise which is to perform the three other parts and this one.
   3. Verify that all the requirements for the exercise have been completed and update the compliance matrix accordingly.
5. *The solution*: consists of working through each of the problems. A partial solution is presented below.

5.8.3.1.4.2   The Compliance Matrix   The partially completed compliance matrix is shown in Table 5.6.

5.8.3.1.4.3   Lessons Learned   Lessons learned included:
- The same lessons as for the previous exercises.
- Before starting the exercise:
  - Read and understand the instructions.
  - Create the compliance matrix.

This is an important lesson because the requirement for the three previous exercises was to format the problem posed by the exercises, not actually perform the exercises. Accordingly, of all work performed in:

- Section 5.8.3.1.1: the only required work was to perform Section 5.8.3.1.1.2.
- Section 5.8.3.1.2: the only required work was to perform Section 5.8.3.1.2.1.
- Section 5.8.3.1.3: the only required work was to perform Section 5.8.3.1.3.1.

TABLE 5.6

Compliance Matrix for the Exercise in Section 5.8.3.1.4

ID	Instruction	Deliverable	Section	Completed
1	Brainstorming or active brainstorming	No	All	Yes
2	Research	No	All	Yes
3	Use imagination	No	All	Yes
4	Use lists and charts	No	All	Yes
5	Spend less than 60 minutes	No	All	Yes
5	The preliminary compliance matrix	No	5.8.3.1.4.2	Yes
6	Format the problem posed by the Engaporian drought relief policy in Section 2.7.3	Yes	5.8.3.1.1.2	Yes
7	Format the problem posed by the Engaporian drought relief policy in Section 3.10.3	Yes	5.8.3.1.2.1	Yes
8	Format the problem posed by the Engaporian drought relief policy in Section 4.10.3	Yes	5.8.3.1.3.1	Yes
9	Format the problem posed by this exercise	Yes	5.8.3.1.4.1	Yes
10	Lessons learned	Yes	5.8.3.1.4.3	Yes
11	Updated compliance matrix	Yes	5.8.3.1.4.2	Yes

Accordingly, the rest of the sections were not required and resources were wasted completing them. In this situation, the reasons for performing the extra work were:

1. The compliance matrix was completed after the work had been done, so the requirement to only do the problem formatting was missed. This is a common real world situation.
2. The work wasn't wasted in this instance since it provided sample answers to the exercises in the previous sections.
3. It demonstrates the importance of the compliance matrix in not only verifying that the required work was performed but also helping to make sure that work not required is not performed an important factor in controlling costs and schedules.

5.8.4 THE OFF-WORLD MINING POLICY

In addition to meeting the generic requirements in Section 2.7:

1. Format the problem posed by the off-world mining policy in Section 2.7.4 in the problem formulation template (Section 5.1.7.2).
2. Format the problem posed by the off-world mining policy in Section 3.10.3 in the problem formulation template (Section 5.1.7.2).
3. Format the problem posed by the off-world mining policy in Section 4.10.4 in the problem formulation template (Section 5.1.7.2).
4. Format the problem posed by this exercise in the problem formulation template (Section 5.1.7.2).

5. For each of the four formatting problems:
   1. State the level of difficulty of the problem (Section 5.1.3.1) with the reasons for the choice.
   2. State the level of objective complexity of the problem (Section 5.2.2.2.2) with the reasons for the choice.
   3. State the type of problem (Section 5.1.4.4) with the reasons for the choice.
   4. State the structure of the problem (Section 5.1.4.5) with the reasons for the choice.

5.9 SUMMARY

The risk management process is riddled with changes, problems and solutions because managing risks poses problems which need to be solved or remedied and change is an outcome, mostly desired but sometimes undesired. A risk mitigation strategy is a solution to the problem of managing the risk but poses a problem to the people who will have to implement the strategy. Accordingly, this chapter discussed perceptions of the problem-solving process from a number of HTPs, then discussed the structure of problems and the levels of difficulty posed by problems and the need to evolve solutions using an iterative approach. After showing that problem-solving is really an iterative causal loop rather than a linear process, the chapter discussed complexity and how to use the systems approach to manage complexity. The chapter then showed how to remedy well-structured problems and how to deal with ill-structured, wicked and complex problems using iterations of a sequential two-part problem-solving process.

Reflecting on this chapter, it seems that iteration is a common element in remedying any kind of problem other than easy well-structured ones irrespective of their structure.

REFERENCES

Ackoff, Russel L. 1978. *The Art of Problem Solving*. New York: John Wiley & Sons.

Ackoff, Russel L., and Herber J. Addison. 2006. *A Little Book of f-Laws 13 Common Sins of Management*. Axminster: Triarchy Press Limited.

Alexander, J. Eugene, and John Milton Bailey. 1962. *Systems Engineering Mathematics*. Englewood Cliff, NJ: Prentice-Hall, Inc.

Allison, James T. 2004. *Complex System Optimization: A Review of Analytical Target Cascading, Collaborative Optimization, and Other Formulations*. Ann Arbor: The University of Michigan.

APSC. 2007. *Tackling Wicked Problems: A Public Policy Perspective*. Canberra: The Australian Public Service Commission (APSC).

Avison, David, and Guy Fitzgerald. 2003. *Information Systems Development: Methodologies, Techniques and Tools*. Maidenhead: McGraw-Hill Education (UK).

Barry, Katie, Ellen Domb, and Michael S. Slocum. 2007. *TRIZ – What is TRIZ? 2007* [cited 31 October 2007]. Available from http://www.triz-journal.com/archives/what_is_triz/.

Bosch-Rekveldt, Marian, Yuri Jongkind, Herman Mooi, and Alexander Verbraeck. 2010. Grasping project complexity in large engineering projects: The TOE (Technical, Organizational and Environmental) framework. *International Journal of Project Management*. doi:10.1016/j.ijproman.2010.07.008.

Chestnut, Harold. 1965. *Systems Engineering Tools*. Edited by Harold Chestnut, *Wiley Series on Systems Engineering and Analysis*. New York: John Wiley & Sons, Inc.

Colwell, Bob. 2005. Complexity in design. *IEEE Computer* no. 38 (10):10–12.

Dery, David. 1984. *Problem Definition in Policy Analysis*. Lawrence: University Press of Kansas.

Dunn, William N. 2012. *Public Policy Analysis*. 5th edition. Upper Saddle River, NJ: Pearson Education Inc.

Eisner, Howard. 1988. *Computer Aided Systems Engineering*. Englewood Cliffs, NJ: Prentice Hall.

ElMaraghy, Waguih, Hoda ElMaraghy, Tetsuo Tomiyama, and Laszlo Monostori. 2012. Complexity in engineering design and manufacturing. *CIRP Annals – Manufacturing Technology* no. 61 (2):793–814.

Fischer, Andreas, Samuel Greiff, and Joachim Funke. 2012. The process of solving complex problems. *The Journal of Problem Solving* no. 4 (1):19–42.

Ford, Whit. 2010. *Learning and teaching math 2010* [cited 8 April 2015]. Available from http://mathmaine.wordpress.com/2010/01/09/problems-fall-into-four-categories/.

Hall, Arthur D. 1962. *A Methodology for Systems Engineering*. Princeton, NJ: D. Van Nostrand Company Inc.

Hari, Amihud, Shraga Shoval, and Joseph Eli Kasser. 2008. *Conceptual design to cost: A new systems engineering tool*. In *The* 18th *International Symposium of the INCOSE*. Utrecht, Holland.

Hitchins, Derek K. 2007. *Systems Engineering. A 21st Century Systems Methodology*. Chichester: John Wiley & Sons Ltd.

Howard, Ronald A. 1973. Decision analysis in systems engineering. In *Systems Concepts*, edited by Ralph F. Miles Jr, 51–85. Hoboken, NJ: John Wiley & Son, Inc.

Hughes, Thomas P. 1998. *Rescuing Prometheus*. New York: Random House Inc.

Jackson, Michael C., and Paul Keys. 1984. Towards a system of systems methodologies. *Journal of the Operations Research Society* no. 35 (6):473–486.

Jonassen, David H. 1997. Instructional design model for well-structured and ill-structured problem-solving learning outcomes. *Educational Technology: Research and Development* no. 45 (1):65–95.

Kasser, Joseph Eli. 1995. *Applying Total Quality Management to Systems Engineering*. Boston, MA: Artech House.

Kasser, Joseph Eli. 2002a. Configuration management: The silver bullet for cost and schedule control. In *the IEEE International Engineering Management Conference (IEMC 2002)*. Cambridge, UK.

Kasser, Joseph Eli. 2002b. Isn't the acquisition of a System of Systems just a simple multi-phased time-ordered parallel-processing process? In the *11th International Symposium of the INCOSE*. Las Vegas, NV.

Kasser, Joseph Eli. 2015a. *Holistic Thinking: Creating Innovative Solutions to Complex Problems*. 2nd edition. Vol. 1, *Solution Engineering*. Charleston, SC: Createspace Ltd.

Kasser, Joseph Eli. 2015b. *Perceptions of Systems Engineering*. Vol. 2, *Solution Engineering*. Charleston, SC: Createspace Ltd.

Kasser, Joseph Eli. 2018a. *Systems Thinker's Toolbox: Tools for Managing Complexity*. Boca Raton, FL: CRC Press.

Kasser, Joseph Eli. 2018b. Using the systems thinker's toolbox to tackle complexity (complex problems). In *SSSE Presentation at Roche*. Zurich: Swiss Society of Systems Engineering.

Kasser, Joseph Eli. 2019a. *Systemic and Systematic Project Management*. Boca Raton, FL: CRC Press.

Kasser, Joseph Eli. 2019b. *Systems Engineering: A Systemic and Systematic Methodology for Solving Complex Problems.* Boca Raton, FL: CRC Press.

Kasser, Joseph Eli, and Derek K. Hitchins. 2012. Yes systems engineering, you are a discipline. In *the 22nd International Symposium of the INCOSE.* Rome, Italy.

Kasser, Joseph Eli, and Derek K. Hitchins. 2013. Clarifying the relationships between systems engineering, project management, engineering and problem solving. In *Asia-Pacific Council on Systems Engineering Conference (APCOSEC).* Yokohama, Japan.

Kasser, Joseph Eli, and Kent Palmer. 2005. Reducing and managing complexity by changing the boundaries of the system. In *the Conference on Systems Engineering Research.* Hoboken NJ.

Kasser, Joseph Eli, and Yang Yang Zhao. 2017. The myths and the reality of problem-solving. In *the 27th International Symposium of the INCOSE.* Adelaide, Australia.

Kasser, Joseph Eli, and Yang-Yang Zhao. 2016b. Simplifying solving complex problems. In the *11th International Conference on System of Systems Engineering.* Kongsberg, Norway.

Kasser, Joseph Eli, and Yang-Yang Zhao. 2016c. Wicked problems: Wicked solutions. In *the 11th International Conference on System of Systems Engineering.* Kongsberg, Norway.

Kuhn, Thomas S. 1970. *The Structure of Scientific Revolutions.* Second edition, Enlarged edition. Chicago, IL: The University of Chicago Press.

Lui, Pao Chuen. 2007. An example of large scale systems engineering. Paper read at Keynote presentation at the *1st Asia-Pacific Systems Engineering Conference,* at Singapore.

Maier, Mark K., and Eberhardt Rechtin. 2000. *The Art of Systems Architecting.* 2nd edition: Boca Raton, FL: CRC Press.

Merton, Robert King. 1936. The unanticipated consequences of social action. *American Sociological Review* no. 1 (6):894–904.

Merton, Robert King. 1948. The self-fulfilling prophecy. *The Antioch Review* no. 8 (2):193–210.

Miller, George. 1956. The magical number seven, plus or minus two: Some limits on our capacity for processing information. *The Psychological Review* no. 63:81–97.

Needham, Joseph. 1937. *Integrative Levels: A Revaluation of the Idea of Progress.* Oxford: Clarendon Press.

Quesada, Jose, Walter Kintsch, and Emilio Gomez. 2005. Complex problem solving: A field in search of a definition? *Theoretical Issues in Ergonomic Science* no. 6 (1):5–33.

Rittel, Horst W., and Melvin M. Webber. 1973. Dilemmas in a general theory of planning. *Policy Sciences* no. 4:155–169.

Rumsfeld, Donald. 2019. *DoD news briefing – Secretary Rumsfeld and Gen. Myers,* February 12, 2002. DoD 2002 [cited 18 January 2019]. Available from http://archive.defense.gov/ Transcripts/Transcript.aspx?TranscriptID=2636.

Rush, Benjamin C. 1977. *Cost as an Independent Variable: Concepts and Risks.* Fort Belvoir, VA: Defense Systems Management College.

Russo, J. Edward, and Paul H. Schoemaker. 1989. *Decision Traps.* New York: Simon and Schuster.

Savage, Sam L. 2009. *The Flaw of Averages.* Hoboken, NJ: John Wiley and Sons, Inc.

Schön, Donald A. 1991. *The Reflective Practitioner.* Burlington, VA: Ashgate.

Sillitto, Hillary. 2009. On systems architects and systems architecting: Some thoughts on explaining and improving the art and science of systems architecting. In *the 19th International Symposium of the INCOSE.* Singapore.

Simon, Herbert A. 1973. The structure of ill structured problems. *Artificial Intelligence* no. 4 (3–4):181–201. doi:10.1016/0004-3702(73)90011-8.

Waring, Alan. 1996. *Practical Systems Thinking.* London: International Thompson Business Press.

Wilson, Tom D. 2002. Philosophical foundations and research relevance: Issues for information research (Keynote address). In the *Fourth International Conference on Conceptions of Library and Information Science: Emerging Frameworks and Method.* University of Washington, Seattle, USA.

Wilson, Warren E. 1965. *Concepts of Engineering System Design.* New York: McGraw-Hill Book Company.

Yen, Duen Hsi. 2008. *The blind men and the elephant 2008* [cited 26 October 2010]. Available from http://www.noogenesis.com/pineapple/blind_men_elephant.html.

6 Systems and Systems Engineering

This chapter discusses systems and systems engineering from different perspectives because policies and projects are systems which are realized using systems engineering* and project management working interdependently. Moreover, the project takes place in an environment or context which is a system. So, this chapter provides an overview of systems, the system lifecycle (SLC), systems engineering and the system development process (SDP) to help the policymaker, implementer and project manager understand the important aspects of risk management in systems and systems engineering and the interdependency and overlap between risk management in systems engineering and project management. Specifically, this chapter discusses:

- The nature of systems in Section 6.1.
- Properties of systems in Section 6.2.
- Hierarchies of systems in Section 6.3.
- Supply chains as systems in Section 6.4.
- An introduction to systems engineering in Section 6.5 including the interdependency and overlap between risk management in systems engineering and project management in Section 6.5.8.
- Modelling and simulation in Section 6.6.
- The nine-system model in Section 6.7.
- Risks in systems and systems engineering in Section 6.8.

6.1 THE NATURE OF SYSTEMS

Using perceptions from the holistic thinking perspectives (HTPs) (Section 2.5.1), the nature of systems can be summarized as follows:

6.1.1 BIG PICTURE

Perceptions from the *Big Picture* HTP include:

1. Systems have external observer(s).
2. Systems exist within containing systems or meta-systems.

---

* Even if they don't call it systems engineering.

6.1.2 Operational

Perceptions from the *Operational* HTP include:

- Systems performing their purpose using resources to transform inputs to outputs.

6.1.3 Functional

Perceptions from the *Functional* HTP include:

- Many systems transforming inputs into outputs performing appropriate internal functions.

6.1.4 Structural

Perceptions from the *Structural* HTP include:

- A system:
  1. Can be almost anything, including products, objects, things, processes, methodologies and ways of doing or arranging something such as a betting system or a classification system.
  2. Contains the following minimum set of common elements:
     1. An external boundary.
     2. More than one internal component.
     3. Interactions between the internal components.
  3. Generally, has:
     1. Inputs.
     2. Outputs.
     3. Subsystems.
     4. Architectures: data, physical, logical, etc.
- Man-made systems contain three kinds of components (Ramo 1973: p. 24), namely:
  1. *Technology*: the equipment and material within the system boundary, often organized into subsystems.
  2. *People*: trained to operate, maintain and interact with the system
  3. *Information*: acquired, processed, stored and disseminated to internal and external users. This information tells the people and technology what to do and when and where to do it, all of which makes the system operate properly in its designed context.

The support systems may be considered as subsystems in the mission – support subsystem architecture – or as adjacent systems or a combination as shown in Figure 6.1. For example, in the discussion on the camera (Section 2.6.1), the charger may be an internal subsystem or an external adjacent system depending on the design.

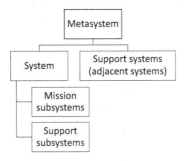

FIGURE 6.1 The generic structure of a system.

System and subsystem boundaries depend on the viewer's perspective. For example, consider the objects shown in Figure 6.2. The top level is an air defence force fighter wing,* and the next level is red leader which consists of two parts: the aircraft and the pilot. The parts of the aircraft include ordnance, an airframe, navigation, propulsion and guidance. Each part can be a system or a subsystem. For example, the aircraft may be a subsystem of red leader, but as far as the airframe or the propulsion parts are concerned, they are subsystems of the system known as the aircraft. This is why the definitions of the system and subsystem are identical and produces the dictum, 'one person's system is another person's subsystem'.

6.1.5 Temporal

Perceptions from the *Temporal* HTP include:

- The past, present and future of systems discussed in Section 6.1.5.1.
- Basic system behaviour over time discussed in Section 6.1.5.2.

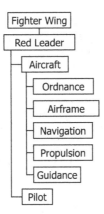

FIGURE 6.2 System or sub-subsystem.

---

* Which is part of a squadron, which is part of an air defence system (ADS).

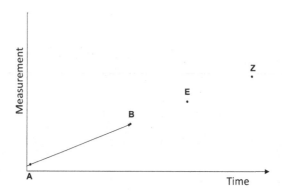

FIGURE 6.3   Predicting the behaviour of a system over time.

- The SLC discussed in Section 6.1.5.3.
- The definitions of a system have changed over the past 40 years and are still changing (Kasser 2013: pp. 178–179).

6.1.5.1   The Past, Present and Future of Systems

Perceptions from the *Temporal* HTP note that systems have a:

- *Past*: which set up the situation (context) and the current state of the system.
- *Present*: which is the current state of the system and its context (meta-system).
- *Future*: which is where the system is evolving at the same time as the adjacent systems are also evolving.

6.1.5.2   Basic System Behaviour

Perceptions from the *Temporal* HTP show system behaviour over time and are often plotted in graphs. Perceive the behaviour of a system between points A and B in Figure 6.3 (Kasser 2015a), draw a line joining points B and E and then extend it to Z. Most people will draw a straight linear line from B to E to join the points and then draw another straight line joining E and Z. Apply some critical thinking and ask why straight lines and why not curves. In general, the lines drawn joining points A, B and E represent the known (observed) behaviour over time. When asked to extrapolate the line beyond point E to point Z, most people continue the line in the same direction. This action is based upon the assumption that the conditions that resulted in the behaviour of the system between points A and B will not change in the near future. This is the assumption people use to predict future behaviour with various degrees of accuracy.

The types of system behaviour include:

- *Cyclic*: the system exhibits repetitive patterns of behaviour or cycles when perceived from the *Temporal* HTP. Examples of such cycles include the annual seasons (spring, summer, autumn and winter) and the 11-year sunspot cycle.* Many people think the same type of patterned behaviour applies to the stock market.

- *Exponential growth*: the system grows by a fixed percentage at regular intervals of time. Exponential growth is often caused by positive feedback in the system.
- *Goal seeking*: The system moves from a starting position to a goal or target value over a period of time. When shown as a graph, the starting position may be above or below the goal. Depending on the rate of change, the system may overshoot the goal and oscillate a few times until a stable state is reached. Perceived from the *Generic* HTP, the multiple-iteration problem-solving process (Section 5.3.2) can be considered as goal-seeking behaviour in which the system is moved from being in an undesirable situation to the goal or feasible conceptual future desirable situation (FCFDS) through a number of iterations.
- *Homeostatic*: the system tends to maintain a stable condition with some small variation about a mean value generally caused by negative feedback in the system responding to a change. Heating and cooling systems exhibit this behaviour when functioning correctly. A classic example of this type of system is a thermostat.
- *Random*: the system changes in a way that is not understood and cannot be predicted.
- *S-shaped growth*: the system grows exponentially for a period of time, then flattens out at the top of the curve in the manner of goal-seeking behaviour. This curve is often used to describe market penetration of new products where the top of the curve represents a saturated market.

6.1.5.3   The SLC of a Man-Made System

The SLC of a man-made system may be considered as the set of states in which a system exists, starting from its conception through development and operation, and ending after its disposal. The SLC contains the following states per the Hitchins-Kasser-Massie framework (HKMF) (Section 3.4.4).

A. The needs identification state discussed in Section 6.1.5.4.3.
B. The system requirements state discussed in Section 6.1.5.4.4.
C. The system and subsystem preliminary states discussed in Section 6.1.5.4.5 which comprises:
   - The preliminary system and subsystem design states discussed in Section 6.1.5.4.6.
   - The detailed system and subsystem design states discussed in Section 6.1.5.4.7.

---

* For example, in 1970 I noted that the average price of gasoline in the northwest suburbs of Detroit seemed to vary between 18 and 32 cents in a cyclic manner. When I plotted the price daily, I saw that the price rose at a rate of about 1 cent every two days until it reached about 32 cents then dropped overnight to 18 cents. Then, after a day or so, the price started to rise again. The cycle repeated for at least 13 more iterations; the graph showed a perfect sawtooth pattern. Knowledge of the cycle allowed the fall in price to be predicted and taken advantage of by waiting for the price to fall to 18 cents a gallon and filling the tank the day before the price rose to 32 cents.

Gantt chart (*Temporal* HTP)

Needs Identification [NI]						
System Requirements [SR]						
System and Subsystem Design [SD]						
Subsystem Construction [SC]						
Subsystem Testing [ST]						
System Integration and System Test [SIT]						

N^2 and Waterfall chart
(*Functional* HTP)

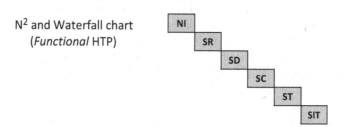

FIGURE 6.4 The SLC: a planning view.

D. The subsystem construction state discussed in Section 6.1.5.4.8.
E. The subsystem testing state discussed in Section 6.1.5.4.9.
F. The system integration and system test states discussed in Section 6.1.5.4.10.
G. The operations and maintenance (O&M) state discussed in Section 6.1.5.4.11.
H. The disposal state discussed in Section 6.1.5.4.14.

While there is no generally accepted set of system states in the SLC, the states are normally thought of as a linear time-ordered sequential process and shown as a Gantt chart or an N^2 chart or waterfall chart (Royce 1987) as shown in Figure 6.4. The waterfall chart is an N^2 chart with the row and column lines blanked out. States A through F can be considered as the SDP. The SDP and SLC map into the problem-solving process shown in Figure 5.11 as shown in Table 6.1. Each state in the SLC may be considered as starting with a problem and ending with a solution.

The activities performed in each state are discussed in *Systems Engineering: A Systemic and Systematic Methodology for Solving Complex Problems* (Kasser 2019) and summarized herein.

6.1.5.4   The SDP

The SDP:

- Often takes place within the context of a project.
- When using the waterfall, is a subset of the SLC (Section 6.1.5.3) consisting of the first seven states of the SLC beginning with the needs identification state and ending when the system becomes operational at the start of the initial O&M state as shown Table 6.2.

TABLE 6.1

Mapping the SDP into Traditional Generic Problem-Solving Process

State in Problem Solving Process	SDP State	Ending Milestone
Defining the problem	Needs identification	OCR
	System requirements	SRR
Conceiving at least two candidate solutions	System design	PDR
Identifying evaluation criteria for choosing between the candidate solutions	System design	PDR
Selecting the preferred solution	System design	CDR
Implementing the solution	System realization	IRR
Verifying the solution remedies the problem	System integration and system test	DRR

- May be:

    1. An acquisition process to purchase commercial off the shelf (COTS) equipment (Section 6.1.5.5).
    2. A development process to design, build, test and integrate a system.
    3. An integration process to integrate equipment (subsystems) acquired from one or more vendors.
    4. A combination of the above.

6.1.5.4.1 The Problem Posed by the SDP

The problem posed by the SDP can be formulated in the problem formulation template (Section 5.1.7.2) as follows.

1. *The undesirable situation*: a situation in which a desired or needed function cannot be achieved. This need may have:
    1. Appeared since the system in the undesirable situation was placed into service.

TABLE 6.2

The SLC, SDP and HKMF States

State	HKMF State	SDP	SLC
Needs identification	A	Yes	Yes
System requirements	B	Yes	Yes
System design	C	Yes	Yes
Subsystem construction	D	Yes	Yes
Subsystem testing	E	Yes	Yes
System integration and system test	F	Yes	Yes
Operations and maintenance	G		Yes
Disposal	H		Yes

2. Arisen because a desired function that was not feasible when the system was originally placed into service has only now become feasible often by the development of new technology.

3. Something is no longer needed and needs to be removed because it is causing the situation to exhibit undesirable behaviour.

4. A combination of the above.

2. *The Assumptions*: specific to the situation.

3. *FCFDS*: the same as undesirable situation but without the undesirable aspects of undesirable situation with perhaps added desirable functionality.

4. *The problem*: to affect a transition from the undesirable situation to the FCFDS. This well-structured complex problem is generally elaborated into several sequential well-structured problems (Section 5.1.4.5.1) including:

   • Determining the root causes of the undesirability in undesirable situation.

   • Determining how to transition from the undesirable situation to the FCFDS by formulating the strategies and plans to realize the solution system.

   • Designing and realizing the solution system in its context in accordance with the plan.

   • Deploying the solution system to complete the transition from the undesirable situation to the future desirable situation.

   • Verifying that the created situation remedies the original and evolved needs and does not seem to contain any undesirable characteristics; this is performed in OT&E.

5. *The solution*: performing the transition process to realize the solution system operating in its context (the FCFDS) sometime in the future by remedying the set of sequential well-structured problems.

6.1.5.4.2 Milestones

Each state starts and ends at a major milestone. While there is also no generally accepted set of major project milestones, the major project milestones used in most system and software development project processes can be mapped into the following template:

• Start-up meeting (which formally starts the SDP).
• Operations concept review (OCR).
• System requirement review (SRR).
• Preliminary design review (PDR).
• Critical design review (CDR).
• Subsystem test readiness review (TRR).
• Integration readiness review (IRR).
• Delivery readiness review (DRR).
• Acceptance test review (ATR) which formally terminates the project.

TABLE 6.3
The Notional States in the SDP

HKMF State	State	Start	End	Produces
A	Needs identification	Start	OCR	Common vision of the system in operation, system architecture
B	System requirements	OCR	SRR	System requirements specifications, project plan
C	System design	SRR	CDR	System design
D	Subsystem construction	CDR	TRR	Subsystems
E	Subsystem test	TRR	IRR	Tested subsystems
F	System integration and system test	IRR	DRR	The tested system
G	Delivery and handover	DRR	ATR	The products or services for which the system was created in operation
G	Closeout	ATR	End	Final reports and lessons learned

Each milestone review covers two sets of work:

1. Work accomplished prior to the milestone.
2. Work to be accomplished before the next milestone review.

The relationship between the states, milestones and major products produced during each state for delivery at the milestones in the 'A' paradigm (Section 6.5.2.1) is shown in Table 6.3.

6.1.5.4.3 The Needs Identification State

This section summarizes the needs identification state from the problem-solving perspective. The needs identification state:

- Is one of the two most complex states in the SLC.
- Begins when the entity experiencing the undesirable situation decides to determine if the undesirable situation can be remedied in an affordable and timely manner.
- Contains the early state systems engineering activities addressing the problem and determining the conceptual solution (Brill 1998, Gelbwaks 1967, Hall 1962, Hitchins 1992).
- Ends at the close of the OCR.

6.1.5.4.3.1 The Problem Posed by the Needs Identification State Using the problem formulation template (Section 5.1.7.2), the needs identification state can be formulated as follows:

1. *The undesirable situation*: may be:
    1. A system providing the current capability is in the in-service state but cannot provide the needed capability. For example, because it has either

reached the point where it can no longer perform its mission or is about to do so in the foreseeable future. Accordingly, the system needs to be upgraded or replaced.

2. Some functionality is needed, but there is no existing system that can provide it.

2. *The assumptions*: are situation specific.
3. *The FCFDS*: (outcome) stakeholder consensus that:
   1. The CONOPS of a conceptual system* operating in its context† will constitute a conceptual future situation without any of the current undesirable characteristics.
   2. The conceptual system is affordable and feasible, and can be realized in a timely manner according to the preliminary project plan (Section 7.4.2.1).
4. *The problem*: is to:
   1. Gain an understanding of the causes of undesirability of the undesirable situation.
   2. Gain consensus on those causes by the stakeholders (desirable but not necessary).
   3. Articulate a proposed conceptual solution system that, when operating in its context, will remedy the undesirable aspects of the situation.
   4. Gain stakeholder consensus that the conceptual solution system, when operating in its context, will remedy the undesirable aspects of the situation.
   5. Determine that there is at least one affordable and feasible way to turn the conceptual solution system into reality.
5. *The solution*: create a new, upgraded or replaced system by following the steps of the problem-solving process to produce an articulation of the proposed conceptual system and a way to make it happen in the form of an approved CONOPS (Section 6.5.9.1) and feasibility study (Section 7.4.1.1) or AoA study (Section 7.4.1.2) and a systems engineering management plan (SEMP) (Section 6.5.9.2) or project plan (Section 7.4.2.1).

6.1.5.4.3.2   Generic Risks in the Needs Identification State   Generic risks in the needs identification state include:

- The state is skipped and the project starts in the system requirements state, namely the project uses the 'B' paradigm which is inherently flawed (Section 6.5.2.2).
- The feasibility study or AoA is not performed often resulting in a doomed project.
- The common vision of the purpose and performance of the solution system among the stakeholders is not developed resulting in the wrong system being produced and wasted efforts leading to cost and schedule overruns.

---

* S6 in the nine-system model (Section 6.7).
† S7 in the nine-system model (Section 6.7).

Ways of mitigating or preventing these risks include:

- Perform the feasibility study or AoA.
- Don't use the 'B' paradigm; use the 'A' paradigm.
- Create that common vision.

6.1.5.4.4 The System Requirements State of the SDP

This section summarizes the system requirements state from the problem-solving perspective. The system requirements state:

- Begins at the close of the OCR and ends at the close of the SRR.
- Has two components:
  1. The systems engineering component traditionally taught in classes on systems engineering.
  2. The project management component traditionally taught in classes on project management.

The systems engineering component traditionally focuses on writing the right requirements during this state of the SDP to produce the system requirements document or specification that is signed off at the SRR. However, during this state of the SDP, systems engineering management or system engineers and project managers working interdependently also architect the process (Kasser 2005) to produce the version of the SEMP (Section 6.5.9.2) or project plan (Section 7.4.2.1) that is signed off at the SRR.

6.1.5.4.4.1 The Problem Posed by the System Requirements State Using the problem formulation template (Section 5.1.7.2), the system requirements state can be formulated as follows:

1. *The undesirable situation* at the start of the state is the lack of:
   1. The complete set of matched specifications for the conceptual solution system.
   2. The detailed strategies and plans to implement the transition from the undesirable situation to the future situation without the undesirable characteristics.
2. *The assumptions*: include:
   1. Ideally, the stakeholders have signed off on the CONOPS (Section 6.5.9.1) and have a common vision of the solution system's mission (namely the 'A' paradigm (Section 6.5.2.1)).
   2. If the stakeholders don't have this vision (namely the 'B' paradigm (Section 6.5.2.2)), then it will have to be created before or while eliciting, elucidating and validating the requirements.
3. *The FCFDS* is a complete set of:
   1. Matched specifications for the solution system.
   2. Detailed strategies and plans for the process to implement the transition from the undesirable situation to the future situation without the

undesirable characteristics, namely accepted versions of the SEMP (Section 6.5.9.2) or project plan (Section 7.4.2.1), test and evaluation master plan (TEMP) (Section 6.5.9.2) and the Shmemp (Section 6.5.9.5), etc. as appropriate.

4. *The problem* is to create the FCFDS from the CONOPS and other sources such as the feasibility study and relevant laws and regulations as well as incorporating any changes that occur during the system requirements state.

5. *The solution*: follow variations of the problem-solving process to produce the FCFDS depending on the paradigm using methods discussed in this chapter or any other appropriate method.

6.1.5.4.4.2   Generic Risks in the System Requirements State   Generic risks in the system requirements state include the perennial problem of poor requirements (Section 6.8.3) due to two basic causes:

1. Production of poorly written requirements.
2. Lack of ways of ensuring completeness of the requirements.

Ways of mitigating or preventing these risks include:

- Using the 'A' paradigm.
- Training in writing requirements at the start of the state.
- Tagging requirements with acceptance criteria to clarify the meaning of the requirement when it is written.

6.1.5.4.5   The System and Subsystem Design State
This section summarizes the system and subsystem design state from the problem-solving perspective. The system and subsystem design state:

- Begins at the close of the SRR and ends at the close of the CDR.
- Is the state in which the system and subsystem designs are created.
- Converts the conceptual system documented in the matched set of specifications into a realizable design which remedies the undesirable situation as it exists at the end of the state (and is robust enough to cope with all foreseen high probability future changes).
- Is split into two parts:
  1. The preliminary system design sub-state (SRR to PDR).
  2. The detailed system design sub-state (PDR-CDR).

6.1.5.4.6   The Preliminary System Design Sub-state (SRR–PDR)
The preliminary system design sub-state:

- Begins at the close of the SRR and ends at the close of the PDR.
- Is the sub-state in which several conceptual designs are created, but only one is selected in accordance with the problem-solving process shown in Figure 5.3.

6.1.5.4.6.1 The Problem Posed by the Preliminary System Design Sub-state The problem posed by the preliminary system design sub-state (SRR to PDR) can be formulated as follows:

1. *The undesirable situation*: at the end of the SRR is the lack of preliminary designs for the solution system that meets the matched set of specifications accepted at SRR.
2. *The assumptions*: are:
    1. Consensus that the matched set of requirements specified a system that when operating in its context will remedy the undesirable situation as it exists at the SRR. Note the undesirable situation may have evolved since the SDP began.
    2. Authorization to proceed with the SDP was received during the SRR.
3. *The FCFDS*: is consensus that a feasible preliminary design for the solution system:
    1. Meets the matched set of specifications accepted at SRR.
    2. Remedies the original and evolved undesirable situation.
4. *The problem*: is to create the FCFDS by:
    1. Converting the matched set of specifications to a preliminary design.
    2. Gaining the consensus that the preliminary design represents a system that will meet the need of remedying the evolved undesirable situation at the time of the PDR.
5. *The solution*: the creation, presentation and acceptance of the preliminary design at the PDR by following an appropriately customized version of the problem-solving process, namely the SDP.

6.1.5.4.6.2 Generic Risks in the Preliminary System Design Sub-state Generic risks in the preliminary system design sub-state include:

- Errors of omission (Section 5.1.4.3.1) in creating the preliminary design.
- Errors of commission in converting the requirements into the preliminary design.
- Failure to perform failure mode analyses on the design.
- Failure to develop the maintenance concepts for the system in operation.

Ways of mitigating or preventing these risks include holding the PDR which provides consensus that the attendees or their staff have examined the preliminary design and to the best of their knowledge have not found any errors of commission or errors of omission (Section 5.1.4.3.1).

6.1.5.4.7 The Detailed System Design Sub-state (PDR–CDR)
The detailed system design sub-state:

- Begins immediately after authorization to proceed has been received at the end of the PDR and ends at the CDR.
- Follows the problem-solving process to convert the preliminary design into the final design that will be created during the remainder of the SDP.

6.1.5.4.7.1   The Problem Posed by the Detailed System Design Sub-state   The problem posed by the detailed system design sub-state (PDR to CDR) can be formulated as follows:

1. *The undesirable situation*: at the end of the PDR is the lack of a final design for the solution system that meets the matched set of specifications accepted at SRR.
2. *The assumptions*:
    1. Consensus that the preliminary design represents a system that will meet the need of remedying the evolved undesirable situation at the time of the PDR.
    2. Authorization to proceed to the CDR has been received.
3. *The FCFDS*: is consensus that a final design for the solution system:
    1. Meets the matched set of specifications accepted at SRR and updated during the state.
    2. Remedies the original and evolved undesirable situations.
    3. Is feasible.
4. *The problem*: is to create the FCFDS by:
    1. Converting the preliminary design to the feasible documented critical or final design.
    2. Gaining the consensus that the design represents a system that will meet the need of remedying the evolved undesirable situation at the time of the CDR.
5. *The solution*: the creation, presentation and acceptance of the final design at the CDR by following the appropriate customized version of the problem-solving process.

6.1.5.4.7.2   Generic Risks in the Detailed System Design Sub-state   Generic risks in the detailed system design sub-state include:

- Errors of omission (Section 5.1.4.3.1) in creating the final design.
- Errors of commission in converting the preliminary design into the final design.

Ways of mitigating or preventing these risks include holding the CDR which provides consensus that the attendees or their staff have examined the final design and to the best of their knowledge have not found any errors of commission or errors of omission (Section 5.1.4.3.1).

6.1.5.4.8   The Subsystem Construction State
This section summarizes the subsystem construction state from the problem-solving perspective. The subsystem construction state:

- Begins at the close of the CDR and ends at the close of the TRR.
- Is where the bulk of the technical activities move from systems engineering to engineering following the SDP created in the system requirements state as documented in the SEMP (Section 6.5.9.2) or project plan (Section 7.4.2.1).

6.1.5.4.8.1 The Problem Posed by the Subsystem Construction State The problem posed by the subsystem construction state can be formulated as follows:

1. *The undesirable situation* at the start of the state is the need to construct each subsystem, often in isolation, according to the final design approved at the CDR.
2. *The assumptions*: the design specifications are feasible with cost and schedule constraints.
3. *The FCFDS*: each subsystem, constructed in isolation, operating according to the final design approved at the CDR.
4. *The problem*: to construct each subsystem in isolation according to the final design approved at the CDR in such a manner that the subsystem should meet all its specifications when tested.
5. *The solution*: follow the SDP.

Each subsystem passes through its own individual SDP in parallel to all the others.

6.1.5.4.8.2 Generic Risks in the Subsystem Construction State Generic risks in the subsystem construction state include:

- Individual subsystems not meeting their performance specifications due to errors of commission or errors of omission (Section 5.1.4.3.1).
- Individual subsystems exceeding their weight, size, power and other non-functional requirements.

Ways of mitigating or preventing these risks include an active role by the systems engineer who monitors the SDP of each subsystem and performs system wide trade-offs in the event problems arise (Kasser 2019: Section 12.3).

6.1.5.4.9 The Subsystem Test State
This section summarizes the subsystem test state from the problem-solving perspective. The subsystem test state:

- Begins at the close of the subsystem TRR and ends at the close of the IRR.
- Is the state in which the subsystems are tested in isolation following the process created in the system requirements state as documented in the SEMP (Section 6.5.9.2) or project plan (Section 7.4.2.1).

6.1.5.4.9.1 The Problem Posed by the Subsystem Test State The problem posed by the subsystem test state can be formulated as follows:

1. *The undesirable situation*: the need to validate each of the subsystems, in isolation, as being compliant to its requirements.
2. *The FCFDS*: the complete set of subsystems has been validated in isolation as being defect free which is usually interpreted as being compliant to their requirements.

3. *The assumptions*: each subsystem has been constructed according to its design specifications.
4. *The problem* is to ensure the set of subsystem tests:
    1. Validates that each of the subsystems, in isolation, is compliant to its requirements.
    2. Is performed in parallel or in a sequential order that will facilitate system integration.
5. *The solution* is the FCFDS. Note that subsystem testing may continue after the IRR should the integration be phased, as long as the subsystem testing for a subsystem is completed before that subsystem is scheduled to be integrated in the system integration state.

6.1.5.4.9.2 Generic Risks in the Subsystem Test State   Generic risks in the subsystem test state include:

- Failure to perform adequate testing of the subsystem resulting in defects not being found. Defects are defined as:
  - The subsystem not being compliant to requirements.
  - Component failures in the subsystem due to installation of bad parts or damage during the subsystem construction state.

Ways of mitigating or preventing these risks include testing the test procedure to ensure that it is designed to test compliance to the requirements.

6.1.5.4.9.3 The Role of the Systems Engineer in the Subsystem Test State   The role of the systems engineer is basically monitoring the situation. If any subsystem fails part or all of a test or does not meet all of its requirements, the systems engineer, considers the nature of the failure and the schedule impact, and recommends if the subsystem in its current state should be:

- Accepted with defects which will be fixed at some later time. This choice will probably mean replanning the system integration and a possible schedule delay.
- Rejected until fixed. This choice will probably mean a schedule delay.

6.1.5.4.10   The System Integration and System Test States of the SDP
This section summarizes the system integration and system testing states from the problem-solving perspective. System integration and system testing is a broad and complex topic worthy of its own book (e.g. Blokdyk 2018). So, this chapter provides an introduction and overview of the system integration and system test states which begin at the close of the IRR and end at the close of the DRR.

6.1.5.4.10.1 The Problem Posed by the System Integration and System Test States   The problem posed by the system integration and system test states can be formulated as follows:

1. *The undesirable situation*: at the start of the states is:
   1. The combination of the subsystems which have been developed and have passed their stand-alone tests in isolation (hopefully) has not been integrated into the solutions system.
   2. The performance of the whole solution system, with optimum effectiveness, in its operational context, under test conditions, has not been established.
2. *The assumptions*: all subsystems have passed their acceptance tests and are fully operational.
3. *The FCFDS*: (outcome) when the performance of the whole solution system, with optimum effectiveness, in its operational context, under test conditions, has been established and shown to meet or exceed the specifications as they exist at the end of the system integration and system test states.
4. *The problem*: to integrate and validate the solution system according to the approved plans.
5. *The solution*: at the end of the system integration and system test states is the successful completion of the set of activities that:
   1. Combines the parts, subsystems, interactions, etc. to constitute the solution system.
   2. Establishes, under test conditions, the performance of the whole solution system, with optimum effectiveness, in its operational context.

6.1.5.4.10.2 Generic Risks in the System Integration and System Test States   Generic risks in the system integration and system test states include:

- Subsystems delivered for integration that do not meet their subsystem specifications. For example, on one National Aeronautics and Space Administration (NASA) contract in the 1990s, the subsystem software passed the subsystem tests when the software code compiled without errors.[*]
- Ways of mitigating or preventing these risks include:
  - Having project managers who understand the difference between software compiling without errors and logic errors.
  - Performing functional tests on the components during the subsystem test state according to approved test procedures.

6.1.5.4.10.3 The Role of the Systems Engineer in the System Integration and System Test States   The role of the systems engineer is to:

- Ensure the system integration and testing are performed according to approved procedures produced earlier in the SDP.
- Assess the impact of failures and non-conformance to specifications during the integration and testing on the system, personnel, cost and schedule,

---

[*] The test did not test the logic to ensure the subsystem met its performance specifications.

interdependently with the project manager; identify appropriate courses of action and their risks and recommend how to mitigate the impact of failures and non-conformance to specifications.

6.1.5.4.11   The O&M States

This section summarizes the O&M states from the problem-solving perspective. The O&M states:

- Are one of the two most complex states in the SLC.
- Begins following the close of the DRR.
- Contains two sub-states:
  1. The system delivery, installation and acceptance testing sub-states discussed in Section 6.1.5.4.12.
  2. The in-service sub-state discussed in Section 6.1.5.4.13.

6.1.5.4.12   The System Delivery, Installation and
Acceptance Testing Sub-states

These are the sub-states in which the system is delivered to the acquiring organization, installed and tested in the operational environment. The state ends when the acquisition organization accepts the delivery at the close of the ATR. At this time the system may either pass the acceptance test or be conditionally accepted with some defects that will be remedied within a short period of time.

6.1.5.4.12.1   The Problem Posed by the System Delivery, Installation and Acceptance Testing Sub-states   The problem posed by the system delivery, installation and acceptance testing sub-states can be formulated as follows:

1. *The undesirable situation*: the system is supposedly ready but not yet operational.
2. *The assumptions*: *are*:
   1. The system has been tested in the factory or other location in which it was constructed.
   2. The system passed its tests or has one or more known defects that allow it to operate in a degraded manner.
   3. The known defects, if any, will be remedied shortly after installation.
   4. The system is ready for operation.
3. *The FCFDS*: the system is operational and successfully performing its MT.
4. *The problem*: is to figure out how to deliver, install and perform the acceptance test.
5. *The solution*: is to create and follow the plans and procedures to deliver, install and perform the acceptance test.

6.1.5.4.12.2   Generic Risks in the System Delivery, Installation and Acceptance Testing Sub-states   Generic risks in the system delivery, installation and acceptance testing sub-states include:

- Damage during shipment.
- Equipment not able to be installed in the operational environment for various reasons including the equipment will not fit through the door or other access routes.
- Installation procedures that damage the equipment or incorrectly install the equipment.
- Ways of mitigating or preventing these risks include:
  - Proper handling procedures during shipment and installation.
  - A site survey to determine the requirements imposed on the system by the delivery and installation during the system requirements state.

6.1.5.4.13   The In-Service Sub-state

This sub-state contains the activities involved in operating the system, providing support to maintain operations and making improvements to the whole to enhance effectiveness and to accommodate changes in the nature of the problematic or undesirable situation over time.

6.1.5.4.13.1   The Problem Posed by the In-Service Sub-state   The problem posed by the in-service sub-state can be formulated as follows:

1. *The undesirable situation*: the system is doing something it should not be doing, or not doing something it should be doing.
2. *The assumptions*: there is a working change management (CM) process.
3. *The FCFDS*: the system is doing what it should be doing. In the event that the system cannot be modified to meet the need, then the system transits to the system disposal state and a replacement project is initiated.
4. *The problem*: to figure out what needs to be changed, and make the change if it is affordable. If it is not affordable, then the problem is absolved but kept in view until it becomes affordable.
5. *The solution*: making the change if it is affordable. If it is not affordable, then the problem is absolved but kept in view until it becomes affordable.

6.1.5.4.13.2   Generic Risks in the In-Service Sub-state   Generic risks posed in the in-service sub-state include:

- Failures of components.
- Accidents in operation.
- Sabotage.
- Changes not being managed properly.
- Ways of mitigating or preventing these risks include:
  - The use of standard operating procedures to avoid risky situations.
  - Designing reliability and robustness into the system.
  - Testing opponents to eliminate early failures before installing the components in the subsystems.

- A maintenance program that replaces parts before they fail, or within a specified time should they fail depending on the system and the system and the situation (Section 3.8.2.1).
- Safety rules and regulations in the way the system is operated.
- Access control and system design such that there is no single point of failure that when sabotaged will result in the system being shut down or causing damage to its environment.
- Having a CM process (Section 4.9.2).
- The use of standard operating procedures to avoid risky situations is not as good as designing risky situations out of the system in the first place.

6.1.5.4.14   The System Disposal State

A system enters the system disposal state when the system can no longer:

- Meet the need of the users usually because the needs have changed.
- Be maintained, due to the obsolescence or pending obsolescence of components: a situation called diminishing manufacturing sources and material shortages (DMSMS) in the US Department of Defense (DoD). As a consequence, suitable spare parts can no longer be acquired such as in the Pacor system (Kasser 2019: Section 10.13) so a replacement system is to be acquired.

Disposal may be considered as a:

- Change of state from 'a system' to 'no system'.
- Realization of the final change request (Section 4.9.2).
- Project in itself with a full SDP depending on the system.

6.1.5.4.15   The Problem Posed by the System Disposal State

The problem posed by the system disposal state which contains the set of activities that dispose of the system and can be formulated as follows:

1. *The undesirable situation*: the system needs to be disposed of.
2. *The assumptions*: resources are or will be available, and the disposal will conform to environmental and other regulations, and statutes.
3. *The FCFDS*: the situation without the system, often containing a replacement system.
4. *The problem*: how to remove the system from service in an orderly manner with minimal impact on the situation. And if necessary, install the replacement system with minimal impact on operations.
5. *The solution*: determining how to remove the system from service in an orderly manner with minimal impact on the situation and making it happen.

The system disposal state may be considered as an SDP or rather a *system* 'undevelopment' *process* because it is the reverse of the SDP. The SDP starts with no system

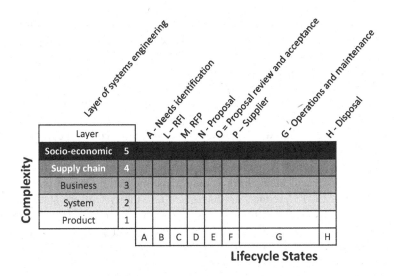

FIGURE 6.5 The modified HKMF for an acquired system.

and ends up with a system, while the disposal state starts with a system and ends up without a system.

6.1.5.4.16 Generic Risks in the System Disposal State

Generic risks in the system disposal state include:

- Failure to dispose of the system safely.
- All the generic risks of the previous states.

Ways of mitigating or preventing these risks include good systems engineering, project management and risk management using the ways mentioned in the previous states.

6.1.5.5 The Modified SLC for an Acquired System

In the event the system is either acquired as a COTS product or constructed to order, the system realization states in the SDP (States C-F) can replaced by system acquisition states (States L-N) and the SLC defined as follows as shown in the modified HKMF (Section 3.4.4) in Figure 6.5:

A. *The needs identification state* discussed in Section 6.1.5.4.3.

L. *The request for information (RFI) state*: an RFI may be issued to potential bidders to clarify certain issues or get advice on certain matters. The sub-state ends on the date and time specified in the RFI. Note an RFI is not always issued.

M. *The request for proposal (RFP) state*: the acquiring organization reviews comments received on the RFI and prepares the RFP. The sub-state ends with the release of the RFP.

N. *The proposal state*: potential bidders review the RFP to finalize their decision to bid, prepare their proposals and submit the proposals according to the instructions in the RFP (Kasser 1995: pp. 187–207). The sub-state ends on the date and time the proposals are due.

O. *The proposal review and acceptance state*: the proposals are reviewed by the acquiring organization, and a winning bid is eventually selected based on predetermined selection criteria. The state ends when the contract/purchase order is signed with the winning bidder who becomes the supplier for the system.

P. *The supplier states*: activities leading up to the system delivery. If the system is COTS and in stock, or is a systems integration project in which a number of COTS products are integrated into the system, the state can be extremely short in duration. If the system is built to order, the bidder may create a project to build the system and the system will pass through the SDP in the supplier's organization. The state ends with a DRR.

G. *The O&M state* discussed in Section 6.1.5.4.11.

H. *The system disposal state* discussed in Section 6.1.5.4.14.

6.1.5.5.1   Generic Risks in Acquiring a System

Generic risks in acquiring a system include:

- The generic risks in developing a system.
- The risk of suppliers not meeting their obligations in a timely manner.
- These risks may be mitigated by:
  - Not using a single supplier where more than one supplier can provide the product or service (Section 8.2) according to the principle of 'no single-point failures in the process or the product'.
  - Inserting penalty clauses into purchase contracts (Section 8.2.4.4).

6.1.6   GENERIC

Perceptions from the *Generic* HTP include:

- The definitions of a system and a subsystem are identical.
- Any system is an instance of a class of systems and:
  1. Inherits the properties of that class (Section 6.2).
  2. Conforms to a functional template for the class (Kasser 2019: Section 4.11).

6.1.7   CONTINUUM

Perceptions from the *Continuum* HTP include:

- The present state of the system may be different to the state or states of the system in the past and also may be different to the state or states of the system in the future.
- Systems exhibit different types of behaviour (Section 6.1.5.2).

- Systems have different properties (Section 6.2).
- Systems produce desired and undesired outputs.
- Systems can be classified in different ways.
- Some of the initial minimum set of common elements (Section 6.1.4) may be incorrect or unknown especially at the time the system boundary is first drawn.
- The word 'system' means different things to different people. For example, *Webster's Dictionary* contains 51 different entries for the word 'system' (Webster 2004).

6.1.8 SCIENTIFIC

Inferences from the *Scientific* HTP include:

- The boundary of a system boundary is crafted by the observer to enclose a section of the real world (Beer 1994: p. 7, Churchman 1979: p. 91, Jackson and Keys 1984) for some purpose known to the observer.
- The reason for the various definitions perceived in Section 6.1.7, namely the various definitions of the word 'system' perceived from the *Continuum* HTP bounded the problems faced by those who made the definitions.
- When a system is created as the solution to a problem, the total solution = functions performed by solution system + functions performed to realize the solution system (Hall 1962).

6.2 PROPERTIES OF SYSTEMS

Systems have various properties made up of (1) the properties of the subsystems and (2) the properties of the interactions between the subsystems. These properties include:

- States discussed in Section 6.2.1.
- Emergence discussed in Section 6.2.2.
- Failures discussed in Section 6.2.3.
- Subsystems discussed in Section 6.2.4.

6.2.1 STATES

Systems, in general, exhibit their behaviour in stable states until a state transition takes place. The system then changes to a new stable state. Examples of various changes in system states include:

- *Ice, water, steam*: if the system (molecules of water) begins at a temperature below zero degrees centigrade and is heated, the behaviour of the ice is homeostatic until the melting point of ice is reached and the system goes through a transition phase changing the state to water. As the liquid is heated further, the water then exhibits homeostatic behaviour until the temperature reaches the boiling point and a further transition phase takes place changing the water to steam.

- *Family, baby, family*[*]: a family exists in one state; a new baby comes along and the family then exists in a changed state.
- *Situation, policy, situation'*: a government perceives an undesirable situation and issues and implements a policy to change the state into a desirable situation. The country then exists in a new state which may or may not be the one the government desired.
- *System, failure, system'*: an operational system suffers a failure and changes the state to a new system. If the failure is total, then the system changes from an operational to a non-operational state. If the failure is partial, then the system changes from an operational state to a partially failed state.
- *System, upgrade, system'*: a system is changed from one state to an upgraded state.

Some system states are known due to experience and can be shown in a graph of system behaviour. Others may be unknown until the system moves beyond point B in Figure 6.3, and an unanticipated state change takes place and the lesson is learnt the hard way. For example, when settlers were moving west in the US in the 19th century, they built towns along the riverbanks not realizing that they were building in flood plains. When a flood came along, the river changed state and the settlers learnt about flood plains the hard way. One way of calculating risk probability is based on history and experience (Section 3.8.3).

6.2.2 EMERGENCE

One definition of emergence is 'the process of coming into existence or prominence' (Oxford 2019), namely the functionality of the system emerges from the components *and the interactions* between the components. For example, when wires, a transparent container and an inert gas or vacuum are combined in the right way, they form an incandescent electric light bulb. When the incandescent light bulb is connected to a working power source, the resulting system produces electric light and heat, which are the properties of the system as a whole, not of any single component. There seem to be two attributes (emergent properties) of emergence:

1. Desired discussed in Section 6.2.2.1.
2. Undesired discussed in Section 6.2.2.2.

The desired and undesired attributes are split as follows between:

- Known (predicted) emergent properties.
- Unknown emergent properties.
- Don't care emergent properties.

---

[*] Family' identifies that the family exists in a different state but is still the same family.

6.2.2.1   Known Emergent Properties

These are the known emergent properties provided by the solution system that are:

1. *Desired*: being the purpose of the system, e.g. the light from an incandescent electric light bulb.
2. *Undesired*: a property of the system known from experience and need to be mitigated in operation or prevented from occurring in the design. Sometimes known as side effects, for example:
   * The incandescent electric light bulb generates heat as well as light. The heat, or rather ways of dissipating the heat, has to be taken into consideration when designing light fixtures, handling, operating and replacing* incandescent light bulbs.
   * Antibiotics commonly used to treat bacterial infections have a side effect of causing diarrhoea because they also kill the beneficial bacteria in the human digestive system.
3. *Don't care*: a property that is known but is irrelevant.

6.2.2.2   Unknown Emergent Properties

These are unknown emergent properties provided by the solution system that are:

* *Undesired*: which have the opposite effect to those intended, or make the solution undesirable or even unable to remedy the problem.
* *Serendipitous*: beneficial and quickly become desired once discovered, but not part of the original specifications. These tend to be accidental discoveries.
* *Don't care*: a property that is unknown until it occurs but is irrelevant.

6.2.3   FAILURES

Sooner or later systems will fail or break down and exhibit abnormal behaviour. Failures may be:

* *Total*: the system stops working.
* *Partial*: some functionality is lost while the system can still operate.

Failures due to wear and tear on parts can be predicted statistically and can often be prevented by a combination of proper maintenance and design for reliability. Failures due to external causes can often be compensated for by designing redundancy into the system in the early states of the SDP (Section 6.1.5.4). Contingency plans for dealing with failures and breakdowns need to be prepared (Section 7.4.2.2) and periodically updated.

---

* For example, railway signals in cold climates may use the heat from the incandescent light bulb to melt snow that accumulates on the signal light container. A replacement upgrade design using light-emitting diodes (LEDs) would need an alternative way of performing the snow-melting function such as a heating element.

6.2.4 SUBSYSTEMS

Systems are made from subsystems. When thinking about systems as a rule of thumb, you need to think about the system and the one that is one level up and those that are one level down in the hierarchy (Machol and Miles Jr 1973: p. 48) as discussed in the nine-system model (Section 6.7).

6.3 HIERARCHIES OF SYSTEMS

Systems are partitioned by the observer for a purpose; understanding how something works, understanding a relationship, arranging things, etc.

Systems are classified in various ways, but each classification generally contains a hierarchy. The hierarchy ranges from an atom at the lowest level to the universe or beyond at the highest level.

6.3.1 PRINCIPLE OF HIERARCHIES

The principle of hierarchies in systems (Spencer 1962) cited by Wilson (2002) is a critical thinking tool that humanity has used to manage complexity for most of its recorded history and is defined in the following three quotations:

> All complex structures and processes of a relatively stable character display hierarchical organisation regardless of whether we consider galactic systems, living organisms and their activities or social organisations.
>
> **(Koestler (1978: p. 31)**

> Once we adopt the general picture of the universe as a series of levels of organisation and complexity, each level having unique properties of structure and behaviour, which, though depending on the properties of the constituent elements, appear only when those are combined into the higher whole,* we see that there are qualitatively different laws holding good at each level.
>
> **Needham (1945) cited by Koestler (1978: p. 32)**

'The English philosopher Herbert Spencer appears to be the first to set out the general idea of increasing complexity in systems (Spencer 1962). The term itself was first used by the English biochemist (and scholar of Chinese science) Joseph Needham (Needham 1937). The following quotation from a Web source provides an insight into the fundamentals of the theory (UIA 2002):

a. The structure of integrative levels rests on a physical foundation. The lowest level of scientific observation would appear to be the mechanics of particles.
b. Each level organizes the level below it plus one or more emergent qualities (or unpredictable novelties). The levels are therefore cumulative upwards, and the emergence of qualities marks the degree of complexity of the conditions prevailing at a given level, as well as giving to that level its relative autonomy.

---

* Namely emergent properties (author).

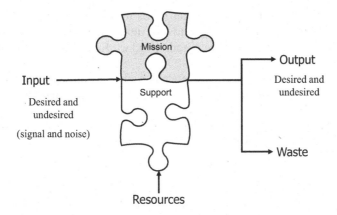

FIGURE 6.6 A generic functional template of a system.

    c. The mechanism of an organization is found at the level below, its purpose at the level above.

    d. Knowledge of the lower level infers an understanding of matters on the higher level; however, qualities emerging on the higher level have no direct reference to the lower-level organization.

    e. The higher the level, the greater its variety of characteristics, but the smaller its population.

    f. The higher level cannot be reduced to the lower, since each level has its own characteristic structure and emergent qualities.

    g. An organization at any level is a distortion of the level below, the higher-level organization representing the figure which emerges from the previously organized ground.

    h. A disturbance introduced into an organization at any one level reverberates at all the levels it covers. The extent and severity of such disturbances are likely to be proportional to the degree of integration of that organization.

    i. Every organization, at whatever level it exists, has some sensitivity and responds in kind' (Wilson 2002).

6.3.2 MISSION AND SUPPORT SUBSYSTEMS

Partitioning a system at the high level between mission and support subsystems reduces the risk of neglecting the support functions in favour of the operational functions as well as providing other benefits. A generic functional view of a system is pictured in Figure 6.6 (Kasser 2015a) as a jigsaw piece because systems can be represented by a number of functions (components or subsystems) fitted together to perform the function of the system. Generically, a function converts inputs to outputs using resources. Hitchins grouped the complete set of functions performed by any system into the following two classes (Hitchins 2007: pp. 128–129):

    1. *Mission*: the functions which the system is designed to perform to remedy the undesirable situation in its operational context under normal and contingency conditions, as and when required (the MT).

2. *Support*: the functions the system needs to perform to ensure that the system is able to perform the mission under normal and contingency conditions, as and when required. Support functions can further be grouped into:
   1. *Resource management*: the functions that acquire, store, distribute, convert and discard excess resources that are utilized in performing the mission.
   2. *Viability management*: the functions that maintain and contribute to the survival of the system in storage, standby and in operation performing the mission.

Ensuring that a design for the solution system is complete constitutes a significant problem or risk. Perceived from the *Generic* HTP, the use of a standard or reference set of functions has a number of benefits including the following:

1. Supports abstract thinking by encouraging problem-solvers to think in abstract terms in the early stages of identifying a problem and providing a solution discussed in Section 6.3.2.1.
2. Maximizes completeness of a system by allowing for the inheritance of generic functions for the class of system discussed in Section 6.3.2.2.
3. Provides for the use of generic functional templates for various types of systems which can help maximize completeness of the resulting system discussed in Section 6.3.2.3.
4. Improves probability of completeness by making it easier to identify missing functions in system functional descriptions than in implementation (physical) descriptions.
5. Allows the system to be modelled in its functional form at design time to determine how well the solution functionality appears to remedy the problem.

6.3.2.1 Supports Abstract Thinking

People tend to use solution language to describe functions. For example, we often use the phrase 'need a car' when we should be saying, 'need transportation'. Using implementation language in the early stages of problem-solving introduces the risk of producing a solution that may not be the best solution to the problem even if it is a complete solution, as well as generally not being an innovative solution. This is because solution language tends to turn examples into solutions with little exploration of alternative solutions. For example, if the need is stated as 'we need a car', the problem-solving process tends to focus on selecting the right car to meet the need. A need should be stated from the *Operational* and *Quantitative* HTPs[*] as 'a transportation function to move N people with B (kg. /cubic M) of baggage K km in H hours over terrain of type T with an operational availability of O'. Creating the

---

[*] The quantitative numbers have to be determined before the system that will meet the need is developed or nobody will know when the need is met.

solution concept in the form of capability or functionality is in accordance with, 'if a problem can be stated as a function, then the total solution is the needed functionality as well as the process to produce that functionality' (Hall 1989). In holistic thinking terms, state the need using functional or problem language, not structural or solution language. Using the language of functions in the early stages of problem-solving will nudge the stakeholders into abstract thinking rather than fixating on an implementation. For example, they might stop saying, 'I need a car', and start saying, 'I need transportation'.

6.3.2.2 Maximizes Completeness of a System

As perceived from the *Generic* HTP, any system is an instance of a class of systems (Section 6.1.5.5), and once the first one has been built, subsequent systems can inherit properties and functions reducing the risk of overlooking little used functions or those outside the designer's knowledge and experience. Consequently, any system can have two types of mission and support functions (Section 6.3.2):

1. *Generic*: to that class of system.
2. *Specific*: to that instance of the system.

For example, when building a spacecraft, the necessary thermal-vacuum properties and functions needed to survive the launch and operate in space can be inherited from previous spacecraft of that type together with the associated risks. Once inherited, the properties and risks need to be examined to determine if they are applicable with or without modification. Other specific properties, risks and functions for that spacecraft must then be examined to determine if they conflict with the generic ones. Any conflicts then need to be resolved.

6.3.2.3 The Use of a Generic Functional Template for a System

Figure 6.6 shows a generic template for a system in which the functionality has been grouped into mission and support functions. Note how Figure 6.6 does not show details of the mission and support functions because these details have been abstracted out since they belong in lower level drawings. An immediate advantage of the *Functional* HTP template is that one can see that there are two system outputs: the desired output labelled 'output' and another undesired output labelled 'waste'. If applied to a system containing an incandescent light bulb, the desired output is light, and the undesired or waste output is heat (Section 6.2.2).

The inheritance concept reduces the risk of incompleteness by inheriting functions from the set of reference functions for the class of system being developed. This seems to be the concept behind Hitchins' generic reference model of any system (Hitchins 2007: pp. 124–142). Problem-solvers and application domain experts working together[*] would assemble the detailed functions to be performed by the solution system from the set of reference functions for the class of system

---

[*] In an integrated team.

being developed, tailoring the functions appropriately. For example, if the solution is a:

- *Spacecraft*: the support functions for surviving launch and the out-of-atmosphere environment would be among those inherited.
- *Information system*: the functions displaying information to ensure that data are not hidden due to colour blindness in the operators would be among those functions inherited.

This method for developing a system design by inheriting from the class of system should lower the risk of missing:

- Risks inherent in the class of system.
- Functions.
- Requirements.

6.4 SUPPLY CHAINS AS SYSTEMS

Supply chains:

- Are systems in layer 4 of the risk framework (Section 1.2).
- Are important support subsystems for all systems in the other layers.

6.4.1 SUPPLY CHAIN FUNCTIONS

Supply chain functions include:

- Producing of materials.
- Transporting materials.
- Storing materials.
- Managing inventory.
- Informing customers and suppliers.
- Measuring inventory levels, rate of consumption, etc.

Each function contains risks which need to be identified and mitigated or prevented. For example, if everything is performed correctly, then supply chain management performs the following seven *rights* (McKeller 2014: p. 14):

- The *right* product or service.
- To the *right* customer.
- In the *right* place.
- At the *right* time.
- In the *right* condition.
- At the *right* quantity.
- For the *right* cost.

Accordingly, the risks are anything that will cause any one or more of the rights to be wrong and need to be mitigated or prevented.

6.4.2 Characteristics of Integrated Supply Management

Some characteristics of integrated supply management (McKeller 2014: p. 68) are:

- All internal supply chain management functions are integrated.
- Suppliers are chosen strategically.
- Common goals exist for both the buying organization and suppliers.
- Proper supplier relationships are in place.
- Integrative performance metrics are in place.
- Cost management tools are actively used.
- Suppliers are actively engaged in all product or services development.
- Redundant, inefficient processes are aggressively eliminated.
- Appropriate technological enablers are integrated.
- Supplier relationship management is ongoing.

Accordingly, the risks and integrated supply chain management are anything that will inhibit these characteristics.

Supply chain performance metrics based on cost, delivery, quality and service can be set up using the same process as management by objectives (MBO) (Section 7.8). Then risk analysis can be performed to identify anything that will negatively affect the metrics. For example, something that will increase the cost is a risk; conversely, something that will decrease the cost provides an opportunity. Once performance metrics have been established, then management by exception (MBE) (Section 7.7) can be used in the management of the supply chain.

6.5 AN INTRODUCTION TO SYSTEMS ENGINEERING

This section helps you to understand the nature of systems engineering by summarizing selected aspects of systems engineering from the HTPs because the content of the literature and courses on systems engineering is confusing. There seems to be agreement on requirements and architectures, but then each book or course covers something different. After reviewing a number of books and discussing systems engineering with practitioners, and developing the concept of pure, applied and domain systems engineering (Section 6.5.3), order appeared out of the confusion.[*] It seemed that each book and course described a part of systems engineering from the author's perspective, in the manner of the fable of the blind men feeling parts of an elephant and each identifying a single and different animal (Yen 2008). This section discusses the following aspects of systems engineering:

1. The SDP in Section 6.1.5.4.
2. The waterfall chart in Section 6.5.1.
3. The two different process paradigms in Section 6.5.2.
4. The difference between systems engineering – the activity (SETA) – and systems engineering – the role (SETR) in Section 6.5.4.

---

[*] Most of the perceptions are summarized in *Perceptions of Systems Engineering* (Kasser 2015b).

5. The eight different camps of systems engineering in Section 6.5.5.
6. The three types of systems engineering in Section 6.5.3.
7. The five layers of systems engineering in Section 6.5.6.
8. Problem-solving and systems engineering in Section 6.5.7.
9. The interdependency and overlap between the systems engineering, project management and other engineering activities in Section 6.5.8.
10. Systems engineering documentation in Section 6.5.9.

6.5.1 THE WATERFALL CHART

The waterfall chart (Royce 1987) shown in Figure 6.4:

- Is a common tool used in systems engineering to incorrectly represent the SDP (6.1.5.4).
- Is a Gantt chart with the rows and columns lines eliminated.
- Is designed to show the relationship between the states in the SDP.
- Was developed as a planning tool for management in a time when the requirements didn't change very quickly.*
- Is easy to teach and understand.
- Copes with changes in requirements:
  1. As long as the time taken to progress through all the states in the waterfall is shorter than the time between changes.
  2. If the project has a working CM system and can adjust the workload to meet the changes in the need. This may require reverting to previous stages in the waterfall depending on the nature the change request (Section 4.9.2).
- Is usually discussed as a single waterfall in a teaching example. However, in the real world there may be multiple iterations of the waterfall which allow the solution system to evolve as in the multiple-iteration problem-solving process (Section 5.3.2) or the cataract methodology (Kasser 2019: Section 10.15.9). Ideally the length of each waterfall SDP should be shorter than the time it takes for the undesirability being remedied by the project to change for the worse. Perceptions from the *Generic* HTP note a similarity to the observe–orient–decide–act (OODA) loop (Boyd 1995). However, perceptions from the *Continuum* HTP note that while the focus of the OODA loop is on the speed of completing the loop to get inside the opponent's decision-making cycle, here the focus is getting inside the rate of change of undesirability.

6.5.2 THE 'A' AND 'B' PARADIGMS IN SYSTEMS ENGINEERING

Perceptions of systems engineering from the *Continuum/Temporal* HTPs note that the SDP has evolved into two process paradigms: the 'A' paradigm and the 'B' paradigm (Kasser 2012) shown in Figure 6.7. Summarizing the two paradigms:

---

* This may have been because microcomputers had not been invented and systems were created in hardware, which was difficult to change.

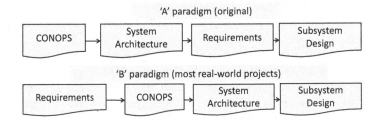

FIGURE 6.7 The 'A' and the 'B' paradigms of systems engineering.

6.5.2.1 The 'A' Paradigm

The 'A' paradigm:

- Begins with the systems engineering activities performed in the needs identification state of the SDP (Section 6.1.5.4.3), the 'A' column in the HKMF (Section 3.4.4).
- Is the original systems engineering paradigm which begins with a focus on converting a problematic or undesirable situation to an FCFDS and creating the CONOPS* (Section 6.5.9.1). Examples of the 'A' paradigm CONOPS include the 'to be' view in business process reengineering and the conceptual model in Checkland's SSM (Checkland and Scholes 1990).

Research into the systems engineering literature found that successful projects such as the NASA Apollo project were characterized by a common vision of the purpose and performance of the solution systems among the customers, users and developers,† namely a paradigm that began in the needs identification state of the SDP (Section 6.1.5.4.3). Moreover, the common vision related to both the mission and support functions performed by the solution system (Section 6.3.2). Perceptions from the *Generic* HTP outside the systems engineering literature support the research with similar findings in the process improvement literature (e.g. Deming 1993, Dolan 2003). In addition, BPR creates and disseminates/communicates a 'to-be' model of the operation of the conceptual reengineered organization (i.e. an FCFDS) before embarking on the change process.

6.5.2.2 The 'B' Paradigm

The 'B' paradigm begins in the systems requirements state of the SDP (Section 6.1.5.4.4), the 'B' column in the HKMF (Section 3.4.4). Many systems and software engineers have been educated to consider the systems engineering activities in column B of the HKMF as the first state of the systems engineering process.

---

* The FCFDS describes the solution system (*Functional* HTP), while the CONOPS (Section 6.5.9.1) describes the context and environment in which the FCFDS will operate and how that operation is anticipated to occur (*Big Picture* and *Operational* HTPs). Depending on the situation, an FCFDS may be associated with several CONOPS or several FCFDS may be associated with a CONOPS.

† The stakeholders.

For example, requirements are one of the inputs to the 'systems engineering process' (e.g. DOD 5000.2-R 2002: pp. 83–84, Eisner 1997: p. 9, Martin 1997: p. 95, Wasson 2006: p. 60).

The 'B' paradigm is inherently flawed. This is because even if systems and software engineers working in a paradigm that begins in HKMF column 'B' (Section 3.4.4) could write perfectly good requirements, they still cannot determine if the requirements and associated information are correct and complete because there is no reference for comparison to test the completeness. Consequently, efforts expended on producing better requirements have not, and will not alleviate the situation. The situation cannot be alleviated because perceptions from the *Generic* HTP perceive the situation as being akin to participating in Deming's red bead experiment, which demonstrates that errors caused by workers operating in a process are caused by the system[*] rather than the fault of the workers in the bead production process (Deming 1993: p. 158). Recognition that the 'B' requirements paradigm is inherently flawed is not a new observation. For example:

- A proposal to reduce human errors in producing requirements by analysing requirements using an approach of creating scenarios as threads of behaviour through a *use case* and adopting an object-oriented approach (Sutcliffe, Galliers, and Minocha 1999), namely they proposed a return to the 'A' paradigm.
- Stand-alone requirements make it difficult for people to understand the context and dependencies among the requirements, especially for large systems, and so *use cases* should be used to define scenarios (Daniels, Bahill, and Botta 2005).
- One of the two underlying concepts of model-based systems engineering (MBSE) is to develop a model of the system to allow various stakeholders to gain a better understanding of how well the conceptual system being modelled could remedy the problem, before starting to write the requirements (Kasser 2019: Chapter 21).

6.5.3 THE THREE TYPES OF SYSTEMS ENGINEERING

Perceptions from the *Continuum* HTP differentiated between three notional types of systems engineering (Kasser and Arnold 2014, 2016) which are related as shown in Figure 6.8:

1. *Pure systems engineering*: cognitive skills, namely thinking, e.g. systems thinking, critical thinking, problem formulation/solving and decision-making (Chapter 2).
2. *Applied systems engineering*: activities traditionally associated with SETA (Section 6.5.4) including:

---

[*] In the bead supply process.

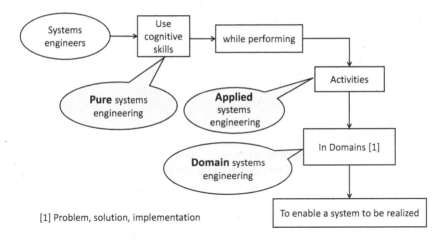

FIGURE 6.8 The relationship between pure, applied and domain systems engineering.

1. Definition of needs/goals/objectives.
2. Operational scenarios analyses.
3. System requirements analysis.
4. System requirements allocation.
5. Functional analysis.
6. Functional allocation.
7. Specification development.
8. System and subsystem design.
9. Trade-off/alternatives evaluation.
10. Development of benchmark tests.
11. Software requirements analysis.
12. Hardware analysis and recommendations.
13. Interface definition and control.
14. Schedule development.
15. Lifecycle costing.
16. Technical performance measurement.
17. Planning.
18. Organizing.
19. Directing junior-level personnel.
20. Program and decision analysis.
21. Risk analysis.
22. Integrated logistics support.
23. Transition planning.
24. Reliability, maintainability and availability (RMA).
25. System integration.
26. Test and evaluation (T&E).
27. Configuration management.
28. Quality assurance.
29. Training.
30. Technical writing.
31. Installation.
32. Operations support.
33. System evaluation and modification.

3. *Domain systems engineering*: pertains to fields such as computers, aerospace, transportation and defence.

6.5.4 The Difference between SETA and SETR

Perceptions from the *Continuum* HTP identified a difference between the activities performed by a systems engineer (the role) and the activities traditionally known as (applied) systems engineering (Section 6.5.3). Namely:

- *SETA*: the activities traditionally associated with applied systems engineering (Section 6.5.3).
- *SETR*: the role or job of the system engineer is to perform a mixture of the activities known as systems engineering (SETA), engineering and project management as well as any other activities in their job description.

6.5.5 The Eight Different Camps of Systems Engineering

Perceptions from the *Continuum* HTP identified eight different somewhat overlapping camps of systems engineering (Kasser and Hitchins 2012) based on sorting the different views of/opinions on/worldviews of systems engineering. Each opinion seems to represent a viewpoint based on the experience of the writer.[*] The somewhat overlapping camps are:

1. *Lifecycle:* this camp is one of the earliest camps, articulated as, 'Despite the difficulties of finding a universally accepted definition of systems engineering, it is fair to say that the systems engineer is the man[†] who is generally responsible for the over-all planning, design, testing, and production of today's automatic and semi-automatic systems'[‡] (Chapanis 1960: p. 357).
2. *Process*: this camp of system engineers, particularly in the International Council on Systems Engineering (INCOSE) and the US DoD, is process-focused (Eisner 1988, Lake 1994) seemingly in accordance with US DoD 5000 Guidebook 4.1.1, which states, 'The successful implementation of proven, disciplined system engineering processes results in a total system solution that is – robust to changing technical, production, and operating environments; adaptive to the needs of the user; and balanced among the multiple requirements, design considerations, design constraints, and program budgets'. The focus is on conforming to the process and not on providing an understanding of the undesirable situation. These campers are often graduates from 'B' paradigm systems engineering courses which focus on the process.
3. *Problem-solving*: members of this camp of system engineers are often old-timers who worked as system engineers before and during the Apollo program. Examples in the literature include:
   - 'System engineers are problem solvers' (Wymore 1993: p. 2). Wymore summed up the philosophy of the principal functions of systems engineering as, 'to develop statements of system problems comprehensively,

---

[*] At least in my case (Kasser 1995).

[†] He was writing in 1960. These days we can write, 'man or woman'.

[‡] Project managers might disagree.

without disastrous oversimplification, precisely without confusing ambiguities, without confusing ends and means, without eliminating the ideal in favour of the merely practical, without confounding the abstract and the concrete, without reference to any particular solutions or methods, to resolve top-level system problems into simpler problems that are solvable by technology: hardware, software, and bioware, to integrate the solutions to the simpler problems into systems to solve the top-level problem' (Wymore 1993: p. 2).

- IEEE 1220 which stated that 'the systems engineering process is a generic problem-solving process' (IEEE Std 1220 1998: Section 4.1) – a statement ignored or forgotten by the process camp.

4. *Discipline and meta-discipline*: Wymore defined systems engineering as a discipline (Wymore 1994), and systems engineering meets Kline's requirements for a discipline (Kline 1995). However, all the elements of the current mainstream SETR approach to systems engineering overlap those of project management and other disciplines which make it difficult to identify systems engineering as a distinct discipline. The discipline camp tends to account for the overlap by viewing systems engineering as a meta-discipline incorporating all the other disciplines and hold that systems engineering needs to widen its span to take over the other disciplines (Kasser and Hitchins 2012).

5. *Systems thinking*: this camp of system engineers tends to be system engineers who can view an issue from more than one perspective (e.g. Beasley and Partridge 2011, Evans 1996, Martin 2005, McConnell 2002, Rhodes 2002, Selby 2006).

6. *Non-systems thinking*: this camp of system engineers tends to have a single viewpoint of systems engineering and generally exhibits the 'biased jumper' level of critical thinking (Section 2.2).

7. *Domain*: this camp of system engineers tags the role of systems engineers to the system being system engineered. Examples are network system engineers/engineering, control system engineers/engineering, communications system engineers/engineering, hydraulic system engineers/engineering and transportation system engineers/systems engineering. However, the name of the type of system is dropped from the role.[*]

8. *Enabler*: this camp of system engineers evolved from the problems-solving camp. In the enabler camp, systems engineering is the application of systems thinking and beyond (Section 2.5) to problem-solving. Moreover, it can be, and is, used in all disciplines for tackling certain types of complex and non-complex problems; see '[systems engineering] is a philosophy and a way of life' (Hitchins 1998).

---

[*] For example, my first job as a systems engineer was as an Apollo Lunar Surface Experiment Package (ALSEP) Control System systems engineer. Each experiment in the ASLEP had its own systems engineer, and there was a systems engineer for the ALSEP itself.

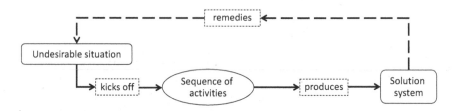

FIGURE 6.9   From an undesirable situation to a solution system.

6.5.6   THE FIVE LAYERS OF SYSTEMS ENGINEERING

There are differences between systems engineering as performed on products, systems and large-scale systems. Hitchins proposed a five-layer model for systems engineering (Hitchins 2000) (Section 1.2) which can be extended downwards to add a sixth layer (Kasser 2015b: p. 69):

- *Layer 0*: the component layer where many components make a product.

6.5.7   PROBLEM-SOLVING AND SYSTEMS ENGINEERING

Systems engineering begins in a problematic or undesirable situation because 'problems do not present themselves as givens; they must be constructed by someone from problematic situations which are puzzling, troubling and uncertain' (Schön 1991). Perceived from the *Big Picture* HTP, the context for systems engineering (the sequence of activities) is shown in Figure 6.9 and begins with the existence of a problematic or undesirable situation and ends with a solution system which remedies the undesirable situation. Figure 6.9 can be expanded into Figure 6.10 (Kasser 2013: p. 261) which converts the undesirable situation into a solution system operating in the context of an FCFDS. From this perspective, the observer becomes aware of an undesirable situation that is made up of a number of related factors. A project is authorized to do something about the undesirable situation. The problem-solver tries to understand the situation, determine what makes the situation undesirable and then create a vision of an FCFDS. The problem then becomes one of how to move from the undesirable situation to the FCFDS. Once the problem is identified, the remedial action is taken to create and transition to the solution system which will operate in the context of the FCFDS. The so-called systems engineering process is a version of the problem-solving process (IEEE Std 1220 1998). Inferences from the *Scientific* HTP note that the confusion between the 'systems engineering process' and the problem-solving process can be resolved by recognizing when the problem is:

- *Small or non-complex*: the sequence of activities in the remedial action is known as the 'problem-solving process'.
- *Large or complex*: the sequence of activities in the remedial action is known as the 'systems engineering process' instead of the SDP. This is in

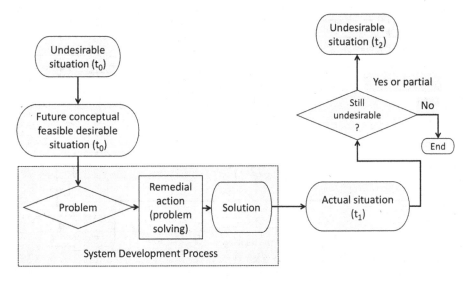

FIGURE 6.10 A holistic approach to managing problems and solutions.

accordance with Jenkins' definition of systems engineering as 'the science of designing complex systems in their totality to ensure that the component subsystems making up the system are designed, fitted together, checked and operated in the most efficient way' (Jenkins 1969).

This relationship is shown in Figure 6.11 (Kasser and Hitchins 2013).

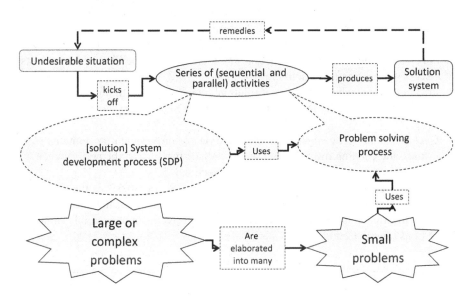

FIGURE 6.11 The SDP and the problem-solving process.

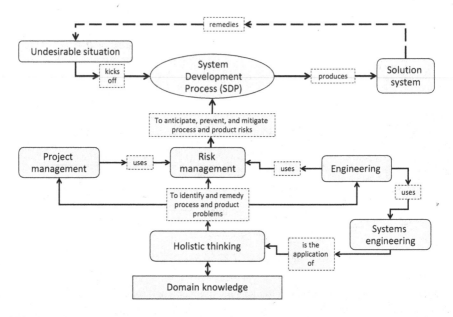

FIGURE 6.12   The relationships between the activities in the SDP in the SETA paradigm.

6.5.8   The Interdependency and Overlap between the Systems Engineering, Project Management and Other Engineering Activities

The relationships between the interdependent SETA (Section 6.5.4), project management and engineering activities (functions) performed to realize a solution system
are shown in Figure 6.12* (Kasser and Hitchins 2013). The top part of the figure
is the problem-solving process causal loop in Figure 6.9 providing the context to
the SDP. The first iteration of the loop begins when someone with the appropriate
authority associated with the undesirable situation kicks off the SDP which consists
of a set of three streams of activities (Section 7.3.1.1.2) performed in series and in
parallel (a process) which produces a solution system designed to remedy the undesirable situation. The interdependent activities in the SDP can be separated where:

- *Domain knowledge* in the problem, solution and implementation domains is
  the underpinning information used by holistic thinking in the performance
  of the activities performed in the SDP (Section 6.5.3).
- *Pure systems engineering* or *holistic thinking* (Section 6.5.3) is the use of
  the systems thinking and beyond conceptual tools that use the knowledge
  in all three domains to identify and remedy problems in undesirable situations (Section 2.5).

---

* This is a figure with high subjective complexity and is broken out into a series of simpler figures in
  accordance with the principle of hierarchies (Section 6.3.1) which mask the non-pertinent aspects
  when explaining the aspects of the figure.

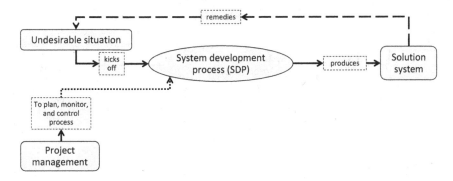

FIGURE 6.13 Project management in the SDP.

- *Risk management* is the set of activities that anticipate, prevent and mitigate risks in the problem, solution and implementation domains.
- *Project management* is the set of activities known as planning, organizing, directing and controlling (Fayol 1949: p. 8) shown in Figure 6.13 (Kasser and Hitchins 2013). Project management incorporates risk management to manage process risks.
- *Applied systems engineering* (SETA) (Section 6.5.4) is the set of activities known as designing, integrating and testing the overall system/interacting subsystems shown in Figure 6.14 (Kasser and Hitchins 2013).
- *Engineering* is the set of non-SETA engineering activities that create subsystems.

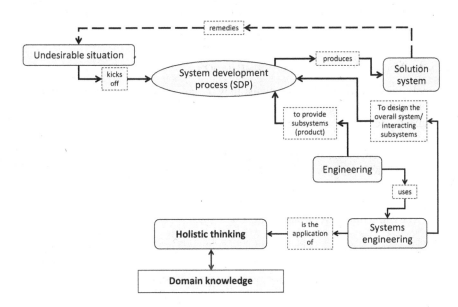

FIGURE 6.14 Engineering and systems engineering in the SDP.

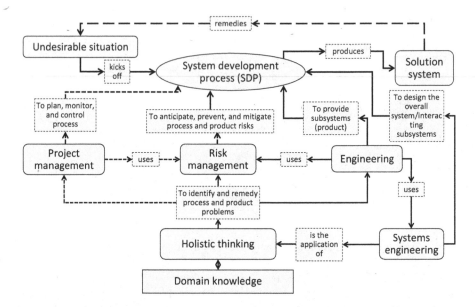

FIGURE 6.15 The non-overlapping activity paradigm (SETA).

Project management, systems engineering and engineering all perform risk management as shown in Figure 6.15 (Kasser and Hitchins 2013). The complex concept map shown in Figure 6.15 shows a non-overlapping relationship between the functions of SETA, project management and engineering. In Figure 6.15, engineering activities may 'provide' by creating, by building, by purchasing COTS products, by changing a process, by reorganizing human activities or by a combination of all or some of the above.

6.5.9 SYSTEMS ENGINEERING DOCUMENTATION

Systems engineers produce documents; they do not produce systems although they do help in the realization of these systems. The most important systems engineering documents are:

- The CONOPS summarized in Section 6.5.9.1.
- The SEMP summarized in Section 6.5.9.2.
- The TEMP summarized in Section 6.5.9.3.
- The interface control document (ICD) summarized in Section 6.5.9.4.
- The Shmemp summarized in Section 6.5.9.5.

Systems engineering documents:

- Must be tailored to the scope of the project.
- Are products and must be delivered to customers who will use the documents. There is no point in producing a document for the sake of producing

documents or because standards mandate that such and such documents be produced if the document will not be used.
- Must be produced in a timely manner.
- May be combined or even written as presentations if the scope is small enough.

6.5.9.1   The CONOPS

The CONOPS is one of the most important system documents (Kasser 2019: Section 4.4). Written in the needs identification state (Section 6.1.5.4.3), it:

- Explains how the system would be created, deployed and employed; who will use it and where, when, why and how they will use it.
- Contains the scenarios which describe what the system will do – scenarios to be converted to requirements in the system requirements state (Section 6.1.5.4.4).

Gabb summarizes the purpose of the operations concept document (OCD)* as describing the operation of a system in the terminology of its users, stating that it may include identification and discussion of the following (Gabb 2001):

- Why the system is needed and an overview of the system itself.
- The full SLC from deployment through disposal.
- Different aspects of system use including operations, maintenance, support and disposal.
- The different classes of user, including operators, maintainers, supporters and their skills and limitations.
- Other important stakeholders in the system.
- The environments in which the system is used and supported.
- The boundaries of the system and its interfaces and relationships with other systems and its environments.
- When the system will be used and under what circumstances.
- How and how well the needed capability is currently being met (typically by existing systems).
- How the system will be used, including operations, maintenance and support.

6.5.9.2   The SEMP

The SEMP (Kasser 2019: Section 4.5):

- Is also one of the most important system documents.
- Is drafted in the needs identification state (Section 6.1.5.4.3) and finalized in the systems requirement state (Section 6.1.5.4.4).†
- Describes the resources and technology needed to implement the system and when they will be needed (schedule).

---

* An alternative name for the document.
† And updated before each major milestone.

- Is a comprehensive work plan and describes how the fully integrated program engineering effort will be managed and conducted.
- Is also often known as a project plan (Section 7.4.2.1).
- Has a focus on the management stream of activities (Section 7.3.1.1.2).
- Includes descriptions of (Kasser and Schermerhorn 1994):
  - The process.
  - Technical program planning and control.
  - Engineering specialty integration.

6.5.9.3   The TEMP

The TEMP is the guiding document for the prevention and testing activities in the SDP (Section 6.1.5.4). The conceptual TEMP is developed during the needs identification state (Section 6.1.5.4.3) and finalized during the system requirements state (Section 6.1.5.4.4).* The activities documented in the TEMP are those of the quality/test stream of activities (Section 7.3.1.1.2).

The purpose and use of a TEMP are:

1. To facilitate the technical tasks of testing.
2. To improve communication about testing tasks and the testing process.
3. To provide a structure for organizing, scheduling and managing the testing activities during each state of the SDP (Section 6.1.5.4).
4. To maximize the yield (number of errors found) in the testing investment.

6.5.9.4   Interface Control Documents

An ICD describes and defines interfaces at all levels in the system. ICDs also contain intrinsic and extrinsic information. The purpose of an ICD is to minimize the risk of interface problems by communicating between the people on each side of an interface. Information in the ICD shall be relevant to the interface or to an understanding of the interface. The ICD shall contain a:

- Description of where the interface exists within the system.
- Brief description of the system on each side of the interface as seen from the interface.
- Complete specification of everything that crosses the interface.

6.5.9.5   The Shmemp

The Shmemp is a collective noun for the remainder of the systems engineering and engineering specialty plans and documentation including:

- Configuration management.
- Engineering specialties.
- Human factors
- Human systems integration.
- Integrated logistics support (ILS).

---

* And updated before each major milestone.

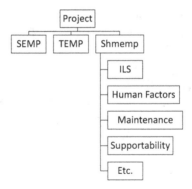

FIGURE 6.16 The SEMP, TEMP and Shmemp.

- RMA.
- Risk management.
- Safety.
- Sustainability.
- CM.
- Others as appropriate to the situation.

Production of each of these plans may require specialists in the relevant discipline and domain, and it is the systems engineer's role to coordinate with the specialist producing these plans to ensure the system-level specifications are met.

Shmemp is a Yiddish-based term similar in meaning to 'bumph'.* It is used to:

- Simplify and provide a template for the documentation hierarchical breakdown structure as shown in Figure 6.16 (Kasser 2010).
- Rhyme, 'SEMP, TEMP and Shmemp' when referring to the complete set of systems engineering documentation (Kasser 2010).

6.6 MODELLING AND SIMULATION

Essentially, all models are wrong, but some are useful.

(Box and Draper 1987: p. 424)

Modelling and simulation:

- Are based on underlying assumptions.
- Are tools used to study situations and proposed solutions.
- Are generally wrong but good enough to use.
- Represent some aspects of the real world.

---

* Documents containing information which you may not need or find interesting (Collinsdictionary 2019).

- Allow us to engineer things, e.g. wave (Huygens 1690) and particle (Newton 1675) theories of electromagnetic propagation explain different attributes of and allow us to engineer radio communications systems.

Systems engineers use models and simulations especially in:

- The needs identification state (Section 6.1.5.4.3) to analyse the system and situation from the *Operational* and *Functional* HTPs to determine the numerical values associated with a scenario. For example, queueing theory is used in models pertaining to traffic flow situations and dataflows in networks.
- The system and subsystem design states (Section 6.1.5.4.5) to model a design and ensure the design is compliant to trequirements.

Perceptions from the *Continuum* HTP distinguish between operational and functional models and simulations.

- *Operational models and simulations*: focus on actual and conceptual views of the 'what', where:
  - Actual views include models of *what* the system is doing.
  - Conceptual models include models of *what* the system can do, *what* the system should do and *what* the system needs to do. These models can be used to gain consensus on the 'what' aspect of a system.
- *Functional models and simulations*: focus on 'how' it is being done. These are useful when the underlying mechanisms are well-understood and the functionality can be expressed mathematically.

The generic risks of using models include:

- The model or simulation is only as good as its underlying assumptions, and when they are wrong, people can be killed.
- Reliance on functional models and simulations to save the costs of testing and evaluation can be dangerous since there is a major difference between operational and functional models and simulations.
- When the underlying mechanisms in an unprecedented system such as a new aircraft are unknown, then using simulations as training tools can be downright dangerous.

6.7   THE NINE-SYSTEM MODEL

This section summarizes the nine-system model (Kasser 2019: Chapter 6) which:

- Is a tool for managing stakeholders.
- Is a framework for perceiving where the parts of systems engineering are performed and how they fit together as well as a tool for use by system engineers.

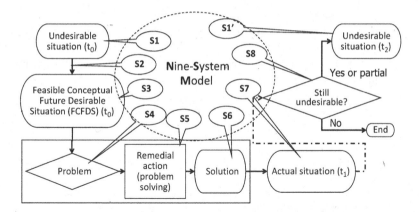

FIGURE 6.17 Mapping the nine systems to the extended holistic problem-solving process.

- Provides a way to manage complexity when creating the system (Kasser 2019: Section 9.6.1).
- Is based on the problem-solving approach to systems engineering in accordance with IEEE 1220 which stated, 'the systems engineering process is a generic problem-solving process' (IEEE Std 1220 1998: Section 4.1).
- Maps into the extended problem-solving process (Section 5.1.6.2) shown in Figure 5.7 annotated as shown in Figure 6.17 (Kasser 2019).
- Manages complexity by abstracting out all information about the system that is not pertinent to the issue at hand (Kasser, Zhao, and Mirchandani 2014).
- Is an application of the theory that complexity can be managed (but not reduced) by applying a set of rules for grouping/aggregation/synthesis.
- Is a self-similar framework model usable in any level of the hierarchy.
- Incorporates much of the content of the MIL-STD-499 (MIL-STD-499A 1974), EIA 632 (EIA 632 1994) and IEEE 1220 (IEEE Std 1220 1998).
- Incorporates the seven principles for systems engineered solution systems (Kasser 2019: Section 4.12).
- Provides a template incorporating built-in best practices that conform to the 'A' paradigm of systems engineering (Section 6.5.2.1).
- Is a conceptual model since as perceptions from the *Temporal* HTP show, all the systems do not coexist at the same point in time.
- Comprises the following nine situations, processes and socio-technical systems in a clearly defined interdependent manner:
  S1. The undesirable or problematic situation.
  S2. The process to create the FCFDS.
  S3. The FCFDS that remedies the undesirable situation.
  S4. The process to plan the transition from the undesirable or problematic situation (S1) to the FCFDS (S3).
  S5. The process to perform the transition from the undesirable or problematic situation (S1) to the FCFDS (S3) by providing the solution system (S6) according to the plan developed in the planning process (S4). S5 could be the SDP or an acquisition process if a suitable COTS system is available.

S6. The solution system that will operate within the FCFDS.

S7. The actual or created situation. S3 evolves into S7 during the time taken to perform S4 and S5.

S8. The process to determine that the realized solution system (S6 operating in the context of S7) remedies the evolved undesirable situation.

S9. The organization(s) containing the processes and providing the resources for the operation and maintenance of the processes. S9 is also often known as the enterprise. Each organization can be perceived as comprising two major subsystems:

1. *The production (mission) subsystem*: produces the products from which the organization makes its profits.

2. *The support subsystem*: provides support such as maintenance, purchasing, human resource supply and finance to the production subsystem.

Each of the nine systems must be viewed from each of the HTPs as appropriate. The nine-system model is not shown in a single figure, instead perceptions of the model from the following HTPs are provided:

- *Operational*: the nine systems map directly into the extended problem-solving process (Section 5.1.6.2) shown in Figure 5.7 annotated as shown in Figure 6.17 kicking off at time t_0. S1 is the undesirable situation. S2 is the process that produces an understanding of the *undesirable or problematic situation* (S1) and develops the FCFDS (F3). Once the FCFDS is approved, S4, the planning process creates the realization process (S5) and the solution system (S6) begins. S4 terminates at the SRR. The realization process (S5) realizes the solution system (S6). Once realized, the *solution system* (S6) is tested in operation in the *actual situation* existing at time t_1 (S7) to determine if it remedies the *undesirable situation*. However, since the solution realization process takes time, the *undesirable situation* may change from that at t_0 to a new *undesirable situation* existing at t_2. If the *undesirable situation* at t_2 is remedied, then the process ends; if not, the process iterates from the undesirable situation at t_2 and the actual situation (S7) becomes the new undesirable situation in the next iteration of the process (S1′).

- *Functional*: shown in Figure 6.18 (Kasser 2019) shows the relationships between the situations, systems and processes. The *process* to plan the transition from the *undesirable or problematic situation* (S1) to the FCFDS (S3) and the *process* to realize the transition from the *undesirable or problematic situation* (S1) to the *FCFDS* (S3), S4 and S5, constitute the two interdependent sequential systems engineering processes (Section 5.3.2).

- *Structural*: shown in Figure 6.19 (Kasser 2019) shows the relationship between the process systems and the *solution system* and the organization(s) containing the process systems and *solution system*. For example, this perspective provides the:

  - Organization charts in S9 for staffing the process systems (S2, S4, S5 and S8).

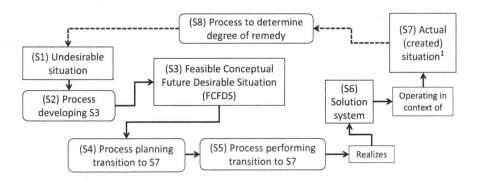

1. The solution systems and the adjacent systems are subsystems in the actual situation

FIGURE 6.18   The nine-system model (*Functional* HTP).

- Product breakdown structure for the solution system (S6).
- *Temporal*: shown in Figure 6.20 (Kasser 2019) uses a Gantt chart to show how the systems relate in time. The nine systems do not coexist at the same point in time; the relationship follows the problem-solving process shown in Figure 6.17, kicking off at time t_0. S2 is the process that develops the FCFDS (F3). Once the FCFDS is approved, S4, the planning process to create the realization process (S5) and solution system (S6) commences. S4 terminates at the SRR. The realization process (S5) realizes the solution system (S6). S1 exists in parallel with S2, S3, S4 and S5 lasting until S6 commences. Once realized, the *solution system* (S6) is tested in operation in the *actual situation* existing at time t_1 (S7) to determine if it remedies the *undesirable situation*. However, since the solution realization process takes time, the *undesirable situation* may change from that at t_0 to a new *undesirable situation* existing at t_2. If the *undesirable situation* at t_2 is remedied, then the process ends; if not, the process iterates from the undesirable situation at t_2 and the actual situation becomes the new *undesirable situation* (S1′).

For a further detailed explanation, see *Systems Engineering: A Systemic and Systematic Methodology for Solving Complex Problems* (Kasser 2019: Chapter 6).

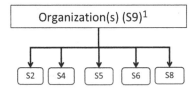

1. Considered as one [class of] system but generally is at least two organizations

FIGURE 6.19   The nine-system model (*Structural* HTP).

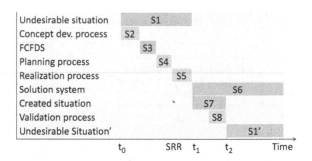

FIGURE 6.20 The nine-system model (*Temporal* HTP).

6.8 RISKS IN SYSTEMS AND SYSTEMS ENGINEERING

This section discusses some additional risks in systems and systems engineering. Having identified these risks, they may be prevented or mitigated by the appropriate actions earlier in the SDP.

6.8.1 RISKS IN SYSTEMS

Generic risks in systems generally show up in the in-service state (Section 6.1.5.4.13.2) and include:

- Failures of components.
- Accidents in operation.
- Sabotage.

Component failures and accidents may also occur in the previous states of the SLC and need to be mitigated and prevented in those states.

6.8.2 RISKS IN SYSTEMS ENGINEERING

Generic risks in systems engineering are mainly people-based rather than technical.[*] For example, (Kasser and Schermerhorn 1994)

- Failing to understand the customer's requirements discussed in Section 6.8.2.1.
- Failing to communicate and maintain the vision discussed in Section 6.8.2.2.
- Failing to perform systems engineering with skilled systems engineers discussed in Section 6.8.2.3.
- Failing to perform adequate specialty engineering discussed in Section 6.8.2.4.

---

[*] It is the engineers who work with the technology.

- Failing to apply lessons learned from previous projects discussed in Section 6.8.2.5.
- Failing to plan ahead to ensure resources are available when needed discussed in Section 6.8.2.6.
- Failing to document the reasons for decisions discussed in Section 6.8.2.7.
- Failing to use a systems engineering methodology that seamlessly interfaces to the software development methodology discussed in Section 6.8.2.8
- Failing to control changes discussed in Section 6.8.2.9.

6.8.2.1 Failing to Understand the Customer's Requirements

The major consequence of failing to understand the customer's requirements is an unhappy customer.

6.8.2.2 Failing to Communicate and Maintain the Vision

The major consequences of failing to communicate and maintain the vision are complete project failure and cancellation, or initial failure then recovery. Without a clear vision, such as the one in the CONOPS (Section 6.5.9.1), different sections of the project proceed in different directions at different rates.

6.8.2.3 Failing to Perform System Engineering with Skilled Systems Engineers

The major consequence of failing to perform system engineering with skilled systems engineers is that the problem tends to be posed in terms of a solution optimized for the expertise of the people responsible for the system design. For example, when hardware expertise is used, it tends to purchase some brand-new hardware that looks like a neat thing on which to base a project, but is later found to be unsuitable. When software expertise is used, the system becomes software-based. Moreover, since it is so easy to make changes in software that there is a propensity for the following to occur:

- Failure to recognize the need to plan ahead, and thus
- failing to set up objectives, and consequently,
- ending up in a continual crisis management situation.

6.8.2.4 Failing to Perform Adequate Specialty Engineering

The major consequences of failing to perform adequate specialty engineering are unexpected problems that cause schedule delays and potential costly redesigns.

6.8.2.5 Failing to Apply Lessons Learned from Previous Projects

The major consequence of failing to apply lessons learned from previous projects is that mistakes are repeated and paid for on the current project, with accompanying cost escalation and schedule delays.

6.8.2.6 Failing to Plan Ahead

The major consequences of failing to plan ahead to ensure resources are available when needed are schedule delays, high pressure and crisis management, which leads to undesirable high corporate visibility.

6.8.2.7 Failing to Document the Reasons for Decisions

The major consequence of failing to document the reasons for decisions is that decisions are questioned throughout the system design, implementation and test states.

6.8.2.8 Failing to Use a Systems Engineering Methodology That Seamlessly Interfaces to the Software Development Methodology

The major consequences of failing to use a systems engineering methodology that seamlessly interfaces to the software development methodology are delays and errors in translation from the system requirements to the design state.

6.8.2.9 Failing to Control Changes

The major consequences of failing to control changes are moving baselines and confusion leading to cost escalation and schedule delays.

The consequences of ineffective systems engineering in any state of a project are cost escalations and schedule delays throughout the remainder of the project. Eliminating ineffective systems engineering is akin to fixing defects in a manufacturing process.

The major products produced by systems engineers on their way to delivering the system are presentations and documentation. Ineffective systems engineers produce flawed or defective documentation. As an example of the cost of defective documentation, consider just the formal and informal meetings on a typical large project resulting from trying to interpret a single defective document. Just multiply the time spent in these meetings by the number of meetings and the numbers of attendees; *the unplanned labour cost of these meetings can very quickly reach $500,000 or so* over the course of the project. Now multiply that by the number of defective documents in a large engineering project (LEP) or even a simpler project, and think about the effect on the project budget and schedule.

6.8.3 THE PERENNIAL PROBLEM OF POOR REQUIREMENTS

The perennial problem of poor requirements may be formatted according to the problem formulation template (Section 5.1.7.2) as:

1. *The undesirable situation*: *research* has shown that there is an ongoing consensus that (Kasser 2012):
   1. Good (well-written) requirements are critical to the success of a project (Larson 2014).
   2. The current requirements paradigm produces poorly written requirements (e.g. Alexander and Stevens 2002, Jorgensen 1998, Lee and Park 2004).
   3. Suggestions for producing good requirements have been around for more than 25 years (Hooks 1993).
2. *Assumptions*: none.
3. *The FCFDS*: systems and software engineers provide customers' needs to system providers using well-written requirements or an alternative method of communication.

4. *The problem*: how to create the FCFDS.
5. *The solution*: to be determined.

This section perceives the undesirable perennial problem of poor requirements from different HTPs to gain an understanding of the undesirable situation.

6.8.3.1 Big Picture

Perceptions of requirements from the *Big Picture* HTP include:

* Systems and software engineers continue to produce poor requirements when ways to write good requirements have been documented in conference papers and textbooks.
* Contemporary requirements management practice irrespective of the process used to generate the requirements is far from ideal, producing:
  * *Vague and unverifiable requirements*: due to poor phrasing of the written text.
  * *Incompletely articulated requirements*: due to a poor requirements elicitation process.
  * *Incomplete requirements*: due to various factors including domain inexperience, and the lack of expertise in eliciting and writing requirements by technical staff.
  * *Poor management of the effect of changing user needs during the time that the system is under construction*: due to lack of the understanding of the need for CM, and use of appropriate tools to do the function in an effective manner.
* Recognition that a requirement is more than just the imperative statement. For example, adding additional properties to the text-based requirement (e.g. priority and traceability) (Alexander and Stevens 2002, Hull, Jackson, and Dick 2002). However, in practice, there is difficulty in adding these additional properties to the traditional requirement document or database and then managing them. This is because the current systems and software development paradigm generally divides the work in a project into three independent streams of activities (Section 7.3.1.1.2). Thus, requirements engineering tools contain information related to the development and test streams (the requirements), and the additional properties tend to be separated into several different incompatible tools (e.g. requirements management, project management, work breakdown structures, configuration control and cost estimation).

6.8.3.2 Operational

System engineers eliciting, elucidating and validating requirements.

6.8.3.3 Functional

System engineers communicating with stakeholders, writing requirements, clarifying requirements, etc.

1. ID number
2. Subject
3. Verb
4. Object
5. Qualifiers

FIGURE 6.21   The five parts of a requirement statement.

6.8.3.4   Temporal

Requirements evolve or change during the SDP and once the system is in-service. Accordingly, requirements are a representation (snap shot) of an evolving need caught at a specific point in time.

6.8.3.5   Structural

Perceptions from the *Structural* HTP noted:

- The parts of a requirement shown in Figure 6.21. A text-mode requirement should just be a simple sentence. Yet there are problems in the way requirement sentences are structured (Scott, Kasser, and Tran 2006).

6.8.3.6   Generic

Text-mode requirements are but one way to communicate information. Fanmuy clarifies the definition of a requirement statement by adding, 'this statement is written in a language which can take the form of a natural language or a mathematical, arithmetic, geometrical or graphical expression' (Fanmuy 2004). Timing and state diagrams are often used in requirements documents. Thus, the concept of stating user needs (under certain circumstances) via diagrams is already in use in systems engineering (Kasser 2002b). Thus, from this perspective, requirements are but one of a number of communications tools. Other ways of communicating all or part of the same information include models, simulations, photographs, schematics, drawings and prototypes. That the focus should be on user needs, not on requirements, has already been recognized, 'we don't perform system engineering to get requirements', and 'we perform system engineering to get systems that meet specific needs and expectations' (Van Gaasbeek and Martin 2001).

6.8.3.7   Continuum

Perceptions from this perspective note ranges and ambiguities. For example:

- The information in a requirements document can be looked at as both a solution and a problem. The matched set of specifications documents a conceptual solution system that should remedy the problem. Thus, they document a solution as far as the customer is concerned, but at the same time, they also document the problem faced by the designers who have to design the solution system.
- The word 'requirement' may have different meanings in different states of the acquisition life cycle. In the early state systems engineering activities

in column A of the HKMF, 'requirement' and 'need' may also be used interchangeably, but have slightly different meanings in the worldviews of the customer and contractor. These differences in meanings show up in the IEEE definition of a requirement (IEEE Std 610 1990).

6.8.3.8   Scientific

Inferences from this perspective note that there seem to be a number of reasons for systems and software engineers to continue to produce poor requirements when ways to write good requirements have been documented in conference papers and textbooks, including (in no particular order):

- Lack of time to write the requirements due to schedule constraints, which results in poorly drafted and incomplete requirements.
- Failure of the stakeholders to articulate the requirements, which results in incomplete and sometimes results in incorrect requirements.
- Lack of training for writing good requirements, which results in poorly written requirements.
- Fundamental lack of management understanding of the need for, and the purpose served by, requirements, which results in lack of sufficient time for the requirements elicitation and elucidation process.
- Lack of implementation and solution domain knowledge (Section 5.1.4.6) in the systems and software engineers eliciting and elucidating the requirements, which tends to result in incomplete and sometimes unachievable requirements.
- Lack of functionality in commercial requirements tools that can call attention to poorly written requirements.
- They are working in the 'B' paradigm.

These reasons can be aggregated into two issues:

1. *Production* of poorly written requirements.
2. *Lack* of ways of ensuring completeness of the requirements.

Ways of mitigating or preventing these risks include:

- Using the 'A' paradigm.
- Training in writing requirements at the start of the system requirements state.
- Tagging requirements with acceptance criteria (Kasser 2019) to clarify the meaning of the requirement when it is written.

6.9   EXERCISES

This section provides a number of exercises as opportunities to think about systems and the systems lifecycle. A partial acceptable solution is provided for the off-world mining policy (Section 6.9.4). It is up to the reader to use the sample as a template when performing the remaining exercises.

6.9.1   THE ENU TRAFFIC LIGHT UPGRADE

In addition to meeting the generic requirements in Section 2.7:

1. Format the problem posed by this exercise in the problem formulation template (Section 5.1.7.2).
2. Define the boundary of the system in operation and explain the reason for the boundary.
3. Estimate the levels of objective complexity and subjective complexity (Section 5.2.2.2) of the system in operation.
4. State, and justify the selection, the layer in the risk framework (Section 1.2) and HKMF (Section 3.4.4) in which the system in operation exists.
5. Describe the system which develops the system in operation in each state of the SLC (Section 6.1.5.3).
6. Describe the relationship between the system which develops the system in operation and the system in operation from the perspective of risk management.

6.9.2   THE ENGAPOREAN MAID REDUCTION POLICY

In addition to meeting the generic requirements in Section 2.7:

1. Format the problem posed by this exercise in the problem formulation template (Section 5.1.7.2).
2. Define the boundary of the system in operation and explain the reason for the boundary.
3. Estimate the levels of objective complexity and subjective complexity (Section 5.2.2.2) of the system in operation.
4. State, and justify the selection, the layer in the risk framework (Section 1.2) and HKMF (Section 3.4.4) in which the system in operation exists.
5. Describe the system which develops the system in operation in each state of the SLC (Section 6.1.5.3).
6. Describe the relationship between the system which develops the system in operation and the system in operation from the perspective of risk management.

6.9.3   THE ENGAPORIAN DROUGHT RELIEF POLICY

In addition to meeting the generic requirements in Section 2.7:

1. Format the problem posed by this exercise in the problem formulation template (Section 5.1.7.2).
2. Define the boundary of the system in operation and explain the reason for the boundary.
3. Estimate the levels of objective complexity and subjective complexity (Section 5.2.2.2) of the system in operation.

4. State, and justify the selection, the layer in the risk framework (Section 1.2) and HKMF (Section 3.4.4) in which the system in operation exists.
5. Describe the system which develops the system in operation in each state of the SLC (Section 6.1.5.3).
6. Describe the relationship between the system which develops the system in operation and the system in operation from the perspective of risk management.

6.9.4 THE OFF-WORLD MINING POLICY

In addition to meeting the generic requirements in Section 2.7:

1. Format the problem posed by this exercise in the problem formulation template (Section 5.1.7.2).
2. Define the boundary of the system in operation and explain the reason for the boundary.
3. Estimate the levels of objective complexity and subjective complexity (Section 5.2.2.2) of the system in operation.
4. State, and justify the selection, the layer in the risk framework (Section 1.2) and HKMF (Section 3.4.4) in which the system in operation exists.
5. Describe the system which develops the system in operation in each state of the SLC (Section 6.1.5.3).
6. Describe the relationship between the system which develops the system in operation and the system in operation from the perspective of risk management.

6.9.4.1 One Acceptable Solution

A completed exercise could take the form shown in this section. Note it provides one acceptable solution, not the single correct solution as there will be other acceptable solutions (Section 5.1.1.2.2).

6.9.4.1.1 The Assumptions

The assumptions include:

1. The question asking for the relationship between the systems does not specify which systems and is open to interpretation. Accordingly, the system is interpreted as being the interface between the SDP and the system in operation (S5 and S6 in the nine-system model) (Section 6.7).

6.9.4.1.2 The Formatted Problem

The problem formatted according to the problem formulation template (Section 5.1.7.2) is:

1. *The undesirable situation*: is the need to:
   1. Comply with the generic requirements in Section 2.7.
   2. Format the problem posed by this exercise in the problem formulation template (Section 5.1.7.2).

3. Define the boundary of the system in operation and explain the reason for the boundary.
4. Estimate the levels of objective complexity and subjective complexity (Section 5.2.2.2) of the system in operation.
5. State, and justify the selection, the layer in the risk framework (Section 1.2) and HKMF (Section 3.4.4) in which the system in operation exists.
6. Describe the system which develops the system in operation in each state of the SLC (Section 6.1.5.3).
7. Describe the relationship between the system which develops the system in operation and the system in operation from the perspective of risk management.

2. *The assumptions*: are:
1. The exercise can be completed within 60 minutes.
2. As listed in Section 6.9.4.1.1.

3. *The FCFDS*: the exercise has been completed and the desired grade has been achieved.

4. *The problem*: how to do the exercise, namely:
1. Meet the generic requirements in Section 2.7.
2. Format the problem posed by this exercise in the problem formulation template (Section 5.1.7.2).
3. Define the boundary of the system in operation and explain the reason for the boundary.
4. Estimate the levels of objective complexity and subjective complexity (Section 5.2.2.2) of the system in operation.
5. State, and justify the selection, the layer in the risk framework (Section 1.2) and HKMF (Section 3.4.4) in which the system in operation exists.
6. Describe the system which develops the system in operation in each state of the SLC (Section 6.1.5.3).
7. Describe the relationship between the system which develops the system in operation and the system in operation from the perspective of risk management.
8. Verify that all the requirements for the exercise have been completed and update the compliance matrix accordingly.

5. *The solution*: consists of performing each of the problems. A partial solution is presented below.

6.9.4.1.3   The Compliance Matrix

The partially completed compliance matrix is shown in Table 6.4.*

---

* Placing the section number in the completed column not only shows that the requirement has been met but shows where the response can be found, mitigating the risk of the instructor failing to find and grade the response.

TABLE 6.4

Compliance Matrix for the Exercise in Section 6.9.4

ID	Instruction	Deliverable	Section	Completed
1	Brainstorming or active brainstorming	No	All	Yes
2	Research	No	All	Yes
3	Use imagination	No	All	Yes
4	Use lists and charts	No	All	Yes
5	Spend less than 60 minutes	No	All	Yes
6	The formatted problem	Yes	6.9.4.1.2	Yes
7	Assumptions	Yes	6.9.4.1.1	Yes
8	Preliminary compliance matrix	No	6.9.4.1.3	Yes
9	The boundary of the system	Yes	6.9.4.1.4	Yes
10	Level of objective complexity	Yes	6.9.4.1.5	Yes
11	Level of subjective complexity	Yes	6.9.4.1.6	Yes
12	Layer in the risk framework	Yes	6.9.4.1.7	Yes
13	Layer in the HKMF	Yes	6.9.4.1.8	Yes
14	System which develops the system	Yes	6.9.4.1.9	Yes
15	Relationship between the systems	Yes	6.9.4.1.10	Yes
16	Lessons learned	Yes	6.9.4.1.11	Yes
17	Updated compliance matrix	Yes	6.9.4.1.3	Yes

6.9.4.1.4 The Boundary of the System

The goal of the policy is to mine the asteroids and the atmosphere of Jupiter and deliver the materials down to the Earth. The CONOPS contains scenarios for:

- Moving people and supplies from the surface of the Earth to low earth orbit (LEO).
- Activities in LEO.
- Transit from LEO to a point in or near the asteroids.
- Mining the asteroids.
- Mining the atmosphere of Jupiter.
- Activities in the asteroid belt other than mining such as recreation and law enforcement.

Accordingly, the justification for drawing the system boundary system boundary is to include all the activities listed above. The external interfaces will be to:

- The spaceport on the ground.
- The atmosphere of Jupiter.
- The asteroids.

The subsystems are:

- The LEO system.
- The asteroid transfer subsystem.
- The asteroid subsystem

The subsystems are partitioned to optimize the interfaces between (Kasser 2019: Section 9.6.1.8):

- The Earth to LEO subsystem.
- The LEO to the asteroid transfer subsystem.
- The asteroid subsystem.

The asteroid subsystem is further divided into its own subsystems, namely:

- The mining subsystem.
- The recreation subsystem.
- The manufacturing subsystem
- The shipment subsystem that interfaces to the LEO to the asteroid transfer subsystem.
- Each of these subsystems will also have their own subsystems.

Each system or subsystem will be partitioned into mission and support subsystems (Section 6.3.2). Since the system and subsystem partitioning has been described, there is no need for a graphic depiction.

6.9.4.1.5   The Level of Objective Complexity
The level of objective complexity is extremely high because:

- There are a large number of entities in the system.
- The approximate number can only be guessed at this time.

6.9.4.1.6   The Level of Subjective Complexity
The level of subjective complexity is medium to low because:

- The subsystems and the interfaces between them are very well defined.
- Their operation and the operation of the interfaces between them are understood.
- The system architecture optimized the interfaces for minimum coupling and maximum cohesion according to the rule for performing aggregation (Kasser 2019: Section 9.6.1.6.1).

6.9.4.1.7   The Layers in the Risk-Framework
The policy begins in layer 5 in the policy initiation state and spreads out down the layers as it moves through the planning and performance states. During the performance states, the system is complex enough to have parts in all five layers.

The problems in the higher layers are generally more research problems than intervention problems, whereas in the lower layers the problems tend to be more technical than people-related intervention problems.

6.9.4.1.8   The Layers in the HKMF
The policy begins in layer 5 in the HKMF (Section 3.4.4) and spreads out down the layers as it moves through the planning and performance states. During the

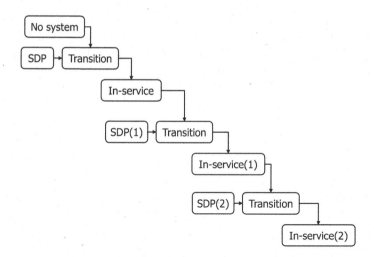

FIGURE 6.22 The generic extended SLC.

performance states, the system is complex enough to have parts in all five layers. However, systems engineering in layer 5 is very people-focused, whereas in layers 2 and 3 it is very technologically focused.

The problems in the higher layers are generally more research problems than intervention problems, whereas in the lower layers the problems tend to be more technical than people-related intervention problems.

6.9.4.1.9 The System Which Develops the System in Operation

The system which develops the system is the SDP (Section 6.1.5.4), S5 in the nine-system model (Section 6.7) in the context of the organization and S9 in the nine-system model.

6.9.4.1.10 Relationship between the Systems

The relationship between the SDP and the system in operation at the highest level is that the SDP produces the system in operation. However, since this is going to be a phased development in which there will be pilot projects taking place in different organizations all marching to their own drummers' pace, the development will be a time-ordered sequential and parallel set of activities (Kasser 2002a). Accordingly, some systems will be under development at the same time as other systems will be in operation or upgraded, a concept similar to the one shown in the generic extended SLC in Figure 6.22 (Kasser 2019)* but with an undetermined number of parallel systems under development. Perceptions from the *Generic* and *Temporal* HTPs note the similarity to the development of the Semiautomatic Ground Environment (SAGE) system (Hughes 1998: p. 15); however, perceptions from the *Continuum* HTP note

---

* Including cited sources is a good way to demonstrate compliance with the requirement to do research and impart expertise to the information.

that the development of the SAGE system was managed by a single organization, whereas the development of the system that this policy will produce will not be managed by any single organization. However, there probably will be a degree of coordination and cooperation since it's expected that the benefits of a degree of coordination and cooperation will make that happen (enlightened self-interest).

6.9.4.1.11 Lessons Learned

Lessons learned include:

- The compliance matrix is very handy for keeping track of what needs to be done and making sure that it is being done.
- People well-read in science fiction are an asset to the team due to their domain knowledge.

6.10 SUMMARY

This chapter discussed systems and systems engineering from different perspectives because projects are systems and create products which are systems using systems engineering.* The chapter provided an overview of systems, the SLC, systems engineering and the SDP to help the policymaker, implementer and project manager understand the important aspects of risk management in systems and systems engineering. Specifically, this chapter discussed the nature of systems, properties of systems, hierarchies of systems, supply chains as systems, an introduction to systems engineering, modelling and simulation, the nine-system model and risks in systems and systems engineering.

REFERENCES

Alexander, Ian. F., and Richard Stevens. 2002. *Writing Better Requirements*. London: Addison-Wesley.

Beasley, Richard, and Richard Partridge. 2011. The three Ts of systems engineering – Trading, tailoring and thinking. In *the 21st International Symposium of the INCOSE*. Denver, CO.

Beer, Stafford. 1994. *The Heart of Enterprise*. Chichester: John Wiley & Sons, Stafford Beer Classic Edition.

Blokdyk, Gerardus. 2018. *System Integration Testing: The Ultimate Step-By-Step Guide*. Brendale: The Art of Service.

Box, George Edward Pelham, and Norman R. Draper. 1987. *Empirical Model Building and Response Surfaces*. New York: John Wiley & Sons.

Boyd, John. 2017. *The essence of winning and losing 1995* [cited 13 November 2017]. Available from www.pogoarchives.com.

Brill, James H. 1998. Systems engineering – A retrospective view. *Systems Engineering* no. 1 (4):258–266.

Chapanis, Alphonse. 1960. Human engineering. In *Operations Research and Systems Engineering*, edited by Charles D. Flagle, William H. Huggins and Robert H. Roy, 534–582. Baltimore, MD: Johns Hopkins Press.

---

* Even if they don't call it systems engineering.

Checkland, Peter, and Jim Scholes. 1990. *Soft Systems Methodology in Action*. Chicester: John Wiley & Sons.

Churchman, C. West. 1979. *The Systems Approach and Its Enemies*. New York: Basic Books, Inc.

Collinsdictionary. 2019. *Definition of 'bumf'* [cited 25 March 2019]. Available from https://www.collinsdictionary.com/dictionary/english/bumf.

Daniels, Jesse., Terry Bahill, and Rick Botta. 2005. A hybrid requirements capture process. In *the 15th International Symposium of the INCOSE*. Rochester, NY.

Deming, W. Edwards. 1993. *The New Economics for Industry, Government, Education*. Cambridge: MIT Center for Advanced Engineering Study.

DOD 5000.2-R. 2002. Mandatory procedures for major defense acquisition programs (MDAPS) and major automated information system (MAIS) acquisition programs. Washington, DC: US Department of Defense.

Dolan, Tom. 2003. Best practices in process improvement. Quality Progress 36:23–28.

EIA 632. 1994. EIA 632 Standard: Processes for engineering a system.

Eisner, Howard. 1988. *Computer Aided Systems Engineering*. Englewood Cliffs, NJ: Prentice Hall.

Eisner, Howard. 1997. *Essentials of Project and Systems Engineering Management*. New York: John Wiley & Sons, Inc.

Evans, Richard P. 1996. Engineering of computer-based systems (ECBS) three new methodologies – Three new paradigms. In *the 6th International Symposium of the INCOSE*. Boston, MA.

Fanmuy, Gauthier. 2004. Best practices for drawing up a requirements baseline. In *the 14th International Symposium of the INCOSE*. Toulouse, France.

Fayol, Henri. 1949. *General and Industrial Management*. London: Sir Isaac Pitman and Sons, Ltd.

Gabb, Andrew. 2001. Front-end operational concepts – Starting from the top. In the *11th Annual Symposium of the INCOSE*. Melbourne, Australia.

Gelbwaks, Norman L. 1967. AFSCM 375-5 as a methodology for system engineering. *IEEE Transactions on Systems Science and Cybernetics* no. 3 (1):6–10.

Hall, Arthur D. 1962. *A Methodology for Systems Engineering*. Princeton, NJ: D. Van Nostrand Company Inc.

Hall, Arthur D. 1989. *Metasystems Methodology. A New Synthesis and Unification*. Oxford: Pergamon Press.

Hitchins, Derek K. 1992. *Putting Systems to Work*. Chichester: John Wiley & Sons.

Hitchins, Derek K. 1998. Systems engineering … In search of the elusive optimum. In *the 4th Annual Symposium of the INCOSE-UK*, Warwick.

Hitchins, Derek K. 2000 *World class systems engineering – The five layer model* [Web site] [cited 3 November 2006]. Available from http://www.hitchins.net/5layer.html.

Hitchins, Derek K. 2007. *Systems Engineering. A 21st Century Systems Methodology*. Chichester: John Wiley & Sons Ltd.

Hooks, Ivy. 1993. Writing good requirements. In *the 3rd International Symposium of the NCOSE*. Crystal City, VA.

Hughes, Thomas P. 1998. *Rescuing Prometheus*. New York: Random House Inc.

Hull, Elizabeth, Ken Jackson, and Jeremy Dick. 2002. *Requirements Engineering*. London: Springer.

Huygens, Christiaan. 1690. *Treatise on Light*. New York: Dover.

IEEE Std 610. 1990. IEEE standard glossary of software engineering terminology. IEEE.

IEEE Std 1220. 1998. Standard 1220 IEEE standard for application and management of the systems engineering process.

Jackson, Michael C., and Paul Keys. 1984. Towards a system of systems methodologies. *Journal of the Operations Research Society* no. 35 (6):473–486.

Jenkins, Gwyn Morgan. 1969. The systems approach. In *Systems Behaviour*, edited by John Beishon and Geoff Peters, 82. London: Harper and Row.

Jorgensen, Raymond W. 1998. Untangling the twists in requirements analysis. In *the 8th INCOSE International Symposium*. Vancouver, BC.

Kasser, Joseph Eli. 1995. *Applying Total Quality Management to Systems Engineering*. Boston, MA: Artech House.

Kasser, Joseph Eli. 2002a. The acquisition of a system of systems is just a simple multi-phased parallel-processing paradigm. In *the International Engineering Management Conference*. Cambridge, UK.

Kasser, Joseph Eli. 2002b. Does object-oriented system engineering eliminate the need for requirements? In *the 12th International Symposium of the INCOSE*. Las Vegas, NV.

Kasser, Joseph Eli. 2005. Introducing the role of process architecting. In *the 15th International Symposium of the INCOSE*. Rochester, NY.

Kasser, Joseph Eli. 2010. SEMP, TEMP and SHMEMP! It's time to stop the Mishigas. In the *Researches and Development Directions in Systems Engineering*, at The Gordon Center, Technion, Haifa, Israel.

Kasser, Joseph Eli. 2012. Getting the right requirements right. In *the 22nd International Symposium of the INCOSE*. Rome, Italy.

Kasser, Joseph Eli. 2013. *Holistic Thinking: Creating Innovative Solutions to Complex Problems*. Charleston, SC: Createspace Ltd.

Kasser, Joseph Eli. 2015a. *Holistic Thinking: Creating Innovative Solutions to Complex Problems*. 2nd edition. Vol. 1, Solution engineering. Charleston, SC: Createspace Ltd.

Kasser, Joseph Eli. 2015b. *Perceptions of Systems Engineering*. Vol. 2, Solution engineering. Charleston, SC: Createspace Ltd.

Kasser, Joseph Eli. 2019. *Systems Engineering: A Systemic and Systematic Methodology for Solving Complex Problems*. Boca Raton, FL: CRC Press.

Kasser, Joseph Eli, and Eileen Arnold. 2014. Academia is not teaching the right things in systems engineering Master's courses. In *the 24th International Symposium of the INCOSE*. Las Vegas, NV.

Kasser, Joseph Eli, and Eileen Arnold. 2016. Benchmarking the content of Master's degrees in systems engineering in 2013. In *the 26th International Symposium of the INCOSE*. Edinburgh, Scotland.

Kasser, Joseph Eli, and Derek K. Hitchins. 2012. Yes systems engineering, you are a discipline. In *the 22nd International Symposium of the INCOSE*. Rome, Italy.

Kasser, Joseph Eli, and Derek K. Hitchins. 2013. Clarifying the relationships between systems engineering, project management, engineering and problem solving. In *Asia-Pacific Council on Systems Engineering Conference (APCOSEC)*. Yokohama, Japan.

Kasser, Joseph Eli, and Robin Schermerhorn. 1994. Gaining the competitive edge through effective systems engineering. In *the 4th International Symposium of the NCOSE*. San Jose, CA.

Kasser, Joseph Eli, Yang-Yang Zhao, and Chandru J. Mirchandani. 2014. Simplifying managing stakeholder expectations using the nine-system model and the holistic thinking perspectives. In *the 24th International Symposium of the INCOSE*. Las Vegas, NV.

Kline, Stephen Jay. 1995. *Conceptual Foundations for Multidisciplinary Thinking*. Stanford, CA: Stanford University Press.

Koestler, Arthur. 1978. *JANUS A Summing Up*. New York: Random House.

Lake, Jerome G. 1994. Axioms for systems engineering. *Systems Engineering. The Journal of the NCOSE* no. 1 (1):17–28.

Larson, Elizabeth. 2014. I still don't have time to manage requirements: My project is later than ever. In *PMI® Global Congress 2014-North America*. Phoenix, AZ: Project Management Institute.

Lee, Joong Yoon, and Young Won Park. 2004. Requirement architecture framework (RAF). In *the 14th International Symposium of the INCOSE*. Toulouse, France.

Machol, Robert E., and Ralph F. Miles Jr. 1973. The engineering of large scale systems. In Systems Concepts, edited by Ralph F. Miles Jr, 33–50. New York: John Wiley & Son, Inc.

Martin, James N. 1997. *Systems Engineering Guidebook: A Process for Developing Systems and Products*. Boca Raton, FL: CRC Press.

Martin, Nathaniel G. 2005. Work practice in research: A case study. In *the 14th International Symposium of the INCOSE*. Toulouse, France.

McConnell, George R. 2002. Emergence: A partial history of systems thinking. In *the 12th International Symposium of the INCOSE*. Las Vegas, NV.

McKeller, John M. 2014. *Supply Chain Management Demistified*. New York: McGraw Hill Education.

MIL-STD-499A. 1974. Mil-STD-499A engineering management: United States Department of Defense (USAF).

Needham, Joseph. 1937. *Integrative Levels: A Revaluation of the Idea of Progress*. Oxford: Clarendon Press.

Newton, Isaac. 1675. *Hypothesis of Light*. http://www.newtonproject.ox.ac.uk/view/texts/normalized/NATP00002.

Oxford. 2019. *Oxford Dictionaries*. Oxford: Oxford University Press [cited 22 January 2019]. Available from https://en.oxforddictionaries.com/definition/emergence.

Ramo, Simon. 1973. The systems approach. In *Systems Concepts*, edited by R.F. Miles Jr, 13–32. New York: John Wiley & Son, Inc.

Rhodes, Donna H. 2002. Systems engineering on the dark side of the moon. In *the 12th International Symposium of the INCOSE*. Las Vegas, NV.

Royce, Winston. W. 1987. Managing the development of large software systems: Concepts and techniques. In *ICSE '87 the 9th International Conference on Software Engineering*. Los Alamitos, CA.

Schön, Donald A. 1991. *The Reflective Practitioner*. Aldershot: Ashgate.

Scott, William, Joseph Eli Kasser, and Xuan-Linh Tran. 2006. Improving the structure and content of the requirement statement. In *the 16th International Symposium of the INCOSE*. Orlando, FL.

Selby, Richard W. 2006. Enabling measurement-driven system development by analyzing testing strategy tradeoffs. In *the 16th International Symposium of the INCOSE*. Orlando, FL.

Spencer, Herbert. 1962. First principles. In *A System of Synthetic Philosophy*, edited by Spencer Herbert. London. Williams and Norgate

Sutcliffe, Alistair, Julia Galliers, and Shailey Minocha. 1999. Human errors and system requirements. In *the IEEE International Symposium on Requirements Engineering*. Limerick, Ireland.

UIA. 2002. *Integrative knowledge project: Levels of organization*. Union of International Associations 2002 [cited 28 May 2002]. Available from http://www.uia.org/uialists/kon/c0841.htm.

Van Gaasbeek, James R., and Judith N. Martin. 2001. Getting to requirements: The W5H challenge. In *the 11th International Symposium of the INCOSE*. Melbourne, Australia.

Wasson, Charles S. 2006. *System Analysis, Design, and Development Concepts, Principles and Practices*. Hoboken, NJ: Wiley-Interscience.

Webster. 2004. *Merriam-Webster online dictionary* [cited 12 January 2004]. Available from http://www.webster.com.

Wilson, Tom D. 2002. Philosophical foundations and research relevance: Issues for information research (Keynote address). In *the Fourth International Conference on Conceptions of Library and Information Science: Emerging Frameworks and Method*. University of Washington, Seattle.

Wymore, A. Wayne. 1993. *Model-Based Systems Engineering, Systems Engineering Series*. Boca Raton, FL: CRC Press.

Wymore, A. Wayne. 1994. Model-based systems engineering. *Systems Engineering: The Journal of INCOSE* no. 1 (1):83–92.

Yen, Duen Hsi. 2008. *The blind men and the elephant* [cited 26 October 2010]. Available from http://www.noogenesis.com/pineapple/blind_men_elephant.html.

7 Basic Project Management

Policies and projects are systems which are realized using systems engineering and project management working interdependently. Project management is the heart of effecting any change, caused by the implementation of a policy, a large engineering project (LEP) or any type of project. Chapter 6 provided an overview of systems, the system lifecycle (SLC), systems engineering and the system development process (SDP) to help the policymaker, implementer and project manager understand the important aspects of risk management in systems and systems engineering and the interdependency and overlap between risk management in systems engineering and project management. Accordingly, this chapter provides an overview of the systems approach to, and the risks in, projects and project management to help the policymaker, implementer and systems engineer understand the important aspects of risk management in systems and systems engineering and the interdependency and overlap between risk management in systems engineering and project management. Specifically, this chapter discusses:

1. Pure and applied project management in Section 7.1
2. The top seven risk-indicators or a project in trouble in Section 7.2.
3. Projects in Section 7.3.
4. The project lifecycle in Section 7.4.
5. The project plan in Section 7.4.2.1.
6. Generic planning in Section 7.4.2.1.1.
7. Specific planning in Section 7.4.2.1.2.
8. Contingencies and contingency planning in Section 7.4.2.2.
9. Categorized requirements in process (CRIP) charts in Section 7.5.
10. Enhanced traffic light (ETL) charts in Section 7.6.
11. Management by exception (MBE) in Section 7.7.
12. Management by objectives (MBO) in Section 7.8.
13. Using prevention to lower project completion risk in Section 7.9.
14. Stakeholders and stakeholder management in Section 7.10.
15. Software project risks
16. Generic project and project management risks in Section 7.12.

7.1 PURE AND APPLIED PROJECT MANAGEMENT

Perceptions from the *Generic* HTP note that just as there are three types of systems engineering, (Section 6.5.3) there are three types of project management, namely:

1. *Pure project management*: cognitive skills, namely thinking, e.g. systems thinking (Section 2.4), critical thinking (Section 2.2), problem formulation/ solving (Section 5.1) and decision-making (Section 5.1.5.1).
2. *Applied project management*: activities traditionally associated with management; planning, organizing, directing and controlling (Fayol 1949, Kezsbom, Schilling, and Edward 1989).
3. *Domain project management*: pertains to the domain in which the project is taking place, e.g. organizations, computers, law enforcement, etc.

7.2 THE TOP SEVEN RISK-INDICATORS OR A PROJECT IN TROUBLE

Anecdotal evidence suggests that most projects do not fail due to the non-mitigation of technical risks. Rather, they fail as a result of poor management of the human element (Deming 1986, Harrington 1995). In addition, while the Standish Group identified ten major causes for project failure along with their solutions, they also stated that it was unclear if those solutions could be implemented (VOYAGES 1996). While the situation has improved somewhat in the past 25 years, there is still a lot of room for improvement.

The SDP for large systems can take several years to complete. During this time, the:

- *Customer* makes periodic progress payments to the supplier. In this situation, since the acceptance tests are only made at the end of the SDP, the suitability of the product for its mission is unknown for the time in which the bulk of the payments are made.
- *Supplier (contractor)* provides the customer with information to reassure the customer that the requirements in the statement of work are being implemented. The information is provided in the form of:
  - *Management information*: e.g. budget (estimated and actual), Gantt and program evaluation review technique (PERT) charts, conformance to 'best practices'.
  - *Intermediate products*: e.g. documents, lines of code produced, defects found, number of requirements satisfied.
  - *Process*: e.g. degree of compliance to pertinent documentation, regulations and standards.

The reports containing this intermediate information are produced to demonstrate a low risk of non-delivery and non-compliance to the requirements to the customer. These measurements provide post facto information, namely they report on what has already happened. This causes management to be reactive instead of being proactive.

In addition, in spite of all the measurements being made, the supplier is often unable to tell the customer:

- The exact percentage of completeness of the system under construction anytime during the SDP.
- The probability of successful completion within budget and according to schedule.

The methodology for developing metrics for predicting risks of project failures has its origins in a class on software independent verification and validation (IV&V) in the Graduate School of Management and Technology at University of Maryland University College* in 1997† and 1998 (Kasser and Williams 1998). Students in those classes wrote and presented term papers describing their experiences in projects that were in trouble. The term 'papers' adhered to the following instructions:

1. Document a case study based on personal experience.
2. Analyse the scenario.
3. Document the reasons the project succeeded or ran into trouble.
4. List and comment on the lessons learned from the analysis.
5. Identify a better way with 20/20 hindsight.
6. List a number of situational indicators that could be used to identify a project in trouble or a successful project while the project is in progress.

Once the papers had been graded, the methodology for developing metrics for predicting risks of project failures:

- Summarized the student papers to identify common elements called 'risk-indicators'.
- Surveyed systems and software development personnel via the Internet to determine if they agreed or disagreed with the risk-indicators.

One hundred and forty-eight responses were received. The raw findings are summarized in Table 7.1 (Kasser and Williams 1998). The first column contains a number identifying the risk-indicator described in the second column. The third column lists the number of students that identified the risk-indicator. The fourth column contains the percentage of agreement by the survey respondents. The fifth column contains the percentage of disagreement. The sixth column is the ranking of the risk-indicator. Thus, e.g. poor requirements ranked the highest as a cause of project failures with 97% agreement and 3% disagreement.

---

* These part time students were employed in the workforce and were working towards their degree in the evening. Their employment positions ranged from programmers to project managers. Some also had up to 20 years of experience in their respective fields.
† Dated but still very pertinent.

TABLE 7.1
Initial Findings

Risk ID	Risk-Indicators	Students	Survey Agree (%)	Survey Disagree (%)	Rank
1	Poor requirements	19	97	3	1
2	Failure to use experienced people	7	79	1	13
3	Failure to use IV&V	6	38	62	31
4	Lack of process and standards	5	84	16	11
5	Lack of, or, poor plans	4	95	5	2
6	Failure to validate original specification and requirements	3	91	9	3
7	Lack of configuration management	3	66	34	19
8	Low morale	2	51	49	24
9	Management does not understand the system development lifecycle	2	59	41	22
10	Management that does not understand technical issues	2	56	44	23
11	No single person accountable/responsible for project	2	69	31	18
12	Client and development staff fail to attend scheduled meetings	1	42	58	28
13	Coding from high-level requirements without design	1	75	25	14
14	Documentation is not produced	1	63	38	21
15	Failure to collect performance and process metrics and report them to management	1	48	52	25
16	Failure to communicate with the customer	1	88	12	5
17	Failure to consider existing relationships when replacing systems	1	85	15	10
18	Failure to reuse code	1	27	73	34
19	Failure to stress test the software	1	75	25	15
20	Failure to use problem language	1	34	66	30
21	High staff turnover	1	71	29	16
22	Key activities are discontinued	1	74	26	17
23	Lack of a Requirements Traceability Matrix (RTM)	1	67	33	19
24	Lack of clearly defined organizational (responsibility and accountability) structure	1	82	18	11
25	Lack of management support	1	87	13	6
26	Lack of priorities	1	85	15	8
27	Lack of understanding that demo software is only good for demos	1	47	53	26
28	Management expects a CASE tool to be a silver bullet	1	45	55	27
29	Political considerations outweigh technical factors	1	86	14	9

(Continued)

TABLE 7.1 (*Continued*)
Initial Findings

Risk ID	Risk-Indicators	Students	Survey Agree (%)	Survey Disagree (%)	Rank
30	Resources are not allocated well	1	92	8	4
31	The QA team is not responsible for the quality of the software	1	40	60	29
32	There are too many people working on the project	1	36	64	32
33	Unrealistic deadlines hence schedule slips	1	86	14	7
34	Hostility between developer and IV&V	1	33	67	33

The top seven (high priority) risk-indicators were identified using the following approaches:

1. *The Tally* discussed in Section 7.2.1.
2. *The priorities* discussed in Section 7.2.2.
3. *The top seven* discussed in Section 7.2.3.

7.2.1 THE TALLY

An 'agree' was allocated a value of +1 and a 'disagree' a value of −1. The answers to each survey statement were then tallied. The seven risk-indicators that received the highest positive values (most agreement) as causes of project failure are shown in Table 7.2 (Kasser and Williams 1998).

7.2.2 THE PRIORITIES

The survey asked respondents to rank the top seven risk-indicators in order of priority. The weighted results are shown in Table 7.3 (Kasser and Williams 1998) (top priority first).

TABLE 7.2
Top Seven Causes of Project Failures (Tally)

Risk ID	Risk-Indicators	Responses
1	Poor requirements	134
5	Lack of, or, poor plans	125
6	Failure to validate original specification and requirements	113
30	Resources are not allocated well	109
16	Failure to communicate with the customer	106
25	Lack of management support	98
33	Unrealistic deadlines hence schedule slips	97

TABLE 7.3

Priority Causes of Project Failure

Risk ID	Risk-Indicators	Weight
1	Poor requirements	864
16	Failure to communicate with the customer	683
5	Lack of, or, poor plans	574
4	Lack of process and standards	361
25	Lack of management support	350
6	Failure to validate original specification and requirements	329
29	Political considerations outweigh technical factors	304

7.2.3 THE TOP SEVEN

Since the actual position may be subjective, the number of times a risk-indicator showed up in any position in the top seven priority list was also counted. The results for the top seven showing items are as shown in Table 7.4 (Kasser and Williams 1998). The results show a high degree of consensus on these risk-indicators as causes of project failures.

7.2.4 SENSITIVITY ANALYSIS ON THE PROJECT MANAGEMENT RISK-INDICATORS

The sample size for respondents without management experience was 99. The raw tallies for the risk-indicators associated with project management shown in Table 7.5 (Kasser and Williams 1998) were examined to see if there was a difference between non-managers and managers with various years of experience. No differences of more than 10% were noted.

7.2.5 THE 'OTHER' CATEGORY

The 'other' category was added to perform risk management and avoid Simpson's paradox (Kasser 2018a: Section 10.1.5.6.12, Savage 2009). Several respondents

TABLE 7.4

Top Seven Causes

Risk ID	Risk-Indicators	Count
1	Poor requirements	99
16	Failure to communicate with the customer	86
5	Lack of, or, poor plans	77
4	Lack of process and standards	51
25	Lack of management support	51
29	Political considerations outweigh technical factors	45
6	Failure to validate original specification and requirements	44

TABLE 7.5

Project Management Related Risk-Indicators

Risk ID	Risk-Indicators
5	Lack of, or, poor plans
8	Low morale
15	Failure to collect performance and process metrics and report them to management
25	Lack of management support
27	Lack of understanding that demo software is only good for demos
29	Political considerations outweigh technical factors
32	There are too many people working on the project
33	Unrealistic deadlines hence schedule slips

added one or two additional risk-indicators in the 'other' category of the questionnaire. These were:

- Failure to control change.
- Rapid rate of change of technology.
- Low bidding.
- Poor management.
- Lack of a technical leader.

Thus, the small student sample size of 19 seems to have identified most of the important risk-indicators.

7.2.6 VALIDATING THE SURVEY RESULTS

Applying some critical thinking, the question remains as to what these other items would have scored had they been in a list of 39 risk-indicators. This is why the survey results needed validating by comparison with an independent similar study. The approach used to validate the survey results was to use the Chaos study as a reference (CHAOS 1995). The study had recently (at that time) identified a number of major contributors to project failure. Five risk-indicators in this study that were chosen as the most important causes for project failure also appear on the Chaos list of major reasons for project failure. The correlation between the findings of this study and the Chaos study is shown in Table 7.6 (Kasser and Williams 1998). While 'resources are not allocated well' did not show up in the top seven lists of this study, it was fourth in the tally. 'Changing requirements and specifications' which showed up in the Chaos study as a contributor to project failure was not identified by the students as such but was written into the survey results as an 'other' by a few of the professional respondents.* Accordingly, the findings of this study support the findings of the Chaos study.

---

* Which brings us back to the question posed in Section 4.6.3.5.

TABLE 7.6

Comparison of Results with Chaos Study

Risk ID	This Study	Chaos Study
1	Poor requirements	Incomplete requirements
16	Failure to communicate with the customer	Lack of user involvement
30	Resources are not allocated well	Lack of resources
–	No equivalent	Unrealistic expectations
25	Lack of management support	Lack of executive management support
–	No equivalent	Changing requirements and specifications
5	Lack of, or, poor plans	Lack of planning

7.2.7 CONCLUSIONS

The conclusions are:

1. These risk indicators constitute generic risks in projects
2. A periodic project audit for these risk-indicators by an external independent entity should pick these up if they exist in time to take corrective action to prevent project failure.

7.3 PROJECTS

A project is a temporary endeavor undertaken to create a unique product, service or result.

PMI (2017)

Projects are systems that span a range of purposive activities such as cooking a meal, arranging a meal, planning and enjoying a vacation, moving house, developing software, deploying disaster relief teams and fighting a war. From conception to completion, a project begins when someone in an undesirable situation decides to remove the undesirability and faces the problem of turning the undesirable situation into a desirable situation. The project then passes through several states, each of which begins with a problem and is completed when a solution to the problem is developed, and the solution becomes a problem for the next state.

The project ends when the solution is provided to the problem posed by the last state, and someone verifies that the solution actually remedies the original problem for which the project was commissioned.

7.3.1 PROJECT ATTRIBUTES

All projects have the following attributes:

1. Activities discussed in Section 7.3.1.1.
2. Conflict discussed in Section 7.3.1.2.
3. Customers discussed in Section 7.3.1.3.

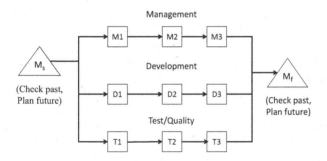

FIGURE 7.1   The three streams of activities.

4. Funding discussed in Section 7.3.1.4.
5. Meetings discussed in Section 7.3.1.5.
6. Milestones discussed in Section 7.3.1.6.
7. Need discussed in Section 7.3.1.7.
8. Outcomes discussed in Section 7.3.1.8.
9. People discussed in Section 7.3.1.9.
10. Politics discussed in Section 7.3.1.10.
11. Priority discussed in Section 7.3.1.11.
12. Purpose discussed in Section 7.3.1.12.
13. Return on investment discussed in Section 7.3.1.13.
14. Risks discussed in Section 7.3.1.14.
15. Sponsor discussed in Section 7.3.1.15.
16. Stakeholders discussed in Section 7.3.1.16.
17. Timeline discussed in Section 7.3.1.17.

7.3.1.1   Activities

Perceptions of activities from the HTPs include:

7.3.1.1.1   Operational

Purposive activities are the basic building blocks of projects. All work in the project is done by activities that take time to produce a product using resources. Activities may take place in series or in parallel.

7.3.1.1.2   Functional

There are three streams of activities (Kasser 1995) management, development and test as shown in Figure 7.1 where:

1. *Management*: the set of activities which include:
   - Monitoring and controlling the development and test stream activities to ensure performance in the state in accordance with the systems engineering management plan (SEMP) (Section 6.5.9.2) or project plan (Section 7.4.2.1).
   - Updating the SEMP or project plan to elaborate the work packages (WPs) (Section 7.3.1.1.4) for the three streams of activities in the subsequent state in more detail.

- Endeavouring to ensure that needed resources in the subsequent state will be available on schedule.
- Providing periodic reports on the condition of the project to the customer and other stakeholders.
- Being the contractual interface with the customer.
- Ensuring the appropriate risk management activities on the project management process.

2. *Development*: the set of activities which produce the products appropriate to the state by performing the design and construction tasks and appropriate risk management activities on the product.
3. *Test*: the set of activities known as quality control (QC) or quality assurance (QA), test and evaluation (T&E) and IV&V which include:
   - Identifying defects in products.
   - Verifying the degree of conformance to specifications of the products produced by the development stream by performing appropriate tests or analyses.

Most activities contain risks – undesirable events that will increase the cost or schedule to the detriment of the project. The time the activity takes is shown in a schedule in the form of a Gantt chart. Each activity has:

- *Customer(s)*: to whom the product produced by the activity is delivered.
- *Prerequisite(s)*: something that has to happen or be created before the activity can commence.
- *Resources*: people, materiel, time, money, etc.
- *Supplier (s)*: who supply something that is transformed by the activity into products.
- *Value*: by transforming the input from the supplier into a product that is used by a customer.

7.3.1.1.3 Continuum
Each activity uses the resources to produce:

1. *Product(s)*: the desired output.
2. *Waste*: the undesired output, e.g. defective products and parts, leftover parts and transformed materials, etc. Waste must be:
   1. Minimized because production of waste incurs an unnecessary cost.
   2. Disposed of carefully.

7.3.1.1.4 Structural
Each activity can be documented in a WP (Kasser 2018a: Section 8.19) which is:

- A tool used in the systems approach to project management to plan the work and identify and prevent risks from happening.
- A set of information associated with an activity.

The WP is based on the quality system elements (Kasser 2000) and contains the following information in a set of linked spreadsheets[*] or database:

1. *Unique WP identification number*: the key to tracking.
2. *Name of activity*: a succinct summary statement of the purposive activity or activities.
3. *Priority*: the priority associated with the activity linked back to the source requirement.
4. *Narrative description of the activity.* A brief description of the activity or activities.
5. *Estimated schedule for the activity*: the estimated time taken to perform the activities. This item contributes to the baseline schedule for the project.
6. *Accuracy of schedule estimate*: median, shortest and longest times to complete. This item is used to create the initial estimate of the critical path and shows up in the PERT charts (Kasser 2018a: Section 8.10).
7. *Actual schedule*: filled in after the activity is performed. This item is compared with the estimated schedule during the performance of the project.
8. *Products*: (outputs) produced by the activities in the WP.
9. *Acceptance criteria for products*: the response to the question, '*how will we know that the product is working as specified*', as agreed between the supplier and the customer.
10. *Estimated cost*: the estimated fixed and variable costs of the activities and materiel. This item contributes to the baseline budget for the project.
11. *The level of confidence in the cost estimate*: or the accuracy.
12. *Actual cost*: filled in as and when the activities are performed. This item is compared with the estimated schedule during the performance of the project as part of earned value analysis (EVA).
13. *Reason activity is being done*: important because personnel leave and join and the reasons may be forgotten. Some activities are preventative and the reason may not be immediately obvious. This information is also useful when considering change requests (Section 4.9.2) during the operations and maintenance (O&M) states of the SLC (Section 6.1.5.3).
14. *Traceability*: (source of work) to requirements, the concept of operations (CONOPS), laws, regulations, etc.
15. *Prerequisites*: the products that must be ready before the activity can commence.
16. *Resources*: the people, equipment, materiel, etc.
17. *Name of person responsible for the activity.*
18. *Risks*: description of risks associated with activities and products in the WP.
19. *Risk probabilities*: estimates and basis for estimates of probabilities of occurrence.
20. *Risk severities*: estimates and basis for estimates of severity consequences should the risks materialize as events.

---

[*] One spreadsheet or tab per WP.

21. *Risk mitigation information*: mitigation WP ID(s) for each risk.
22. *Lower level work package IDs*: if any.
23. *Decision points* (if any).
24. *Internal key milestones*: if the activity is broken out into lower-level WPs.
25. *Assumptions not stated elsewhere*: so as to be available for checking by cognizant personnel.

For examples of using the WPs in projects, see *Systemic and Systematic Project Management* (Kasser 2019a). The benefits of using WPs include:

- Facilitating risk management by incorporating risk identification and mitigation/prevention into the process of constructing the WP during the project planning and systems requirements states. For example, the risk mitigation plan would be based on an abstracted view of the risk elements in the WP database.
- Collecting information that is interdependent between product and process provides the ability to create and use attribute profiles (Kasser 2018a: Section 9.1).
- New perspectives on the system based on the attribute profiles.
- The project information resides in a single integrated database that contains information from the three streams of activities (Section 7.3.1.1.2), which dissolves the problem of updating separate databases to keep them current in the non-systems approach to project management.

7.3.1.2 Conflict

Conflict is an active disagreement between people:

1. With opposing opinions.
2. And the rules, culture or other factors that govern behaviour in various situations.

Project managers need to manage conflict effectively because conflict is inevitable and exists on all projects. Sometimes conflict can even be a force for good because it can result in innovation.

Once a conflict or a potential for a conflict in a situation is recognized, it should be identified as a risk. This will define the urgency of failing to resolve the conflict. Perceptions from the *Continuum* HTP indicate that importance and urgency may be different (Covey 1989).

Conflict arises for various reasons including:

- Ambiguous personnel roles.
- Different opinions.
- Inconsistent and/or incompatible goals.
- Lack of and poor critical thinking.
- Miscommunications.
- Need for joint decision-making.

- Need for consensus.
- Overlapping personnel responsibilities.
- Personality clashes.
- Procedures and regulations that hinder achieving project goals.
- Unresolved prior conflict.

Conflict on a project is generally an undesirable situation. When stated in these terms, the response to conflict is a variation of the traditional generic problem-solving process (Section 5.1.6.1) where the undesirable situation is the conflict and the feasible conceptual future desirable situation (FCFDS) is the lack of conflict.

The first step in resolving conflict is gaining an understanding of the situation, namely understand the:

- Nature of the conflict: e.g. logical (technical), emotional (personal).
- People involved in the conflict.

A useful systems thinking tool for reducing risks by not taking precipitative action and gaining an understanding of the situation when everyone is panicking is STALL (Kasser 2018a: Section 3.2.9). STALL is an acronym and a mnemonic tool to help you remember how to deal with situations that range from being asked a question to avoiding panicking in a crisis. STALL stands for:

- *Stay calm*: don't panic; wait until you understand what's going on. Then you can either panic or deal with the situation in a logical manner (*Continuum* HTP).
- *Think*: think about what you're hearing, experiencing, being told or seeing; you're generally receiving symptoms; you need to understand the cause.
- *Analyze and ask questions*: gain an understanding what's going on. This can be considered as the first stage of the problem-solving process. And when you're asking questions and thinking, use idea-generating tools such as active brainstorming (Section 2.5.2) to examine the situation.
- *Listen*: you learn more from listening than from talking.
- *Listen*: you learn more from listening than from talking.

Listening, analysing and asking questions are done iteratively until you feel you understand the situation. You have two ears and one mouth – that's the minimum ratio in which they should be used. Namely that means do at least twice as much listening as talking.

Perceiving conflict resolution from various HTPs:

- *Big Picture*: conflict takes place in a situation.

- *Continuum*: ways of resolving conflicts are situational. This means that there isn't a single way of resolving conflicts that is the optimal way of resolving every conflict. The project manager needs to recognize the type of situation in which the conflict exists and determine and use the appropriate way of resolving the conflict which include:
  - *Accommodating*: allowing parties to give up their needs and wishes to accommodate the other party.
  - *Collaborating*: working together with the parties involved to find a mutually beneficial solution, often a win–win solution.
  - *Competing*: a win–lose approach in which one party puts forward their resolution to the conflict at the expense of other resolutions.
  - *Compromising*: working together with the parties involved to find a middle ground in which each party is partially satisfied: leads to less than optimal solutions and lose–lose solutions.
  - *Delaying*: postponing conflict by introducing a cooling off period to give the parties a chance to work it out for themselves.
  - *Denying*: denying that the conflict even exists and ignoring or avoiding it.
  - *Dictating*: asserting one viewpoint at the potential expense of another, such as majority rule, or a party with authority dictates the resolution and the other parties accept it.
  - *Smoothing*: suppressing the conflict, a short-term approach to buy time for a more complete resolution.
  - *Win-wining*: resolving the conflict so that everyone gets something. Since different parties want different things, what is a desired outcome of the conflict for one party may be a 'don't care' outcome for another party (Chacko 1989, Kasser 2018a: Section 3.2.2).
- *Functional*: negotiation (Section 8.2.6) is a useful tool in resolving conflicts.
- *Quantitative*: the Thomas-Kilmann Conflict Mode Instrument (TKI®) assesses an individual's behaviour in conflict situations in two dimensions: assertiveness and cooperativeness in five conflict-handling modes (accommodating, avoiding, collaborating, competing and compromising) (Thomas and Kilmann 1974).

7.3.1.3 Customers

The most important stakeholder is the one who funds the project. In this book, that stakeholder is known as the customer.

7.3.1.4 Funding

Adequate funding is critical to project success. Underfunded projects tend to fail. However, many projects are knowingly underfunded by the project sponsors (Section 7.3.1.15); sometimes because if the true costs were known, the project would never be approved: politics (Section 7.3.1.9) in action. The project sponsors hope that there will be changes during the performance state of the project that will inflate the cost of the project to the realistic expected amount.

7.3.1.5 Meetings

Project personnel spend a lot of time in meetings, most of which tend to be a waste of time (Augustine 1986).* However, if organized systemically and systematically, meetings can be an effective tool for achieving desired results (Kasser 1995) such as:

- Obtaining consensus for a course of action, e.g. milestone reviews (Section 6.1.5.4.2).
- Transferring information, e.g. staff meetings.

Effective meetings have a (Kasser 1995):

1. *Purpose*: which provides focus and determines who is to attend and why.
2. *Agenda*: which serves several purposes:
   - Published ahead of time; it allows participants to think ahead and come to the meeting having thought about the issues.
   - Helps keep the meeting on track. People tend to respect and abide by written guidelines more than they do for verbal guidelines.
3. *Restricted attendance*: each person attending a meeting is a cost to the project. Minimize the cost by restricting attendance to those who have a need to be there as:
   - *Contributors*: people who will make a contribution.
   - *Recipients*: people who need to receive information and their attendance at the meeting is the optimal way for them to receive it.
4. *Time limit*: people have limited attention spans. They tend to be more active at the start of a meeting, so the effectiveness of the meeting decreases over time (Mills 1953). After an hour or so, terminate the meeting or take a break. If the meeting continues without the break, after an hour and half or so, there is a good probability that at least one person will need to answer the call of nature. If they are counting down the seconds till the break, because they do not wish to disturb the meeting, they are not participating in the meeting.
5. *Prompt beginning*: there is a cost associated with each person attending. If people wait around for a latecomer, they are being paid to waste time. Unless there is a line item in the budget for wasted time, there is going to be a cost overrun, or some other activity will not be performed as well it should be. People learn by observing. If they see meetings starting late, they will estimate the real start time and arrive accordingly. If they see the meetings start on time, they will arrive on time.
6. *Leader*: facilitates the meeting by:
   - Guiding it through the agenda.
   - Encouraging discussion without deviation in a tactful manner.
   - Ensuring the meeting does not digress or adjourn without reaching its goal, a conclusion or assigning an action item to achieve a conclusion.
   - Managing disruptors.
   - Merging the official agenda with each person's hidden agenda.

---

* This section is a modified version of Kasser (2019a: Section 2.1.5).

7. *Action items*: make sure the action item:
- Is relevant.
- Can be concluded without cost and schedule impact to the project.
- Is assigned to one person to simplify accountability. If the action item must be carried out by more than one person, then assign it to the leader.
- Is completed on time. Use the just-in-time (JIT) approach to assigning completion dates.
- Is followed up on, to ensure timely completion. This may be done by briefly reviewing its status at a project progress meeting or sending out periodic progress reports as appropriate.
- Is not arbitrarily assigned a completion date; people soon learn which managers 'cry wolf' and act accordingly.

8. *Summary*: review of what was achieved or agreed to at the meeting. This can be done effectively when reviewing the action items assigned during the meeting.

9. *Timely termination*: people have other things to do and want to do them; consequently, they lose interest in the proceedings as the meeting stretches out.

10. *Metric to determine the degree of success of the meeting*: define the criteria for success at the same time as setting the objective of the meeting. Make the measurement and determine the effectiveness of the meeting.

7.3.1.6 Milestones

Projects incorporate major and minor milestones. The number of milestones depends on the duration of the project. Major milestones may be the milestones in the SDP (Section 6.1.5.3); minor milestones may be reporting meetings that take place periodically, e.g. weekly or monthly depending on the scope and direction of the project.

7.3.1.7 Need

All projects shall have a need. If there is no need to deliver the output of the project, the effort put into the project is wasted and should be usefully applied elsewhere.*

7.3.1.8 Outcomes

In the ideal world, projects have two kinds of outcomes: successful and unsuccessful.

7.3.1.8.1 Successful Projects

The keys to project success generally include:

- Clear and concise communications.
- A common vision of what the project is going to achieve such as in a CONOPS (Section 6.5.9.1).
- Adequate funding.
- Good time management.
- The absence of the top seven risk-indicators (Section 7.2) portending a project failure.

Conversely, the lack of these attributes introduces generic risks into the project.

---

* On the other hand, there might be a need to keep people busy even if the product is not going to be used.

7.3.1.8.2 Unsuccessful or Failed Projects
The characteristics of unsuccessful or failed projects in the:

- *People dimension*: are summed up in the top seven risk-indicators for predicating project failure (Section 7.2). They are well known, yet continue to manifest themselves – an instance of Cobb's paradox (VOYAGES 1996). Cobb's Paradox states, 'We know why projects fail, we know how to prevent their failure; so why do they still fail?' Now a paradox is a symptom of a flaw in understanding the underlying paradigm. Perhaps Juran and Deming provided the remedy to this paradox. Juran as quoted by Harrington (1995) stated that management causes 80%–85% of all organizational problems. Deming stated that 94% of the problems belong to the system (i.e. were the responsibility of management) (Deming 1993). More than 25 years later, the same paradigm still produces the same results as satirized by the Dilbert cartoons (Adams 1996) and Figure 1.1.

  One definition of project management is 'the planning, organizing, directing, and controlling of company resources (i.e. money, materials, time and people) for a relatively short-term objective. It is established to accomplish a set of specific goals and objectives by utilizing a fluid, systems approach to management by having functional personnel (the traditional line-staff hierarchy) assigned to a specific project (the horizontal hierarchy)' (Kezsbom, Schilling, and Edward 1989). Poor performance in any of these activities puts a project at risk. By adding the people element to the equation, the systems approach has the potential to dissolve Cobb's paradox (Section 7.2.8.8.2).
- Process dimension: include
  - The wrong management style for the type of technical project (Section 8.4).
  - Assuming a success-oriented schedule.
  - Planning insufficient iterations of the SDP for the project.

7.3.1.9 People
Experience has shown that the right people can make a project and the wrong people can break a project (Augustine 1986), and this factor needs to be considered during the life of the project since:

- People who make decisions need to be competent and effective and knowledgeable in the appropriate problem, solution and implementation domains (Section 5.1.4.6).
- Competent people can lower costs and shorten schedules by preventing problems and taking advantage of opportunities and minimizing errors of omission and commission (Section 5.1.4.3.1).
- Incompetent people can increase costs and schedules by causing problems or not mitigating problems (risks) in a proactive manner and by making errors of omission and commission.
- The reliability of the people can also affect the project (Section 3.8.2.3).

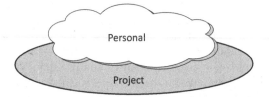

FIGURE 7.2   The need for overlapping goals. (Republished with permission of Artech House, from Applying total quality management to systems engineering, Joe Kasser, 1995; permission conveyed through Copyright Clearance Center, Inc.)

Individuals working in a project generally have their own goals. The goals of the organization are to complete the project in a cost-effective manner within the applicable constraints. The goals will be met in the optimal manner if the individual and organizational goals can be made to overlap to the maximum possible extent as illustrated in Figure 7.2. Even self-starters may need to be provided with incentives to do what the project/organization wants them to do.

Making the goals overlap is a process in itself. The process contains three steps, namely:

- Find out why people have goals in the first place (what's in it for me?)
- Find out why people behave the way they do in organizations.
- Provide incentives to motivate the employee to adopt the goals of the organization.

Finding out why people have goals means that we have to understand why people behave the way they do, then assume each one of them asks, 'what's in it for me?'

The answer to the question 'what's in it for me?' provides the key to determining how to ensure a project can be staffed by people whose needs and goals match those of the project. See *Systemic and Systematic Project Management* (Kasser 2019a: Section 2.1.4) for a brief summary of theories of motivation. Although these theories are complex, there are enough common threads to make some practical use of them (Kast and Rosenzweig 1979). The key point in motivation is that people motivate themselves based on an attempt to satisfy the need that is *most important* to them at that point in time. For example, when faced with an unexpected bill for a large amount of money, and a low bank balance, a person will experience a dramatic change in the relative order of needs. Maslow states that the hierarchy is not a rigid fixed order, which is the same for all individuals; the order varies somewhat from person to person especially near the middle of the hierarchy (Maslow 1968). Maslow described the situation in terms of a dynamic hierarchy of needs. Consider the category level in Maslow's hierarchy as a weighting factor defining the urgency of a need. Then, if instead of a hierarchy or pyramid, consider the situation at a point in time as a pie chart (Kasser 2018a: Section 2.12), then the answer to the question 'what's in it for me?' is 'a slice of the pie'.

The leader's problem then becomes to use the theories to figure out which slice, and its size (the incentive), to offer the person being motivated.

7.3.1.10 Politics

Successful project management is directly linked to the ability of project managers and other key players to understand the importance of organizational politics and how to make them work for project success. While most of us view politics with distaste, there is no denying that effective managers are often those who are willing and able to employ appropriate political tactics to further their project goals.

Pinto (2000)

Perceptions of politics from the HTPs include:

1. *Structural*: definitions of politics include:
   1. 'The process of making decisions that apply to members of a group' (Hague and Harrop 2013).
   2. 'The use of intrigue or strategy in obtaining any position of power or control, as in business, university, etc.' (Dictionary.com 2013).
   3. 'The processes by which differing interests reach accommodation' (Checkland and Scholes 1990).
2. *Big Picture*: politics is:
   1. A major part of stakeholder management (Section 7.10).
   2. A way of interacting with people to influence the achievement of goals by facilitating or impeding the achievement.
   3. Distained by many people who have only been exposed to its negative use.

   Project management is generally performed in a political context because projects tend to exist outside the traditional organizational structure. Projects borrow people from functional departments so those people report to the managers of the functional departments who perform their revaluations and can reward good performance in various ways. The project manager who doesn't have this ability has to influence the project team to perform. Accordingly, one of the roles of the project manager is to use positive politics to facilitate the achievement of the project goals via negotiation (Section 8.2.6) and conflict management (Section 7.3.1.2).
3. *Operational*: note that projects produce changes which tend to be resisted (Section 4.5.1). This resistance can be mitigated by giving stakeholders a personal reason for the change that will be to their benefit such as:
   • Reducing the amount of work they will have to do once the change is implemented which will give them increased leisure or extra time to accomplish other tasks.
   • Increasing their income.

In other words, find out what is undesirable about the current situation or what problem people are having, then using positive politics, show them how the change will help them (Crosby 1981: p. 92), namely what's in it for them (Section 7.3.1.9).

4. *Continuum*: note that politics is a tool that can be used:
   - *Negatively*: for self-serving purposes, revenge, spite, cover ups, etc.
   - *Positively*: to achieve the goals of a project, to improve the quality of life, etc.
5. *Temporal*: note that even successful project managers can be hurt by negative politics and need to understand the nature of negative politics to prevent it from happening.

7.3.1.10.1   Positive Use of Politics

Some examples of using politics in a positive way are (Reardon 2005: p. 86) cited by Irwin (2007):

- *Creating a positive impression*: assuring that key people find you interesting and approachable.
- *Positioning*: being in the right place at the right time to advance your project.
- *Cultivating mentors*: locating experienced advisors.
- *Lining up the ducks*: making strategic visits to peers, senior people and support staff at which you mention your project and accomplishments and let them know what you can do for them.
- *Developing your favour bank*: favours usually require reciprocation; by agreeing to – or even offering – favours, you are making 'deposits' in anticipation that someday, when you need to call in a chit, you will have the currency to do so.

7.3.1.10.2   Negative Use of Politics

While the positive use of politics will advance your project or your organization, if it does this at the expense of others, it becomes negative politics. Some examples of using organizational politics in a negative manner include (Reardon 2005: p. 87) cited by Irwin (2007):

- *Poisoning the well*: fabricating negative information about others, dropping defaming information into conversation and meetings in the hope of ruining the target's career chances.
- *Faking left while going right*: leading others to believe you will take one action in order to increase the likelihood of succeeding via an entirely different maneuver; allowing or encouraging someone to think one condition exists when in reality another condition holds.
- *Deception*: lying, telling different people different things at the same time.
- *Entrapment*: steering or manipulating someone into a political position or action that results in embarrassment, failure, discipline or job loss.

People use negative politics for personal gain to the detriment of their organizations. They are successful because:

- Of the support given by their management who lack the perceptions from the *Big Picture* HTP to realize that the shenanigans they are supporting are harming the organization as well as the reputation of their organization.
- The organization does not seem to be able to detect and/or mitigate the use of negative politics.

7.3.1.10.3   Seven Steps That a Project Manager Can Take to Become Politically Astute

Seven steps that a project manager[*] can take to become politically astute are (Pinto 2000):

1. *Understand and acknowledge the political nature of most organizations*: politics is a fact of life, exists in all organizations and has an impact on the success of the project. Accordingly, a project manager needs to understand how best to use positive politics (Section 7.3.1.10.1) to achieve a successful project.
2. *Learn to cultivate appropriate political tactics*: developing a reputation for keeping your word will assist negotiation and bargaining because you are often promising a future favour in return something in the present. It is important to keep promises so do not make promises you cannot keep as part of the negotiation (Section 8.2.6) and bargaining. They will come back to haunt you.
3. *Understand and accept 'what's in it for me?'*: the first thing to understand is 'what's in it for you – the project manager'. What's in it for you is politics is a tool that if used properly will facilitate the success of the project and if misused čan alienate influential stakeholders who can and probably will sabotage the project. Cultivate relationships with influential stakeholders *before* you need their aid. Develop positive personal relationships with stakeholders. Build this time into your daily work plan (Kasser 2019a: Section 12.1). When asking for help, you need to give your potential helper a reason as to why they should provide that help. Similarly, when dealing with project personnel, there needs to be a reason for the project personnel to perform. Do not be too blatantly obvious that you are attempting to influence the other person; keep it subtle. It is not wise to develop a reputation as a person playing politics.
4. *Try to level the playing before taking over the project*: many organizations do not give the project manager influence in performance evaluation is of project personnel. Line managers often resist sharing the power of performance evaluation. However, since the project manager is most familiar with the performance of the personnel, it's to the benefit of the organization for the project manager to be involved in the performance evaluation. So, when taking over a project, ask for influence in the performance evaluation of the project

---

[*] Systems engineer, policy maker or implementer, as well as everybody else working with other people.

personnel. It's probably not advisable to ask to take that role away from the line manager, and it may not be possible to share it, but it should be possible to provide input to the performance evaluation. When the performance evaluation is made, use positive politics; let the project team member have a copy of the evaluation before submitting it to the line manager and let the line manager know that the project team member has a copy (fait accompli).

5. *Learn the fine art of influencing*: use some leadership techniques (Kasser 2019a: Chapter 12). Understand the stakeholders you are dealing with, understand their concerns and the best way to influence them and then employ it. As a project manager, you may not have financial reward power (Kasser 2019a: Section 12.2.1), but you can deal with emotion (guilt, pride, shame, etc.). Recognize the type of critical thinker (Section 2.2) you are dealing with and tailor your arguments accordingly. Perceived from the *Generic* HTP, this is similar to understanding what motivates the person and then motivating the person accordingly.

6. *Develop your negotiating skills*: you need to understand the techniques of negotiation (Section 8.2.6) which:
   - Can be used against you.
   - You can use to achieve your goal.*

   The literature encourages the use of win–win techniques, but they can only be achieved with the collaborating style of negotiator. If you can recognize and match the style of the opposing negotiator, you stand a better chance of achieving your goal. Try and set up the negotiation as an example of mutual problem-solving looking for a win–win solution (Kasser 2019a: Section 12.5.2).

7. Recognize that conflict is a natural side effect of project management and can be dealt with constructively (Section 7.3.1.2).

7.3.1.11   Priority

All projects should have a priority within the organization in which the project is performed. This priority allows upper management to determine which projects should be delayed or even cancelled should shared resources not be available when needed.

7.3.1.12   Purpose

The purpose of the project is to create the product, service or result for which it was established (meet a need) on or before the scheduled delivery date within the resource budget. If there is no clear vision of the purpose of the project, it will fail.

7.3.1.13   Return on Investment

The traditional approach to estimating the return on investment is to divide the benefit of doing the project by the cost of doing project. However, perceptions from the *Continuum* HTP indicate that there may also be:

---

* The same techniques used in reverse.

- A *cost of starting late*: starting late means finishing late, which results in missing the opportunity. The missed opportunity cost is there even if the project takes the scheduled amount of time (Malotaux 2010). For example, the late start delays the date of a product introduction into the market allowing a competitor to gain market leadership.
- A *cost of finishing late*: the cost of the resources used after the planned finishing time such as salaries. For example, the project salary budget is $10,000 a month, then for every month the project is late, the cost goes up by $10,000. And this is in addition to any penalty cost clauses in the contract. The systems approach also considers the effect of that month's delay in completing the project (Malotaux 2010). For example, if the project was to develop a new product, and the monthly revenues that the product would bring in are estimated as $500,000, the cost of the delay would be $510,000, per month not $10,000.

7.3.1.14   Risks

Risks are associated with each of the other attributes and are discussed in Chapter 3. Specific risk topics discussed in this book include:

1. Generic risks in each state of the SLC discussed in Section 6.1.5.
2. Risks in systems engineering discussed in Section 6.8.
3. The top seven risk-indicators of a project in trouble discussed in Section 7.2.
4. Software project risks discussed in Section 7.11.
5. Generic project risks discussed in Section 7.12.

7.3.1.15   Sponsor

There are various opinions on what constitutes a sponsor. In some instances, the sponsor represents the customer; in other instances, the activities assigned to the sponsor overlap the activities assigned to the project manager. Perceptions from the *Continuum* HTP show that the there is a difference between the roles of the sponsors in projects that are internal to organizations and the roles of the sponsors in projects that are undertaken under contract for an external customer. The different roles are:

1. *Internal projects*: 'project sponsors champion the project and use their influence to gain approval for the project. Their reputation is tied to the success of the project, and they need to be kept informed of any major developments. They defend the project when it comes under attack and are a key project ally' (Larson and Gray 2011: p. 343). The project sponsor is often the project customer.
2. *External projects*: the customer is the project sponsor.

7.3.1.16   Stakeholders

A stakeholder is an individual, group, or organization who may affect, be affected by, or perceive itself to be affected by a decision, activity, or outcome of a project. Stakeholders may be actively involved in the project or have interests that may be

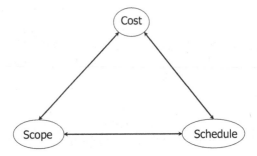

FIGURE 7.3   The traditional triple constraints.

positively or negatively affected by the performance or completion of the project. Different stakeholders may have competing expectations that might create conflicts within the project. Stakeholders may also exert influence over the project, its deliverables, and the project team in order to achieve a set of outcomes that satisfy strategic business objectives or other needs.

PMI (2013: p. 30)

Stakeholders need to be managed over the project lifecycle (Section 7.4).

7.3.1.17   Timeline

A timeline:

1. Extends from the start of the project to the end of the project passing through a number of pre-planned milestones (Section 6.1.5.4.2).
2. Contains the three streams of activities (Section 7.3.1.1.2), each of which consists of a sequence of activities and milestones: in series, parallel or a combination of series and parallel activities.
3. Is often shown from two perspectives:
   - *Gantt charts* (Kasser 2018a: Section 8.4) show the timing and are often called schedules.
   - *PERT charts* (Kasser 2018a: Section 8.10) show the dependency between the activities and identify the critical path.

Projects have fixed timelines or at least they are fixed when the project begins. This facilitates MBO (Mali 1972) (Section 7.8), if the prerequisite for achieving a milestone is the accomplishment of one or more objectives.

If the project is developing a new product, then in general the shorter the timeline, the faster the product will get to market and the greater will be the profit or return on investment.

7.3.2   The Triple and Quadruple Constraints of Project Management

In traditional project management, the effect of a change on a project is said to impact the triple constraints of project management shown in Figure 7.3, namely

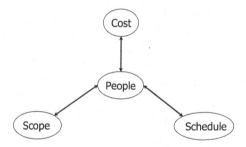

FIGURE 7.4 The systems approach's quadruple constraints.

scope, cost and schedule. This is because a change in any one cause changes in the other two. Accordingly, it is possible to:

- Shorten the schedule of a project but that will increase the cost.
- Lengthen the schedule of a project to reduce the cost.
- Lower the cost of a project, but that will increase the schedule or reduce the scope.
- Change the scope of a project, but that will affect cost or schedule or both.

The systems approach considers the quadruple constraints[*] shown in Figure 7.4 by adding the people element to the triple constraints because the people attribute (Section 7.3.1.9) is one of the seven interdependent P's of a project and closely coupled to the others (Kasser 2019a: Chapter 2). Experience has shown that the right people can make a project and the wrong people can break a project (Augustine 1986), and this factor needs to be considered during the life of the project because selecting the wrong people increases the risk of project failure.

7.3.3 PROJECT ORGANIZATION

Project management and leadership are about making optimal decisions in a timely manner. Consequently, the leadership and/or management of a task can only be accomplished effectively if the leader/manager understands the technical implications of decisions, as well as the people, schedule and budget implications. For this reason, the systems approach to project organization uses situational leadership and allows leaders with the relevant competencies to emerge for specific activities. The functions performed by the leadership and management roles can be classified as administrative and technical. An administrative assistant performs the administration functions for the team,[†] whereas the technical personnel perform the technical functions and the leader/manager takes care of the entire project. This arrangement also reduces the cost of the project by not having the administrative activities performed at project manager salary rates.

---

[*] A change in one will produce a change in the others.
[†] The administrative activities are not spilt between the project manager and the administrative assistance, but they are portioned to avoid duplication.

7.3.3.1   Customer – Project Interfaces

From the contracting perspective, there are two interfaces to the customer, namely:

1. *The contractual interface*: between the customer and the project manager. The project manager can access upper management and has been given a high level of independent authority. Thus, problems and issues that may surface should be resolved quickly. As a result, customer satisfaction should remain at a high level throughout the duration of the project (or off with his/her head!). This interface is for:
   - Managing change requests (Section 4.9.2).
   - Reporting to the customer.
   - Billing the customer.
   - Compliance with legal requirements.
   - Other contract management functions.
2. *The information interface*: between the customer and the project personnel. This interface is primarily for WP (Section 7.3.1.1.4) management in the following areas:
   - Joint or customer resolution of problems in the WP.
   - Ensuring WP performance as an integral part of the project/customer team effort.

The information interface is not for accepting informal changes from the customer bypassing the formal CM process.

7.4   THE PROJECT LIFECYCLE

This section discusses the project lifecycle, the SLC and the differences between them. The SLC may be considered as the states a system exists in, starting from its conception through development and operation, and ending once it has been disposed of. The SDP contains the system development states of the SLC. Projects, being temporary endeavours, may appear in any one state of the SLC or in multiple states crossing state boundaries. For example, if a project is to:

- Develop a new product it will take place over the whole SDP of that product.
- Improve the way requirements are elicited and elucidated it will take place in the system requirements state of the SDP.

The project lifecycle contains the following four states[*]:

1. The project initialization state discussed in Section 7.4.1.
2. The project planning state discussed in Section 7.4.2.
3. The project performance state discussed in Section 7.4.3.
4. The project closeout state discussed in Section 7.4.4.

---

[*] Identical to the states in the policy lifecycle (Section 9.5.1).

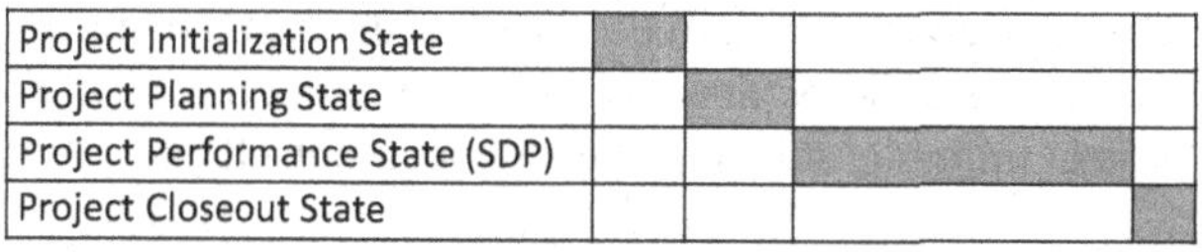

Project Initialization State				
Project Planning State				
Project Performance State (SDP)				
Project Closeout State				

FIGURE 7.5   The project lifecycle.

The project sequences through the lifecycle states as shown in Figure 7.5. Each state:

- Starts and ends at a milestone.
- May take an amount of time that ranges from minutes to months depending on the complexity of the project. For example, if the project is to:
  - *Develop a new government policy*: the project initialization state may take months to determine what the policy should be (and may be a project in itself).
  - *Build an airport*: the project planning state may take months if it includes figuring out where to locate the airport.
  - *Organize a party*: the project planning state may take 20 minutes or less.

7.4.1   THE PROJECT INITIATION STATE

The project begins in the project initiation state as a proposed project. The product produced in the project initiation state is the project initiation document (PID). This document contains the guiding information for the project including:

- The sponsor.
- The business case for the project.
- The undesirable situation that needs to be remedied by the project.
- The priority of the project with respect to other projects taking place in the organization.
- The stakeholders in the project and their order of importance and influence, if known.
- The proposed project organization.
- The customer who funds the project and who must be satisfied with the deliverables.
- An estimate of the budget for the proposed project.
- An estimate of the completion date; the date the product to be produced by the proposed project is needed.
- An estimate of the other types of resources needed by the proposed project.
- A summary of the analysis of alternatives (AoA) (Section 7.4.1.2) or the feasibility study (Section 7.4.1.1) of the project.

The PID may be created using annotated outlines (Kasser 2018a: Section 14.1) and the documentation process (Kasser 2018a: Section 11.4).

The project initiation state terminates at a milestone which reviews the information in the PID and provides consensus to proceed with the project. The review is not the place to adjust the scope of the project to lower the cost, change the schedule or reduce the scope. This should be done as part of creating the PID. If the project is complex, the project initiation state may be a project in itself

7.4.1.1 Feasibility Studies

The feasibility study is a major risk management tool because it can prevent undertaking a doomed project. Perceptions from the *Continuum* and *Structural* HTPs note that definitions of a feasibility study include:

- An analysis used in measuring the ability and likelihood to complete a project successfully including all relevant factors (Investopedia 2018).
- An analysis and evaluation of a proposed project to determine if it (1) is technically feasible, (2) is feasible within the estimated cost and (3) will be profitable (Businessdictionary 2018).
- Is an analysis of the viability of an idea. The feasibility study focuses on helping answer the essential question of 'should we proceed with the proposed project idea?' (Hofstrand and Holz-Clause 2009).

Perceived from the *Continuum* HTP, the focus of the feasibility study is generally to find out if there is at least one way to meet the objective within the constraints, whereas AoA (Section 7.4.1.2) explores the attributes of different ways of meeting the objectives.

7.4.1.1.1 Reasons to Do a Feasibility Study

Reasons to do a feasibility study include (Hofstrand and Holz-Clause 2009):

- Gives focus to the project and outline alternatives.
- Narrows business alternatives
- Identifies new opportunities through the investigative process.
- Identifies reasons not to proceed.
- Enhances the probability of success by addressing and mitigating factors early on that could affect the project.
- Provides quality information for decision-making.
- Provides documentation that the business venture was thoroughly investigated.
- Helps in securing funding from lending institutions and other monetary sources.
- Helps to attract equity investment.

7.4.1.1.2 Reasons Not to Do a Feasibility Study

Project managers may find themselves under pressure to skip the 'feasibility analysis' step. Major risk! Reasons given for not doing a feasibility analysis include (Hofstrand and Holz-Clause 2009):

- We know it's feasible. An existing business is already doing it.
- Why do another feasibility study when one was done just a few years ago?
- Feasibility studies are just a way for consultants to make money.

- The market analysis has already been done by the business that is going to sell us the equipment.
- Why not just hire a general manager who can do the study?
- Feasibility studies are a waste of time. We need to buy the building, tie up the site and bid on the equipment.

The reasons given above should not dissuade you from conducting a meaningful and accurate feasibility study. Once decisions have been made about proceeding with a proposed business, they are often very difficult to change. You may need to live with these decisions for a long time. Perceptions from the *Generic* HTP note similarities between these reasons and the decision traps (Section 5.1.5.3).

7.4.1.1.3 Types of Feasibility

There are various types of feasibility studies including:

- *Technical*: which should answer the following questions (Cleverism 2018):
  1. What is the proposed product or service?
  2. Is the product or service already on sale? If not, how far is it from an existing marketplace and what will the introduction cost?
  3. How can you protect the product or service from the competition?
  4. What are the strengths of the product or service?
  5. What are the main benefits to customers or users?
  6. What resources are required for producing or providing it?
  7. How capable is the organization to acquire these resources?
  8. What are the regulatory standards surrounding the product or service and its use?
- *Market*: which focuses on testing the market for the proposed action or idea. It examines issues like whether the product or service can be sold at reasonable prices or if there's a marketplace for it. Market feasibility should answer the following questions (Cleverism 2018):
  1. What market segments are you targeting?
  2. Why would people buy the product or service?
  3. Who are the potential customers and how many of them are there?
  4. What are the buying patterns of these potential customers?
  5. How will you sell the product or service?
  6. Where will you sell the product or service?
  7. Who are your competitors? Including past, current and future competitors.
  8. What are the strengths and weaknesses of your competitors?
  9. What is your product or service's competitive edge?
- *Commercial or economic*: which focuses on the probability of commercial success. It's mainly focused on studying a new business or a new product or service, and whether your organization can create enough profit with it. The questions that require answering as part of the commercial feasibility study include (Cleverism 2018):
  1. What are the strengths and weaknesses of your business?
  2. What are the potential sales volumes of the product or service?

3. What is the pricing structure you'll use?
4. What are the sensitivity points for your business in terms of sales?
5. What is the return on investment (ROI)?

- *Operational*: which focuses on the following questions.
  1. How can the system meet the operational needs at an affordable price?
  2. How can the system be supported when deployed?
  3. How can the system be maintained?
- *Risks*: which focus on the following questions (Cleverism 2018):
  1. What are the major risks associated with the operation?
  2. What is the survival outlook for each of the above risks?
  3. How sensitive are the profits?
  4. What are the best ways to minimize these risks?

7.4.1.2 Analysis of Alternatives

AoA performed in HKMF sub-state A.1 is a term that was adopted in the US by the Office of Management and Budget (OMB) and the US Department of Defense (DoD) to ensure that the effectiveness, cost and risk associated with multiple alternatives had been analysed prior to making costly investment decisions (Ullman 2009). Perceptions from the *Generic* HTP show that:

- AoA is intended to be an early state risk management activity to prevent wasting funds on doomed projects.
- AoA is almost an instance of a feasibility study (Section 7.4.1.1).
- Comparing multiple alternative solutions in AoA is an adaptation of the generic problem-solving process, namely Steps 3, 4 and 5 in Figure 5.11.

The Office of Aerospace Studies in the Air Force Materiel Command at Kirtland Air Force Base published a practical guide to the use of AoA in the form of a handbook (OAS 2013). This handbook provides an excellent template for not only performing AoA but also providing useful ideas and questions for performing feasibility studies (Section 7.4.1.1). This section contains a summary of the relevant aspects of AoA to risk identification and mitigation/prevention in industry, academia and government. These aspects are also relevant to feasibility studies and the other activities in the needs identification state of the SDP/SLC (Section 6.1.5.4.3).

7.4.1.2.1 The AoA Process

The AoA process includes the following activities (OAS 2013):

1. Planning the analysis discussed in Section 7.4.1.2.2.
2. Performing the analysis discussed in Section 7.4.1.2.3.
3. Performing the risk analysis discussed in Section 7.4.1.2.4
4. Performing the sensitivity analysis discussed in Section 7.4.1.2.5.
5. Performing the cost analysis discussed in Section 7.4.1.2.6.

7.4.1.2.2   Planning the Analysis
Planning the analysis contains the following activities:

1. Scoping the effort discussed in Section 7.4.1.2.2.1.
2. Defining the alternative concepts discussed in Section 7.4.1.2.2.2.
3. Identifying stakeholder community discussed in Section 7.4.1.2.2.3
4. Determining the level of effort discussed in Section 7.4.1.2.2.4.
5. Establishing the study team discussed in Section 7.4.1.2.2.5.

7.4.1.2.2.1   Scoping the Effort
The scope of the effort needs to address the following typical questions for each alternative:

1. How well does it close the capability gaps?
2. How does it compare to the current capability?
3. What are all the support and enabling capabilities?
4. What are the risks that could prevent or delay the development process, e.g. technical, operational, integration, political, etc.?
5. What is the life cycle cost estimate (LCCE)?
6. What are the significant performance parameters?
7. What are the trade-offs between effectiveness, cost, risk and development schedule?

If there have been previous analyses, they need to be reviewed and determine if they are still applicable.

The ground rules, constraints and assumptions help scope the analysis and must be carefully documented and coordinated with senior decision-makers and cognizant personnel. In this context, the specific definitions of the ground rules, constraints and assumptions are:

- *Ground rules*: broadly stated procedures that govern the general process, conduct and scope of the study.
- *Constraints*: imposed limitations that can be physical or programmatic.
- *Assumptions*: conditions that apply to the analysis.

7.4.1.2.2.2   Defining the Alternative Concepts
At a minimum, the AoA must include an analysis of:

1. The existing situation which provides a baseline reference against which to compare other alternatives.
2. Similar systems in use elsewhere.
3. Commercial off the shelf (COTS) implementations.
4. A custom development.

The number of alternatives to be analysed needs to be reduced to a small number of serious contenders. There is no formula for doing this; it is an art whose practice benefits from experience. However, in general, it is prudent to continuously screen the alternatives throughout the AoA process. This has the advantage of eliminating non-viable alternatives before resources are expended on analysing them. The basis for eliminating each alternative from further consideration should be documented at the time it becomes clear that it is non-viable. A summary of this documentation should be included in the final AoA report to provide an audit trail which may be very important in the event the AoA results are questioned.

7.4.1.2.2.3 Identifying the Stakeholder Community   AoA defines a stakeholder as any entity having a vested interest (a stake) in the outcome of the analyses. A stakeholder may contribute directly or indirectly to the activities and is usually affected by decisions made as a result of these activities. Using the nine-system model (Section 6.7) and other tools for identifying and grouping stakeholders (Section 4.1.3) and asking the following questions helps to identify the stakeholders (Section 7.10):

1. Who are the end-users of the capability in S7?
2. What enablers have interdependencies within the solution being analysed in the AoA?
3. How do the other entities in S7 fit into the mission area being explored in the AoA?

7.4.1.2.2.4 Determining the Level of Effort of the Analysis   The level of effort of the analysis will depend on various factors such as the study questions, complexity of the problem, time constraints, manpower and resource constraints and type of analysis methodology. By controlling the scope of the study, the level of effort is more likely to remain manageable over the course of the analysis. Answers to the following questions will aid in determining the level of effort:

- How much analysis has been accomplished to date (as part of the activities in defining the problem in sub-state A.1)?
- What remaining information needs to be learned from the AoA?
- Which stakeholders are available to participate in the effort?
- Are the right experts available and can they participate?
- What data and tools are needed to execute the AoA?
- How much time and funding are available to execute the AoA?
- What level of analytic rigor is required?
- Where and what amount of analytic risk are acceptable to the decision-makers?

There are other risks associated with uncertainties inherent in the study process such as the effectiveness and cost analysis methodologies, funding and resources limitations and insufficient time to conduct the study. For example, a study with limited time and resources may reach different conclusions compared to similar study with less constrained time and resources.

7.4.1.2.2.5 Establishing the Study Team   Organize the study team in a way that meets the study needs. The structure of the study team depends upon the scope of the AoA and the level of effort required. Study teams are not identical; they are tailored in size and skill sets to meet the objectives of the specific AoA. Team membership should be interdisciplinary and should include the appropriate problem, solution and implementation domain expertise (Section 5.1.4.6) such as operators, developers, cost estimators and other specialists.

7.4.1.2.3 Performing the Analysis

The goal of the analysis is to identify alternative solution concepts and then determine the value of the alternative concepts in performing operational scenarios or mission tasks (MTs).[*] MTs must be stated in functional language, not in solution language. For example, an MT might be to provide transportation rather than to provide an aircraft. Each MT shall have at least one measure supporting it.

Once the MTs are well defined and understood, the next step is to identify the necessary attributes of successful MTs. An attribute is essentially a property or characteristic of an entity, e.g. colour, shape, size, power consumption, electromagnetic radiation, cost, etc. An entity may have many attributes. Perceptions from the *Continuum* HTP indicate that attributes may be desired, undesired or not relevant to the study of the characteristics of the entity. Accordingly, attributes included in the study should be problem specific and should be used to enlighten decisionmakers, answer key questions and respond to guidance. The key should be to keep the study logical, identify the desired attributes first and then craft the measures to address them.

The ability to satisfy the MT is determined from estimates of each alternative concept's performance with respect to measures of effectiveness (MOE), measures of performance (MOP) and measures of suitability (MOS). The methodology must:

1. Be systematic and logical.
2. Be at the appropriate level of detail required.
3. Be executable and repeatable.
4. Not be biased for or against any alternative.
5. Determine the effectiveness of each conceptual alternative based on value.

Measures are a central element when conducting an AoA. Without them, there is no way to determine the effectiveness and suitability of an alternative and the ability of the alternative to close capability gaps either partially or completely. Properly formed and explicitly stated, measures will:

- Specify what to measure (what data to collect, e.g. time to deliver message).
- Determine the type of data to collect (e.g. transmit start and stop times).
- Identify the source of the data (e.g. human observation).
- Establish personnel and equipment required to perform data collection.
- Identify how the data can be analysed and interpreted.
- Provide the basis for the assessment and conclusions drawn from the assessment.

---

[*] Tasks a system will be expected to perform; the effectiveness of system alternatives is measured in terms of the degree to which the tasks would be attained.

Examples of measures include attrition rate, quantity (and types) of resources consumed (e.g. fuel, in litres per minute, power consumption in kilowatt/hour) and time to perform various functions such as repair, reinitialize and replace.

Thee four important system attributes in an AoA analysis are as follows:

1. *Effectiveness*: a characteristic of a system, where:
   - *Operational effectiveness*: the overall degree of mission accomplishment of a system when used by representative personnel in the environment planned or expected for operational employment of the system considering organization, doctrine, tactics, survivability, vulnerability and threat.
   - *MOE*: a qualitative or quantitative measure of operational success that must be closely related to the objective of the mission or operation being evaluated. MOEs should normally represent raw quantities such as numbers of something or frequencies of occurrence. Attempts to disguise these quantities through a mathematical transformation (e.g. through normalization), no matter how well meaning, may reduce the information content and might be regarded as tampering with the data. Although ratios are typically used for presenting information such as attrition rates and loss/exchange ratios, use caution as a ratio can also essentially hide both quantities.
2. *Suitability*: a characteristic of a system, where:
   - *Operational suitability*: the degree to which a system can be placed satisfactorily in field use with consideration given to availability, compatibility, transportability, interoperability, reliability, wartime usage rates, maintainability, safety, human systems integration (HSI), manpower supportability, logistics supportability, natural environmental effects and impacts, documentation and training requirements.
   - *MOS*: a measure of a system's operational suitability ability to support mission/task accomplishment with respect to reliability, availability, maintainability, transportability, supportability and training. Suitability issues such as reliability, availability, maintainability and deployability can be significant force effectiveness multipliers. Maintainability issues could dramatically increase the number of maintainers needed to sustain a system. Major HSI issues might increase operator workload. Significant reliability issues could result in low operational availability.
3. *Sustainability*: a system's capability to maintain the necessary level and duration of operations to achieve objectives. Sustainability encompasses a wide range of elements such as systems, spare parts, personnel, facilities, documentation and data. Sustainability performance not only impacts mission capability but is also a major factor that drives the life cycle cost of a system. For example, maintainability issues could considerably increase life cycle costs by increasing the number of maintainers needed to sustain a system in the field. In other situations, significant HSI issues may increase an operator's workload or poor reliability performance could result in low operational availability.

Defining how alternatives will be employed in the operational environment is an essential step in conducting the sustainability analysis in the AoA study. The concept of employment for each alternative should be defined and contain descriptions of the projected maintenance concept and product support strategy. Given that the alternatives are primarily developmental or conceptual at this early state of the SLC, defining the maintenance concept and product support strategy can be challenging and may require the assistance of system engineers and acquisition logistics, maintenance, supply and transportation

The maintenance concept is a general description of the maintenance tasks required in support of a given system or equipment and the designation of the maintenance level for performing each task.

4. *Performance*: is a quantitative measure of physical characteristics where:
   - *MOP*: a measure of the lowest level of physical performance (e.g. range, velocity, throughput, etc.) or physical characteristic (e.g. height, weight, volume, frequency, etc.).

MOPs are chosen to support the assessment of one or more MOEs. MOPs support the MOEs by providing causal explanation for the MOE and/or highlighting high-interest aspects or contributors of the MOE. MOPs may apply universally to all alternatives or, unlike MOEs, they may be system specific in some instances. In order to determine how well an alternative performs, each MOP should have an initial minimally acceptable value of performance (often the 'threshold' value).

7.4.1.2.4   Performing the Risk Analysis

Risks in this context refer to the operational, technical and programmatic risks associated with the alternative solutions. Various factors such as technical maturity, survivability and dependency on other programs shall be considered in determining these risks.

7.4.1.2.5   Performing the Sensitivity Analysis

Alternatives whose effectiveness is stable over a range of conditions provide greater utility and less risk than those lacking such stability. Alternatives in an AoA are typically defined with certain appropriate assumptions made about their performance parameters, e.g. weight, volume, power consumption, speed, accuracy, impact angle, etc. These alternatives are then assessed against AoA-defined threats and scenarios under a set of AoA-defined assumptions. This provides very specific cost and performance estimates but does little to assess the stability of alternative performance to changes in system parameters or AoA threats, scenarios, employment and other assumptions.

Stability can only be investigated through sensitivity analyses in which the most likely critical parameters are varied; for instance, reduced speed or increased weight, greater or less accuracy, different basing options, reduced enemy radar cross section or when overarching assumptions are changed. This form of parametric analysis can often reveal strengths and weaknesses that are valuable in making decisions to keep or eliminate alternatives from further consideration.

7.4.1.2.6   Performing the Cost Analysis

Costing of projects is discussed in *Systemic and Systematic Project Management* (Kasser 2019a: Chapter 8).

7.4.2   THE PROJECT PLANNING STATE

The product produced in the project planning state is the initial draft of the project plan (Section 7.4.2.1). Many projects combine all the information in the PID into the project plan. On the other hand, many projects don't have a PID or even a project plan and so don't have clear goals, and accordingly fail (Section 7.3.1.3.2). The project planning state terminates at a milestone which reviews the information in the draft project plan and provides consensus to proceed with the project.

7.4.2.1   The Project Plan

The project plan is:

- A guide to project execution by providing a reference.
- A communications tool for the present and future.
- The controlling document with which to manage a project.
- A description of the:
  - Interim and final deliverables the project will deliver.
  - Activities in the managerial and technical processes necessary to develop the project deliverables in the three streams of activities (Section 7.3.1.1.2).
  - Resources required in the process of creating, verifying and delivering the project deliverables.
  - Additional plans required to support the project.
  - CM process (Section 4.9.2).
- Consists of two parts:
  1. The generic part applicable to all projects discussed in Section 7.4.2.1.1.
  2. The specific part applicable to the specific project discussed in Section 7.4.2.1.2.
- Is produced by the people performing the planning process.

Perceptions from the *Continuum* HTP note that there are two types of planning: generic and specific.

7.4.2.1.1   Generic Planning

The generic planning process begins with completing the problem formulation template (Section 5.1.7.2) for the project, namely:

1. *The undesirable situation*: the need to create a draft project plan.
2. *The assumption*: the knowledge of what products will be produced and what will be done (process) to produce the products is available in the planning team or can be accessed when needed. Namely the planning team has competence in the problem, solution and implementation domains (Section 5.1.4.6).

3. *The FCFDS*: (outcome) having successfully completed the draft project plan in a timely manner.
4. *The problem*: to figure out how to create and deliver the products and document the information in the draft project plan.
5. *The solution*: use the process for creating a generic project plan discussed in *Systemic and Systematic Project Management* (Kasser 2019a: Section 5.6). One advantage of the waterfall chart is that it provides a template for generic project planning by defining the milestones and deliveries at each milestone.

7.4.2.1.2 Specific Planning

Specific planning for a project customizes the generic template for the scope and type of project as well as adding or deleting WPs as required. This activity requires knowledge of the problem, solution and implementation domains, so if the project planner does not have that knowledge, he or she needs to team with people who have that knowledge. The process for creating a specific project plan is discussed in *Systemic and Systematic Project Management* (Kasser 2019a: Section 5.7.1).

7.4.2.2 Contingencies and Contingency Plans

Contingencies are risks that cannot be prevented and must be mitigated. Contingency plans:

- Are the plans for mitigating the risk should it occur and turn it into an event.
- Are created before the event occurs so that the way to resolve the problem posed by the events is thought-out and documented rather than being developed as a panicked response to the event.

7.4.3 THE PROJECT PERFORMANCE STATE

The project performance state is the longest state in the project. It is the state in which the project is managed and produces the products or product that will remedy the undesirable situation using the SDP (Section 6.1.5.4). If the project takes time, it may not completely remedy the undesirable situation for various reasons that include:

1. The product operating in its context does not remedy the entire original undesirable situation.
2. New undesirable aspects have shown up in the situation during the time taken to develop the product.
3. Unanticipated undesired emergent properties of the product and its interactions with its adjacent systems may produce new undesirable aspects of the situation.

Errors of omission and commission (Section 5.1.4.3.1) made in each part of SDP in the project performance state can produce undesirable outcomes. For example:

- If the wrong cause of the undesirable situation is identified, the wrong problem will be stated, and not only will the solution not remedy the undesirable situation, but the solution may make the situation even more undesirable.

- If the correct problem is identified but the wrong solution conceptualized and realized, then the undesirable situation will not be remedied and may become even more undesirable.
- Even if the correct solution is conceptualized, errors in the realization process may produce an incorrect solution.

Accordingly, the project performance state has to use a methodology that will increase the probability of delivering the product or products that will remedy the undesirable situation as it exists at the time of delivery, rather than as it existed at the time the project began.

7.4.3.1 Build a Little Test a Little

A project's most important resource is time. Projects will be completed faster and at lower cost if waste is eliminated (Malotaux 2010). There are two basic types of waste in the project process: doing something that:

1. Does not yield value.
2. Doing something poorly.

A common vision of the project goals and well-written requirements prevent creating some items that do not yield value, hence the focus on those writing the right requirements and creating the common vision in the literature.

The 'build a little test a little' approach (Kasser 2019b: Section 10.15.5):

- Can keep the project focused and swiftly reprioritize work in the event of problems.
- Has been used by electrical engineers to prototype circuits for as long as there have been electrical engineers.
- Is used by software engineers who release daily* updates of a product under development.
- Organizes the project timeline in chunks of work with specific goals for each chunk.† Each activity in the chunk is extracted from the relevant WPs and summarized in a spreadsheet format. The contents of the spread sheet are:
  1. *Priority*: the priority of the activity from the WP as modified by the chunking process.
  2. *WP item or activity*: the reference to the WP activity for further details.
  3. *Date due*: the latest date for completion of the activity.
  4. *Estimated time to do activity*: from the WP.
  5. *Latest start date*: the date due minus the estimated time to do the activity.
  6. *Date completed*: once completed may be hidden from view. This item is used to improve the accuracy of future estimates by seeing how accurate this one was.

---

* Or periodic.
† The chunks can be days or weeks depending on the length of the schedule.

7. *Notes*: as appropriate; a good place to note lessons learned, risks encountered and how they were overcome.

Each chunk begins and ends in a short meeting. Each item is examined at the meeting. The project is in trouble if the current date is later than the:

1. Latest start date for any activity.
2. Due date for any uncompleted activity.

If people are working on activities not listed in the spreadsheet, then either:

1. They are wasting time,* or
2. The spreadsheet is in error and needs to be updated.

In such a situation, some of the quadruple constraints (Section 7.3.2) need to be adjusted to recover the schedule and the priorities of the activities adjusted. These changes often take the form of completing the activities while delaying lower priority activities. If the project ends before the lower priority requirements are completed, then there will have to be a negotiated agreement about how to proceed.

The effect of information from the 'build a little test a little' meetings should end up in the CRIP charts (Section 7.5) and ETL charts (Section 7.6) in the formal meetings with the customer.

7.4.4 THE PROJECT CLOSEOUT STATE

The project closeout state is the reverse of the project performance state. The project performance state started with nothing and created a project. The project closeout state starts with a project and reduces it to nothing. The project closeout state forms a WP of its own and contains the following process:

1. Delivering the products produced by the project to the customer.
2. Getting delivery acceptance from customer.
3. Shutting down resources and releasing to new users.
4. Rewarding the project team; a certificate of recognition of contributions is a minimum.
5. Reassigning project team members: They should be reassigned according to a published schedule so there is no uncertainty as to where people will go when the project ends. This minimizes reduction in morale. Ideally the people have a place to go within the organization, perhaps returning to their home department or moving on to the next project.
6. Closing accounts and seeing all bills are paid.
7. Arranging the project records in a logical manner and storing them in an accessible but secure location. The records need to be available for technical, procedural, liability and other legal reasons.

---

* And those activities should cease immediately.

8. Updating the project lessons learned into an accessible database.
9. Producing a formal project closure notification to higher management.
10. Creating and delivering a final report to the appropriate stakeholders.

7.5   CRIP CHARTS

The CRIP chart (Kasser 2018a: Section 8.1, 2019a: Section 11.5) is a project management, systems and software engineering tool that:

- Facilitates proactive risk management.
- Is used in monitoring and controlling projects.
- Provides a way to think about and measure technical progress.
- Identify potential problems in near real time so as to be able to prevent or mitigate the problems before they occur (risk management).

While simplistic approaches of tracking the realization activities of all the requirements or features such as feature-driven development (FDD) (Palmer and Felsing 2002) can inform about the state of the realization activities, they cannot be used to estimate the degree of completion since each requirement or feature has a different level of complexity and takes a different amount of effort to realize. The need is for a measurement approach that can:

- Roll up the detailed information into a summary that can be displayed in one or two charts.
- Readily relate to the existing cost and schedule information.

The CRIP approach (Kassér 1999) meets that need by looking at the change in the state of a summary of the realization activities which convert requirements into systems during the SDP from several perspectives (categories). The summary information is presented in a table known as a CRIP chart which:

- Covers the entire SDP (Section 6.1.5.4).
- Uses a technique similar to FDD charts.
- Provides simple summaries suitable for upper management uninterested in details.
- Indicates variances between plan and progress but not the causes of the variance. It is up to management to ask for explanations of the variances.
- Is based on the use of WPs (Section 7.3.1.1.4) in planning a project, hence have to be integrated into the SDP during the project planning state.
- Is designed to be used in association with EVA budget and schedule information as shown in Figure 7.6.

Although written up for use with requirements, CRIP charts can also be used for use cases, scenarios, technical performance measures (TPMs) and any other

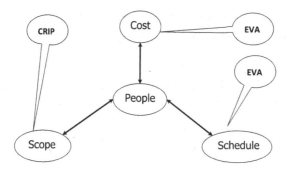

FIGURE 7.6   The relationship between cost, schedule and budget and CRIP and EVA.

technical measurement that can be tracked across the SDP. When the column in the CRIP chart is plotted as a histogram (Kasser 2018b: Section 2.9), it shows an attribute profile (Kasser 2018b: Section 9.1) for that attribute such as the risk profile in Figure 3.2.

7.5.1   THE FIVE-STEP CRIP APPROACH

The five-step CRIP approach is to:

1. Identify categories for the requirements.
2. Quantify each category into ranges.
3. Categorize the requirements.
4. Place each requirement in a category into a range.
5. Monitor the differences in the state of each of the numbers of requirements in a range at the SDP formal and informal reporting milestones.

The first four steps take place before the SRR. The last step which takes place in all the remaining states of the SDP following the SRR is the key element in the CRIP approach because it is a dynamic measure of change rather than a static value. Consider each of the five steps.

7.5.1.1   Step 1: Identify Categories for the Requirements

Identify categories for the requirements. Typical categories include:

- *Complexity*: of the requirement.
- *Estimated cost*: to implement the requirement.
- *Firmness*: the likelihood that the requirement will change during the SDP.
- *Priority*: of the requirement.
- *Risk probability*: of occurrence.
- *Risk severity*: if it occurs.
- *Source*: from a specific source, regulation, etc.

7.5.1.2   Step 2: Quantify Each Category into Ranges

Quantify each category into no more than ten ranges. Thus, for:

- *Priority*: requirements may be allocated priorities between one and ten.
- *Complexity*: requirements may be allocated estimated complexities of implementation between 'A' and 'J'.
- *Estimated cost to implement*: requirements may be allocated estimated costs to implement values. For example, less than $100, between $100 and $500, between $500 and $1000, etc.
- *Risk (probability and severity)*: requirements may be awarded a value between one and five.

The ranges are relative, not absolute. Any of the several quantitative decision-making techniques for sorting items into relative ranges may be used (Kasser 2018a: Section 4.6). The buyer/customer and supplier/contractor determine the range limits in each category.

A requirement may be moved into a different range as more is learned about its effect on the development or the relative importance of the need changes during the SDP. Thus, if the priority of a specific requirement or the estimated cost to implement changes between SDP reporting milestones changes, the requirement may be moved from one range to another in those categories. However, the rules for setting the range limits, and the range limits themselves must not change during the SDP.

7.5.1.3   Step 3: Categorize the Requirements

Every requirement shall be placed in each category.

7.5.1.4   Step 4: Place Each Requirement in Each Category into a Range

Place each requirement into one range slot for each category. The information used to place the requirements into the ranges for the categories comes from the WP (Section 7.3.1.1.4) in the project plan (Section 7.4.2.1). If all the requirements end up in the same range slot, such as all of them having the highest priority, the range limits should be re-examined to spread the requirements across the full set of range slots. If most of them end up in a single range slot, then that slot should be expanded into several slots. There is no need for the ranges to be linear. For example, ranges could be 1–10, 11–20, 20–30, etc., or 1–10, 11–15, 16–20, 21–30, etc. In the second case, it just means that the range 11–20 has been split into two ranges.

Once the requirements have been categorized in ranges, an attribute profile (Kasser 2018a: Section 9.1) for each category can be drawn in the form of a histogram (Kasser 2018a: Section 2.9) as shown in Figure 3.2.

7.5.1.5   Step 5: States of Implementation

At any time during the SDP, each requirement shall be in one, and only one, of the five CRIP states. The CRIP states of implementation of each requirement during the project are:

TABLE 7.7

An Unpopulated CRIP Chart

Range	Identified			In Process			Completed			In Test			Accepted		
	P	E	A	P	E	A	P	E	A	P	E	A	P	E	A
1															
2															
3															
4															
5															
6															
7															
8															
9															
10															
Totals															

1. *Identified*: a requirement has been identified, documented and approved.
2. *In process*: the supplier has begun the development activities to realize the requirement.
3. *Completed*: the supplier has completed development activities on the equipment that will perform the requirement.
4. *In test*: the supplier has started to test compliance to the requirement.
5. *Accepted*: the buyer has accepted delivery of the part of the system (a Build) containing the implementation of the requirement.

The summaries of the number of requirements in each state are reported at project milestones and reporting meetings.

7.5.2 POPULATING AND USING THE CRIP CHART

An unpopulated CRIP chart is shown in Table 7.7 where:

- The vertical axis of the chart is split into the ten ranges in the category (1–10).
- The horizontal axis of the chart is split into five columns representing the CRIP states of a project.
- Each CRIP state contains three cells: planned 'P', expected 'E' and actual 'A', where:
  - *[P] Planned for the next reporting period*: the number of requirements planned to be in the CRIP state before the following reporting milestone.
  - *[E] Expected*: the number of requirements expected to be in the CRIP state based on the number planned in the previous reporting milestone. This is a copy of the 'P' value in the CRIP chart for the previous milestone.
  - *[A] Actual*: the number of requirements actually in the CRIP state.

For the first milestone-reporting period, the values for:

- *'Expected'*: 'E' are derived from the project plan (Section 7.4.2.1) for the time period.
- *'Actual'*: 'A' are the numbers actually measured at the end of the reporting period.
- *'Planned for the next reporting period'*: 'P' are numbers derived from the project plan and the work done during the current reporting period.

As of the first milestone following the start of a project, the numbers in the 'P' column of a CRIP state of the chart at one milestone are always copied into the 'E' column of the same CRIP state in the chart for the next milestone. The 'A' and 'P' values reflect the reality. As work progresses, the numbers flow across the CRIP states from 'Identified' to 'Accepted'.

At each reporting milestone, the changes in the CRIP state of each of the requirements between the milestones are monitored. The numbers of each of the requirements in each of the categories are presented in tabular format in a CRIP chart at reporting milestones (major reviews or monthly progress meetings). Colours can be used to draw attention to the state of a cell in the table. For example, the colours can be allocated* such that:

- *Violet*: shows realization activities for requirements in that range are well ahead of estimates.
- *Blue*: shows realization activities for requirements in that range are ahead of estimates.
- *Green*: shows realization activities for requirements in that range are close (±) to estimates.
- *Yellow*: shows realization activities for requirements in that range are slightly below estimates.
- *Red*: shows realization activities for requirements in that range are well under the estimates.

The range setting for the colours in the CRIP charts should be the same as the range settings using in the ETL charts (Section 7.6) for the project.

CRIP charts show that a problem might exist, namely a risk. Any time there is a deviation between 'E' and 'A' in a CRIP state, the situation needs to be investigated just as a deviation in an EVA chart has to be explained. A comparison of the summaries from different reporting milestones can identify progress and show that problems may exist. On its own, however, the CRIP chart cannot identify the actual problem; its purpose is to trigger questions to determine the nature of the actual problem.

The CRIP charts when viewed over several reporting periods can identify other types of 'situations'. While the CRIP chart can be used as a stand-alone chart, it should really be used together with EVA, budget and schedule information to reduce the risk of realizing a failed project. For example, if there is a change in the number of:

---

* The quantitative numbers for the ranges would be agreed upon between the stakeholders, specified in the contract prior to the commencement of the project, and not changed during the SDP.

TABLE 7.8

The CRIP Chart for a Challenged Project at TRR

Range	Identified			In Process			Completed			In Test			Accepted		
	P	E	A	P	E	A	P	E	A	P	E	A	P	E	A
1	0	0	0	0	0	0	40	81	41	81					
2	0	0	0	0	0	0	40	78	38	78					
3	0	0	0	0	0	0	5	35	30	35					
4	0	0	0	0	0	0	10	30	20	30					
5	0	0	0	0	0	0	10	26	16	26					
6	0	0	0	0	0	0	0	20	20	20					
7	0	0	0	0	0	0	6	8	2	8					
8	0	0	0	0	0	0	4	7	3	7					
9	0	0	0	0	0	0	1	5	4	5					
10	0	0	0	0	0	0	1	2	1	2					
Totals	0	0	0	0	0	0	117	292	175	292					

- Identified requirements and there is no change in the budget or schedule, there is going to be a problem. Thus:
  - If the number of requirements goes up and the budget does not, the risk of failure increases because more work will have to be done without a change in the allocation of funds.
  - If the number of requirements goes down, and the budget does not, there is a financial problem. However, if it is in the context of a fixed price contract, it may show additional profit.
- *Requirements being worked on and there is no change in the number being tested*: there is a potential supplier management or technical problem if this situation is at a major milestone review.
- *Requirements being tested and there is no change in the number accepted*: there may be a problem with the process or a large number of defects have been found and are being reworked.
- *Identified requirements at each reporting milestone*: the project is suffering from requirements creep if the number is increasing. This situation may reflect controlled changes due to the change in the customer's need or uncontrolled changes.

Since projects tend to delay formal milestones until the planned work is completed, the CRIP charts are more useful in the monthly or other periodic meetings between the formal major milestones.

The following example demonstrates the use of the CRIP chart. A CRIP chart for a challenged project at the subsystem TRR shown in Table 7.8 indicates:

- No additional requirements were identified since the CDR.

TABLE 7.9

The CRIP Chart for the Challenged Project at IRR

Range	Identified			In Process			Completed			In Test			Accepted		
	P	E	A	P	E	A	P	E	A	P	E	A	P	E	A
1	0	0	0	0	0	0	0	0	0	40	81	21	81		
2	0	0	0	0	0	0	0	0	0	30	78	48	78		
3	0	0	0	0	0	0	0	0	0	5	35	30	35		
4	0	0	0	0	0	0	0	0	0	8	30	22	30		
5	0	0	0	0	0	0	0	0	0	14	26	12	26		
6	0	0	0	0	0	0	0	0	0	0	20	20	20		
7	0	0	0	0	0	0	0	0	0	7	8	1	8		
8	0	0	0	0	0	0	0	0	0	6	7	1	7		
9	0	0	0	0	0	0	0	0	0	4	5	1	5		
10	0	0	0	0	0	0	0	0	0	0	2	2	2		
Totals	0	0	0	0	0	0	0	0	0	114	292	158	292		

- Development activities in all requirement ranges except Range 6 have not been completed since the 'E' and 'A' values in row 6 of the 'Completed' state do not match.
- The project plans to catch up on the development activities as shown by the numbers in the 'P' column of the 'Completed' state.
- The project is optimistic about commencing testing following the TRR as evidenced by the difference between numbers in the 'P' column of the 'In test' state and the numbers in the corresponding rows of the 'A' column of the 'Completed' state. The customer definitely needs to find out the reason for the optimism.

The CRIP chart for the challenged project at IRR shown in Table 7.9 indicates:

- No additional requirements were identified.
- The project should not have transitioned into the subsystem testing state of the SDP because of the difference between the numbers in the 'E' and 'A' columns in the 'In test' state.
- The project plans to catch up as shown by the numbers in the 'P' column of the 'In test' state.
- The project is still optimistic about completing the testing before the delivery readiness review (DRR) because the 'P' numbers in the 'Accepted' state match the 'E' numbers instead of the 'A' numbers in the 'In test' state. The customer definitely needs to determine the reasons for the optimism.

7.5.3 ADVANTAGES OF THE CRIP APPROACH

The advantages of the CRIP approach include:

- Can be used in both the 'A' and 'B' paradigms of systems engineering (Section 6.5.2).

- Links all work done on a project to the customer's requirements.
- May be used at any level of system decomposition.
- Provides a simple way to show progress or the lack of it, at any reporting milestone. Just compare the 'E' and 'A' numbers and ask for an explanation of the variances.
- Provides a window into the project for upper management (buyer and supplier) to monitor progress (micromanagement by upper management (MBUM)).
- Can indicate if lower priority requirements are being realized before higher priority requirements if priority is a category.
- Identifies the probability of some management and technical problems as they occur, allowing proactive risk containment techniques.
- May be built into requirements management, and other computerized project and development management tools.
- May be incorporated into the progress reporting requirements in system development contracts. Falsifying entries in the CRIP chart to show false progress then constitute fraud.
- Requires a process. Some organizations don't have one, so they will have to develop one to use CRIP charts.
- Requires configuration management which tends to be poorly implemented in many organizations. The use of CRIP charts will enforce good configuration management.

7.5.4 DISADVANTAGES OF THE CRIP APPROACH

The disadvantages of the CRIP approach include:

- Is a different way of viewing project progress.
- Requires categorization of the requirements.
- Requires sorting of the requirements into ranges in each category.
- Requires prioritization of requirements if priority is used as a category, which it should be.

7.5.5 EXAMPLES OF USING CRIP CHARTS IN DIFFERENT TYPES OF PROJECTS

See *Systems Thinker's Toolbox: Tools for Managing Complexity* (Kasser 2018a: Section 8.1) or *Systemic and Systematic Project Management* (Kasser 2019a: Section 11.5) for more examples of how CRIP charts can help manage risks by indicating the technical progress of a project and identifying potential problems (risks) using the following stereotype project examples:

- An ideal project.
- A project with requirements creep.
- A challenged project.
- A 'make up your mind' project.

#	Projects	Last time	Current	Next
1	DMO MPM Degree	Green	Green	Green
2	DMO Certification Contract	Yellow -P	Yellow -P	Red -P
3	DSTO Evaluating Coalition C4ISR Architectures Contract	Yellow -P	Yellow -P	Green
4	DSTO Maritime Support Contract	Red -BS	Red -BS	Red -BS
5	Wedgetail TRDC C4ISR Course	Green	Blue	Green
6	Research student supervision	Green	Green	Green
7	Semester 1 Software T&E Course	Yellow -P	Yellow -P	Green
8	PETS	Green	Green	Green
9	Research Group Meetings	Green	Green	Green
10	SEEC Administrative tasks	Green	Green	Yellow -P
11	INCOSE Region VI Support	Yellow -PB	Yellow -PB	Yellow -PB

FIGURE 7.7   The ETL chart.

7.6   ETL CHARTS

The ETL chart is a tool that:

- Can provide information that is not presently available in most projects.
- Can potentially reduce the time spent in meetings.
- Is based on adding perceptions of the project from the *Temporal* HTP to the basic traffic light chart (Kasser 2018a: Section 8.16.1).

The basic traffic light chart can be enhanced by adding perceptions from the *Temporal* HTP as shown in the ETL chart in Figure 7.7 in the form of two additional columns reminding the viewers of the state of the project at the prior reporting period and showing the expected state of the project at the following future reporting period in a similar manner to the CRIP chart (Section 7.5) (*Generic* HTP). New information that was not previously available can now be seen at the high-level summary. For example:

- Projects 1, 6, 8 and 9 are conforming to the planned schedule and budget and are not facing any problems, namely they are proceeding according to the nominal status.
- Projects 3 and 7 which have been facing minor problems are expected to face a major problem during the next reporting period.
- Project 4 which has been facing minor problems is expected to overcome them and return to a nominal state during the next reporting period.
- Project 5 exceeded expectations during this reporting period and is expected to return to nominal status during the next reporting period.
- Project 11 has been facing problems and the problems are ongoing.

These perceptions from the *Temporal* HTP can be generalized as shown in Figure 7.8 where:

#	Projects	Last time	Current	Next
1	Project Ho-hum	Green	Green	Green
2	Project Oh oh	Yellow -PB	Yellow -P	Red -P
3	Project Catching up	Yellow -P	Yellow -P	Green
4	Project Replace manager	Red -BS	Red -BS	Red -BS
5	Project Very happy customer	Green	Blue	Blue
6	Project Completed	Green	Green	N/A
7	Project Promote manager	Red -P	Yellow -P	Green
8	Project Watch this person	Yellow –BS	Green	Blue
9	Project No risk management	Green	Red -P	Red -P
10	Project Took course in risk management	Green	Green	Yellow -P
11	Project Manager doing risk management	Yellow -P	Yellow -P	Yellow -P

FIGURE 7.8   Project trends using the ETL chart.

- *Project 1* is proceeding according to nominal status.
- *Project 2* has identified that a major problem or risk is expected to arise during the next reporting period. The project may need additional resources to remedy the problem.
- *Project 3* seems to be overcoming its problems and returning nominal status.
- *Project 4* has been facing one or more major problems and no change is expected. This could be due to (a) poor management, (b) being underfunded so there is no funding to complete the remaining work or some other reason to be determined.
- *Project 5* has a very happy customer since it is ahead of the nominal status and is expected to remain so in the next reporting period. Perceptions from the *Continuum* HTP note that this situation may be because the project was overfunded and given more than enough time to complete or has a smart project manager.
- *Project 6* has been nominal and is now complete on schedule and within budget.
- *Project 7* had major problems during the previous reporting period which are being tackled successfully, and the project is expected to return to nominal status by the next reporting period. This project manager may be a candidate for promotion or a candidate to take over other projects in trouble and get them back on track.
- *Project 8* has gone from experiencing budget and schedule problems in the previous reporting period to nominal status, and the project manager is anticipating exceeding the budget and schedule in the next reporting period. This project manager may be a candidate for promotion.
- *Project 9* was nominal in the last reporting period but is currently experiencing one or more major problems which are expected to continue into the next reporting period. If the major problem was not anticipated in the previous status report, then the situation needs to be investigated.

last time'

#	Projects	Last	Current		Next
			E	A	
1	DMO MPM Degree	Green		Green	Green
2	DMO Certification Contract	Yellow -P		P	Red -P
3	DSTO Evaluating Coalition C4ISR Architectures Contract	Yellow -P		P	Green
4	DSTO Maritime Support Contract	Red -BS		BS	Red -BS
5	Wedgetail TRDC C4ISR Course	Green		Blue	Green
6	Research student supervision	Green		Green	Green
7	Semester 1 Software T&E Course	Yellow -P		P	Ended
8	PETS	Green		Green	Green
9	Research Group Meetings	Green		Green	Green
10	SEEC Administrative tasks	Green		Green	Yellow -P
11	INCOSE Region VI Support	Yellow -PB		PB	Yellow -PB

FIGURE 7.9   The final version of the ETL chart.

- *Project 10* which has been nominal since the last reporting period (or earlier) is expecting to experience minor problems in the next reporting period. The project manager may have taken a continuing education course and started to practice risk management.
- *Project 11* which has been experiencing minor problems since the last reporting period (or earlier) is expecting to continue to experience such problems. If there is no corresponding adverse budget and schedule impact (shown in the Gantt (Kasser 2018a: Section 8.4)) and EVA charts accompanying the status report, the project manager may be practicing risk management and bringing it to the attention of senior management.

The chart shown in Figure 7.8 can be improved using an idea from the *Generic* HTP (Section 7.5) to show how the current situation compares with the predicted situation as of the last report presentation as shown in Figure 7.9. The current reporting period has been separated in two parts:

1. *Expected from last time*: copied from the 'next' column in the previous report presentation.
2. *Actual*: as achieved by the work done during this reporting period.

For example:

- *Project 1* was expected to be green and is green.
- *Project 2* was expected to be in the green but is still yellow (problems).
- Differences can be seen in some of the other projects, which require explanation at the reporting presentation.

In summary the ETL chart:

- Facilitates proactive risk management by showing that problems are anticipated.
- Provides information that is not currently available.
- Allows MBUM by delving into the reasons for the deviation from the planned status.
- Is an ideal tool for MBE (Section 7.7).
- Is an ideal tool for MBO (Section 7.8).

7.7 MANAGEMENT BY EXCEPTION

MBE is a tool that:

- Allows the span of control to be widened by allowing the manager to ignore projects that are performing to the planned values (the norm).
- Sets tolerance limits on the parameters of an object (process or product being developed) and allows the manager to ignore the object if the limits are not exceeded.
- Allows delegating decisions to the lowest level, giving employees responsibility, as long as the consequences of the decision do not cause the limits to be exceeded.

7.7.1 THE KEY INGREDIENTS IN MBE

The key ingredients in MBE are (Bittel 1964):

1. *Selection*: pinpoints the criteria that will be measured. In the systems approach, these are the objectives determined by MBO (Section 7.8).
2. *Measurement*: assigns value to past and present performance.
3. *Projection*: analyses the measurements and projects' future performance.
4. *Observation*: informs management of the state of performance.
5. *Comparison*: compares performance with limits or boundaries and reports the variances to management.
6. *Decision-making*: prescribes the action to be taken to:
   1. Bring performance back within the limits.
   2. Adjust the limits or expectations.
   3. Exploit opportunity.

7.7.2 ADVANTAGES AND DISADVANTAGES OF MBE

The advantages include:

- Reduces the amount of MBUM.
- Upper management need not spend time on projects that are progressing normally.
- Undesirable situations are indicated by the limits being exceeded so that upper management can intervene.

The disadvantages include:

- Open to MBUM once a parameter exceeds the limit threshold.
- Mistakes in setting limits may cause problematic situations to be unnoticed until it is too late to take corrective action or may flag too many alarms.
- Can be confused with 'management without caring' resulting in low morale in organizational units that do not raise exceptions as in the C3I Group Case (Kasser 2018a: Section 13.1.2).
- It is based on the assumption that only upper management can deal with a problematic situation. However, this disadvantage may be overcome if the project manager can produce a plan for dealing with the problem that satisfies upper management.

7.7.3 USING MBE

There are two states in MBE, namely:

1. *The project planning state*: in which the limits are set. The objectives of each WP are to produce the product within the estimated cost and schedule. The upper and lower level tolerances for each objective are determined. The cost upper and lower levels may the best- and worst-case cost and schedule estimates or may be assumed by using a number of standard deviations. However, as with most other things there is a trade-off; the smaller the limit, the more exceptions will need to be investigated. The larger the limit, the fewer investigations; but there will be a greater chance that a serious situation will be missed. The MBE upper and lower limits may be added to the WPs.
2. *The monitoring and controlling states*: in which the limits should not be exceeded. Tools used in MBE in this state include process control trend charts (Kasser 2018a: Section 2.15), CRIP charts (Section 7.5) and ETL charts (Section 7.6). Should the limits be exceeded, corrective action may need to be taken.

7.8 MANAGEMENT BY OBJECTIVES

MBO (Mali 1972) is a compound tool:

- For performing risk management.
- For planning a project.
- For monitoring the progress towards completion.
- Which was popular as a management approach in the 1960s.

MBO as introduced was unnecessarily complex and complicated. However, the core concepts in MBO are simple and manageable, namely:

1. Set realistic objectives or goals for every activity.

2. Monitor progress towards those objectives as the project proceeds.
3. Take corrective action if the objectives are not going to be met (risk management).

These core concepts are actually embodied in standard project management and EVA and built into the SDP waterfall chart template (Kasser 2018a: Section 14.9). The SDP is divided into a number of states, each of which has to accomplish predefined objectives by the milestone at the end of the state. For example, the objectives of the system requirements state (Section 6.1.5.4.4) are to produce the system requirements document at the SRR which takes place at the end of the state.

MBO was originally introduced by Peter Drucker as a risk-reducing systems approach to managing an organization (Drucker 1954) where the objectives of one layer in the hierarchy are set to accomplish and support the objectives of the higher level of the organization, a similar concept to the 'Do Statement' (Chacko 1989, Kasser 2018a: Section 3.2.1). In the language of systems, when setting objectives:

1. The objectives of the system shall support those of the meta-system.
2. The objectives of the subsystem shall support the objectives of the system.

7.8.1 MBO in the Project Planning State

MBO relates to activities and objectives; it does not involve schedules and timelines. The process for using MBO in the planning state of a project is as follows:

1. Determine the appropriate states and milestones for the project (Section 6.1.5.4.2). Milestones can be major and minor, and formal and informal.
2. If the project is complex, use the waterfall chart (Kasser 2018a: Section 14.9, Royce 1970) as a template. Since the scope of projects varies, the mixture of major and minor, formal and informal, milestones also vary. Each project will have its own mixture, similar to those of similar projects.
3. Determine the objectives to be completed by the last major milestone (the last state) in the project. These may include events to be achieved and products to be produced.
4. Working back from the last milestone, determine the objectives for each previous state of the SDP.
5. Determine the sub-objectives for each objective. A useful tool for doing this is the 'Do Statement' (Chacko 1989, Kasser 2018a: Section 3.2.1).
6. Create the WPs (Section 7.3.1.1.4) and project network of activities that will produce each objective and sub-objective in a sequential orderly manner. Useful tools for doing this standard planning process are product-activity-milestone (PAM) charts (Kasser 2018a: Section 2.12), PAM network charts (Kasser 2018a: Section 2.12.2) and PERT charts (Kasser 2018a: Section 8.10).

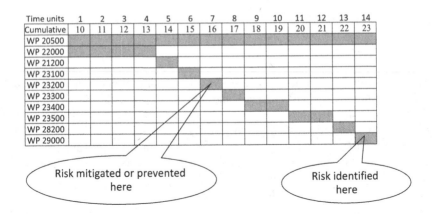

FIGURE 7.10   Example of planned prevention.

7.8.2   MBO IN THE PROJECT PERFORMANCE STATE

The traditional project performance state includes MBO even if the term is not used, since there is a check at each milestone to determine if the predicted objectives from the previous milestone have been met, and if not, why not and what corrective action is planned. Reasons for not meeting objectives constitute risks and include:

- Poor project management.
- Poor systems engineering.
- Poor estimating when setting objectives.
- The impact of unforeseen events.

7.9   USING 'PREVENTION' TO LOWER PROJECT COMPLETION RISK

Preventing risks from turning into events, namely occurring, is reasonably simple when planning is done starting at the end of the project and working back to the beginning. For example, look at the skeleton chart timeline shown in Figure 7.10. Supposing the first risk identified is in the last WP, in this case WP 29000. After identifying the risk, and working out what needs to be done to mitigate or prevent that risk, the WP for the risk mitigation activities earlier in the planned SDP can be created and positioned appropriately or, as shown, added to an existing WP such as WP 23200.

7.10   STAKEHOLDERS AND STAKEHOLDER MANAGEMENT

Stakeholder management is a key area in the management of projects.

Chung and Crawford (2015)

The systems approach to managing stakeholder concerns:

1. Uses perceptions from the *Generic* HTP to note the similarity to the resistance and support of any proposed changes (Section 4.5.1).
2. Begins in the project initiation state (Section 7.4.1) and continues throughout the remainder of the project lifecycle.

7.11 SOFTWARE PROJECT RISKS

Each domain has its generic set of risks. Donaldson and Siegel provide a set of risks in the software domain in three categories based on their experience (Donaldson and Siegel 1997): high, medium and low.

7.11.1 HIGH RISK

A project is at high risk if two or more of the following criteria apply:

- Unique application.
- Lack of up-to-date documentation.
- Inexperienced development staff.
- No schedule slack.
- Uncertain requirements.
- Multiple customers.
- Subcontractor hours are at least 50% of total development hours.
- Software failure results in deaths, injuries or major financial losses.

7.11.2 MEDIUM RISK

A project is at medium risk if the project is not high risk and if two or more of the following criteria apply:

- Application domain is not well understood.
- Lack of up-to-date documentation.
- Some inexperienced development staff.
- Little schedule slack.
- Some major requirements uncertain.
- First-time customer.
- Subcontractor hours are 25%–50% of total development hours.
- Software failure results in high visibility within customer community.

7.11.3 LOW RISK

A project is low risk if the project is not high or medium risk and if three or more of the following criteria apply:

- Application domain is well understood.
- Up-to-date documentation.
- Experienced development staff.
- Flexible schedule.

- All requirements known.
- Established supplier–customer relationship.
- Software failure does not result in deaths, injuries or major financial losses.

7.12 GENERIC PROJECT RISKS

Generic risk can be collected from various sources including those mentioned in this book. For example, Mar provided a sample list of 130 generic project risks (Mar 2016) in the following categories:

1. Approvals and red tape.
2. Architecture.
3. Authority.
4. Change.
5. Commercial.
6. Communication.
7. Cost.
8. Decisions and issue resolution.
9. Design.
10. Executive support.
11. External factors.
12. Integration.
13. Organizational.
14. Procurement.
15. Project management.
16. Resources and team.
17. Requirements.
18. Scope.
19. Secondary risks.
20. Stakeholders.
21. Technical.
22. User acceptance.

However, experience in behavioural psychology has shown that a list of 39 reasons was too long to be practical (Section 1.2) so Mar's 130 generic risks would be even more difficult to use. However, a long list of risks could be built into computer software as an expert system that would 'think' about the situation based on a dialogue and help the problem-solver identify and manage risks pertinent to the situation. The systems approach (Section 1.4) is based on the problem-solving process and the risk framework shown in Table 1.2 and building the expert system into the human via education, training and experience.

7.13 EXERCISES

This section provides a number of exercises as opportunities to create outline project plans using the concept of working backwards from the solution (Kasser 2018a:

Section 18.8), experience creating WPs (Section 7.3.1.1.4) and develop objectives for the states in the process per MBO (Section 7.8). A partial acceptable solution is provided for the ENU traffic light upgrade (Section 7.13.1). It is up to the reader to use the sample as a template when performing the remaining exercises.

7.13.1 The ENU Traffic Light Upgrade

The Ministry of Transport is about to award an acquisition contract to create a project to upgrade the traffic light outside ENU. In addition to meeting the generic requirements in Section 2.7:

1. Format the problem posed by this exercise in the problem formulation template (Section 5.1.7.2).
2. Create an outline project plan for the upgrade documenting the process listing the steps in reverse from the upgrade in operation back to commencing the work including any prerequisites such as purchasing the materiel.
3. Show the objectives for each state in the SDP (Section 6.1.5.4).
4. State at least two generic risks in each state of the SDP.
5. Show how the risks will be mitigated or prevented in earlier states of the SDP.
6. Break the work in any one state into at least three WPs, and make reasonable and realistic assumptions for the information in each WP.
7. Show linkages between the WPs.

7.13.1.1 One Acceptable Solution

A completed exercise could take the form shown in this section. Note it provides one acceptable solution, not the single correct solution as there will be other acceptable solutions (Section 5.1.1.2.2).

7.13.1.1.1 The Assumptions

The assumptions are as follows:

1. Question 5, which asks how the risks will be mitigated or prevented, refers to the two generic risks asked for in question 4.
2. During the project initiation state, a study took place to identify the cause of the complaints. Using perceptions from the *Generic* HTP, the root cause of the complaints was determined not to be the actual waiting line time, but to be the uncertainty in the amount of waiting time.[*] Perceptions from the *Generic* HTP noted that the uncertainty and amount of waiting time could be eliminated by installing a countdown timer and display. This has been done in many other crossings in the country. Accordingly, the decision was made to change the problem from 'how to reduce the waiting time' to 'how to provide information as to the amount of time left before the lights changed at the intersection'. The remedy then became to upgrade the traffic light by installing a countdown timer and display. As a result, the problem was turned

---

[*] Domain knowledge.

over to the Ministry of Transport who will be performing the project using their standard operating procedures for modifying the traffic system.

3. The WPs shown in section are part of a complete set of work packages so that the links in the appropriate elements of the work package can be entered. The assumed links not required by the question are realistic but are greyed out to distinguish them from the links that are required.

4. The Ministry of Transport chose to subcontract the work to one of its pool of contractors who maintain and support the national traffic light system.

7.13.1.1.2   The Formulated Problem

The problem formatted according to the problem formulation template (Section 5.1.7.2) is:

1. The undesirable situation: is the need to:
    1. Comply with the generic requirements in Section 2.7.
    2. Create an outline project plan for the upgrade documenting the process listing the steps in reverse from the upgrade in operation back to commencing the work including any prerequisites such as purchasing the materiel.
    3. Show the objectives for each state in the SDP (Section 6.1.5.4).
    4. State at least two generic risks in each state of the SDP.
    5. Show how the risks will be mitigated or prevented in earlier states of the SDP.
    6. Break the work in any one state into at least three WPs, and make reasonable and realistic assumptions for the information in each WP.
    7. Show linkages between the WPs.
2. The assumptions: are:
    1. The exercise can be completed within the 60 minutes.
    2. As listed in Section 7.13.1.1.1.
3. *The FCFDS*: the exercise has been completed and the desired grade has been achieved.
4. *The problem*: how to do the exercise, namely:
    1. Meet the generic requirements in Section 2.7.
    2. Create an outline project plan for the upgrade documenting the process listing the steps in reverse from the upgrade in operation back to commencing the work including any prerequisites such as purchasing the materiel.
    3. Show the objectives for each state in the SDP (Section 6.1.5.4).
    4. State at least two generic risks in each state of the SDP.
    5. Show how the risks will be mitigated or prevented in earlier states of the SDP.
    6. Break the work in any one state into at least three WPs, and make reasonable and realistic assumptions for the information in each WP.
    7. Show linkages between the WPs.
    8. Verify that all the requirements for the exercise have been completed and update the compliance matrix accordingly.

TABLE 7.10

Compliance Matrix for the Exercise in Section 7.13.1

ID	Instruction	Deliverable	Section	Completed
1	Brainstorming or active brainstorming	No	All	Yes
2	Research	No	All	Yes
3	Use imagination	No	All	Yes
4	Use lists and charts	No	All	Yes
5	Spend less than 60 minutes	No	All	Yes
6	Assumptions	Yes	7.13.1.1.1	Yes
7	Preliminary compliance matrix	No	7.13.1.1.3	Yes
8	Format the problem	Yes	7.13.1.1.2	Yes
9	An outline project plan	Yes	7.13.1.1.4	Yes
10	The objectives for each state in the SDP	Yes	7.13.1.1.5	Yes
11	At least two generic project risks	Yes	7.13.1.1.6	Yes
12	How the risks will be mitigated or prevented	Yes	7.13.1.1.7	Yes
13	At least three WPs	Yes	7.13.1.1.8	Yes
14	Linkages between the WPs	Yes	7.13.1.1.8	Yes
15	Lessons learned	Yes	7.13.1.1.9	Yes
16	Updated compliance matrix	Yes	7.13.1.1.3	Yes

5. *The solution*: consists of remedying each of the problems. A partial solution is presented below.

7.13.1.1.3   The Compliance Matrix

The completed compliance matrix is shown in Table 7.10[*]

7.13.1.1.4   The Outline Project Plan

Due to the constraints on time for this exercise, the outline project plan is limited to the extracts discussed in this section:

7.13.1.1.4.1   Project Purpose   The purpose of the project is to upgrade the traffic light at ENU to eliminate pedestrian complaints about waiting time by installing a countdown timer and display. Moreover, the display needs to be visible to pedestrians on both sides of the street.

7.13.1.1.4.2   Project Activities   The project activities[†] listed in reverse according to the working backwards from the solution (Kasser 2018a: Section 18.8) are:

1. Releasing the upgrade for operation.
2. Testing the upgrade for 24 hours.

---

[*] Placing the section number in the completed that the requirement has been met, but shows where the response can be found, mitigating the risk of the instructor failing to find and grade the response.

[†] Note the use of the present tense for the activities (words ending in -ing)

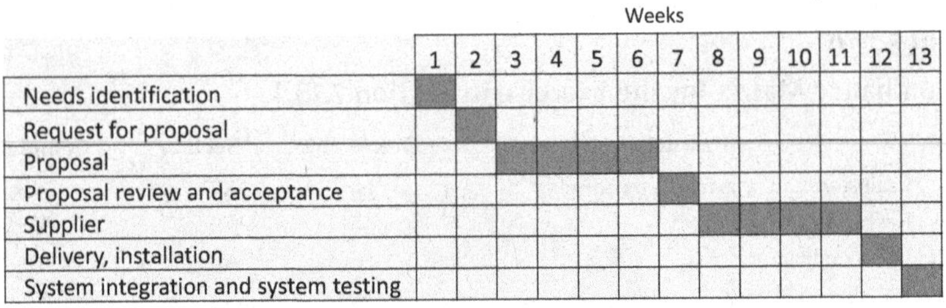

						Weeks							
	1	2	3	4	5	6	7	8	9	10	11	12	13
Needs identification	■												
Request for proposal		■											
Proposal			■	■	■								
Proposal review and acceptance						■	■						
Supplier								■	■	■			
Delivery, installation											■	■	
System integration and system testing													■

FIGURE 7.11   The schedule for the ENU upgrade.

3. Installing the upgrade.
4. Purchasing the equipment for connecting the countdown timer display to the traffic light controller.
5. Modifying the traffic light controller to add the countdown timer.
6. Determining how the traffic light controller could be modified.
7. Selecting the contractor to perform the upgrade via the standard operating procedure for engaging contractors to work on the entire traffic control system.
8. Creating the project team.

7.13.1.1.4.3 Project Deliverables   The project deliverables are the countdown timer, two working displays and the contractually specified documentation.

7.13.1.1.4.4 Project Schedule   The schedule is shown using a Gantt chart in Figure 7.11.[*]

7.13.1.1.4.5 Project Milestones   The project milestones based on the modified SLC for an acquired system (Section 6.1.5.5) are as follows[†]:

1. The project initiation milestone: start of project.
2. The operations concept review (OCR): showing the upgraded traffic light operating with the countdown timer displays at the end of the needs identification state.
3. The requests for proposals (RFPs) are ready to be issued: at the end of the RFP state.
4. The closing date and time for proposals has been reached: at the end of the proposal state.
5. The winning RFP has been selected: at the end of the proposal review and acceptance state.
6. The contract has been signed: with a contractor to provide an upgrade traffic light system.

---

[*] Explanations were not required, so none are given.
[†] Since the question did not ask for an explanation of the milestones, the response to this question takes the form of a list of milestones.

7. The contractor is ready to install the upgrade: to the traffic light at ENU as shown by the DRR at the end of the supplier states.
8. The upgraded traffic light system is operational: at the end of the delivery, installation, system integration and system testing states.

7.13.1.1.4.6   Project Team Roles and Responsibilities   Due to the use of a subcontract, the project will have a minimal team who will be working part time on the project while working on similar projects.

- *Project manager*: has the overall responsibility for managing the project to ensure that it meets the cost and schedule requirements.

   The role of the project manager is to manage the project, ensure the cost and schedule are met and report the state of the project to upper management.
- *System engineer*: responsible for the correctness and completeness of the design by the subcontractor and ensuring that the equipment as delivered meets the requirement of reducing the uncertainty to zero for the pedestrians waiting at the lights, when operating in the traffic light at ENU.

   The role of the systems engineer is to:
   - Monitor the subcontractor's design and construction process to ensure compliance to the requirements, and also ensure that the installed system will be compatible with the overall traffic system.
   - Arrange and monitor the installation in such a way as to minimize disruption to the traffic flow outside ENU.
   - Act as the technical interface between the Ministry of Transport and the contractor's engineering personnel to minimize the risk of interface problems between the upgraded traffic light system at ENU and the overall traffic light system.
   - Serve as a member of the change control board in the event changes need to be made* following the award of the contract.
- *Contract specialist*: responsible for selecting the subcontractor, negotiating the contract, monitoring the performance of the contract and closing out the contract once the traffic light upgrade is operational per its specifications.

   The role of the contract specialist is to handle the contractual interface between the Ministry of Transport and the contractor in the event changes need to be made following the award of the contract.

7.13.1.1.4.7   Risk Management Plan   The risk management section of the project plan includes the activities in the following states of the project:

- *The project initialization state*: ensuring that the requirement to report the project status in the contract includes the use of CRIP charts (Section 7.5) and ETL charts (Section 7.6) at the appropriate milestones to mitigate the risk of running open-loop or not identifying problems in time to take corrective action.

---

* These changes may be requested by the contractor or the Ministry as more details of the situation at ENU become known. As usual, none are expected when the project begins.

- *The project planning state*: incorporating risk identification and mitigation/ prevention planning into the planning process, namely:
  - Working back from the final delivery to create the WPs.
  - Once the set of WPs has been created, work backing from the final delivery to identify the risks in each WP and then determine which WP earlier in the schedule can be modified to include a mitigation or prevention activity. If such a WP cannot be found, modifying the project set of WPs to include a new risk mitigation WP will be done at this time.
- *The project performance state*: monitoring and reporting the cost and schedule using CRIP charts (Section 7.5), ETL charts (Section 7.6) and EVA charts. If the CRIP and the ETL charts show deviation from the plan, asking questions to find out the cause of the deviation, the mitigation activities and the expected date for the return to the planned cost and schedule.
- *The project closeout state*: activities include:
  - Identifying any risks and planning to prevent or mitigate them when the project closeout state begins. These risks include not disposing of hazardous material and safety risks associated with the installed equipment at ENU.
  - Making sure those risks are prevented or mitigated as the project.

7.13.1.1.4.8 CM Plan   The CM plan is basically to how to incorporate change requests for various reasons (Section 4.9.2) into the project. This is not really a large project, so there is no need to develop its own CM plan. Accordingly, this project will perform CM using the Ministry of Transport's standard operating procedures. For example, CM will be performed according to MOT-M-997386.[*]

7.13.1.1.4.9 Supporting Plans   This is a small project for the Ministry of Transport, so it will be supported using the standard operating procedures in the Ministry. For example, ILS will be performed according to MOS-L-738699.[†]

7.13.1.1.5   The Objectives for Each State in the SDP

In this instance, the SDP is part of the modified SLC (Section 6.1.5.5). Accordingly, the states of the SDP and their objectives in the ENU traffic light upgrade project are shown in Table 7.11.

---

[*] A fictitious number for the purpose of showing how such a question should be addressed in a real system because in the real system there would be an ILS plan, or should be, that can be cited.
[†] See Footnote 22.

TABLE 7.11

States of the SDP and Their Objectives in the ENU Traffic Light Upgrade

State in the SDP	Objective
Needs identification (Section 6.1.5.4.3)	A concept of what the upgraded system will be doing documented in a CONOPS (Section 6.5.9.1)
RFP [a]	An RFP ready to be issued
Proposal	Receive at least three proposals
Proposal review and acceptance	A winning bid has been selected
Supplier	The DRR confirms the upgrade is ready to be delivered and installed
Delivery, installation, system integration and system testing states	Demonstrated operational upgraded traffic light system

[a] The was no RFI state in this instance since it is dealing with a well-structured problem which has already been remedied in similar situations.

7.13.1.1.6 The Generic Risks

In addition to the risks listed in Section 7.12, generic project risks include[*]:

- Time and cost estimates are too optimistic (Augustine 1986).
- Stakeholders:
  - Not providing input and feedback in a timely manner.
  - Changing requirements after the project has started.
  - Adding new requirements after the project has started.
- Communications:
  - Unclear roles and responsibilities.
  - Not clearly understanding stakeholder needs.
  - Poor communication resulting in misunderstandings.
- Unexpected budget cuts.

7.13.1.1.7 How the Risks Will Be Mitigated or Prevented

Two generic risks in each state of the modified SDP (Section 6.1.5.5) and how they will be mitigated are shown in Table 7.12. The question did not ask for information about the WPs in which the risks could occur and should be mitigated or prevented so that information is not supplied.

---

[*] The first part of the response to this question, namely generic projects risks include, is typically included in a student answer to the question. However, the question did not ask for this information; the question only asked for two generic risks in each state of the SDP. The way the question is answered does not link it to any state in the SDP. Accordingly, the student wasted time producing that answer, and the instructor grading the answer wasted time reading it and working out to see if points could be awarded for it. A correctly worded compliance matrix carries through the original wording which will help to mitigate the risk of producing answers that are not required when working through exercises. In the same way, in the real world carrying through the same wording all the way through to the compliance matrix and section headings mitigates the risk doing real project work that was not required.

TABLE 7.12

ENU Traffic Light Upgrade Risks and Mitigation Approaches

State in SDP	Risks	Mitigation Approaches
Needs identification	Conceptualizing the wrong solution	Use domain experts to validate the solution
	Incorrect assumptions	Allow domain experts to validate the assumptions
Request for proposal	The proposal is poorly worded and does not communicate the intent	Have the proposal reviewed for clarity and conciseness
	The cost and schedule estimates to be used to when evaluating proposals are wrong	Use Ministry personnel who have successful[a] estimating experience on similar projects
Proposal	Do not receive at least three bids	Publicize the upcoming award to at least ten potential suppliers
	Proposal cost and schedule estimates differ to those developed by the Ministry	Find out the reason for the difference and take remedial action
Supplier sub-states[b]		
Systems requirements	Poorly written requirements	Ask for acceptance criteria to clarify requirements when they are written
	Poor cost and schedule estimation in planning	Use Ministry cost and schedule estimating experts who have experience on many similar projects
Subsystem construction state	Subsystems not completed according to schedule	Monitor progress using CRIP and ETL charts and assist in remedying problems
		Write penalty clauses in contract to discourage delays
	Subsystems when constructed do not meet specifications	Monitor the subsystem SDP and assist in remedying problems
Subsystem testing	Incomplete tests	Verify test procedure tests compliance to all requirements
	Subsystems fail tests	Earlier, monitor the subsystem SDP and assist in remedying problems
		Monitor tests and assist expediting replacements of failed parts if necessary.

[a] Estimates were within 20% of the actual cost.
[b] Not really required by the exercise but included as examples.

(Continued)

TABLE 7.12 (Continued)
ENU Traffic Light Upgrade Risks and Mitigation Approaches

State in SDP	Risks	Mitigation Approaches
Delivery, installation, integration and system testing	Damage during delivery	Package equipment according to the Ministry standard for shipping equipment (MOT-S-8699)[c]
	Failure during testing	Be able to restore system to unmodified state within 5 minutes so as not to impeded traffic flow

[c] In this instance, the number is hypothetical. In a real-world example, the specific standard number and version should be cited. If the entire standard does not apply, then the relevant sections shall be cited. The date is included to ensure the correct version is used.

TABLE 7.13
WP 50000 The Delivery, Installation and Final Testing State of the SDP

1	Unique WP ID	WP 50000	
2	Name of activity	Delivery, installation, integration and final testing	
3	Priority	1	
4	Narrative of the activity	The Ministry is in a monitoring role. The contractor's role is to deliver parts for the upgrade to the traffic light at ENU, install and integrate them into the new system. The upgrade is then tested and the site cleaned up following the successful test. If there is a failure, then the site is restored to its pre-upgrade state and a situational analysis will be made as to how to proceed.[a]	
5	Estimated schedule	The schedule is shown in Figure 7.11.	
6	Accuracy of schedule estimate	80% or ± 1 day	
7	Products	50000-900	The working upgraded system
		50000-260	The delivery report
		50000-261	Documentation that all outstanding bills have been paid
		50000-262	Documentation that all resources have been reallocated to other projects
8	Acceptance criteria for products	The Ministry of Transport signs off as the project has been completed in a satisfactory manner.	
9	Estimated cost	Cost of WP 50600	
		Cost of WP 51600	
		Cost of WP 51700	
		Cost of other WPs not supplied in this section	

[a] Note the use of perceptions from the *Continuum* HTP to cover the failure condition as part of the planning.

(Continued)

TABLE 7.13 (*Continued*)

WP 50000 The Delivery, Installation and Final Testing State of the SDP

10	The level of confidence in the cost estimate	±10% based on experience with similar projects	
11	Reason activity is being done	To complete the Ministry contract for a system to prevent complains about the waiting time at the ENU traffic light.	
12	Traceability	Original complaints endorsed by university provost.	
13	Prerequisites	40000-900	A set of subsystems that have met the subsystem tests and are ready for integration
		40100-700	Validated integration procedures
		40200-710	Validated test procedures
		40500-900	Test equipment specific in the test procedure is available
14	Resources	Mark Time: principal systems engineer Douglas Tomkins: contract specialist	
15	Person responsible for task	Mark Time	
16	Risks	ID 51600-R[b]	See WP 51600 row 17
		ID 51700-R	See WP 51700 row 17
		ID 50000-R	Damage during delivery
17	Risk mitigation	ID 51600-R	See WP 51600 row 17
		ID 51700-R	See WP 51700 row 17
		ID 50000-R	Mitigated in WP 41800
18	Lower level work package IDs (if any)	WP 50600	System installation and integration
		WP 51600	Acceptance test
		WP 51700	System delivery
19	Decision points (if any)	To proceed with test following installation and integration	
		To accept delivery following a successful test	
20	Internal key milestones	End of integration	
		End of acceptance test	
21	Assumptions not stated elsewhere	1. The contractor is delivering upgraded components that will be installed on-site at ENU. 2. Those parts have been tested and are performing according to their subsystem specifications.	
[b]	There is no need to repeat the information since the entire WP is supplied.		

7.13.1.1.8   The WPs in the Delivery, Installation, Integration and System Testing States of the SDP

Three WPs for the delivery, installation, integration and system testing states of the SDP in the supplier state are presented in the section. The WPs based on the information in Section 7.3.1.1.4 are provided in Tables 7.13–7.15. Note:

- The assumed links not required by the question are realistic but are greyed out to distinguish them from the links that are required.

TABLE 7.14
WP 50600 – The System Installation and Integration State

1	Unique WP ID	WP 50600			
2	Name of activity	System installation and integration			
3	Priority	1			
4	Narrative of the activity	The Ministry is in a monitoring role. The role of the contractor is to verify the upgrade is ready to be installed, then install and upgrade the traffic light system. Once integrated, the system will be operated for 3–4 hours with non-rush hour traffic.			
5	Estimated schedule	The schedule is shown in Figure 7.11.			
6	Accuracy of schedule estimate	80% or ± 1 day			
7	Products	50600-900	Installed system		
		50600-910	Integrated system		
		50600-700	Installed system report		
		50600-710	Integrated system report		
8	Acceptance criteria for products	The system has been integrated and operated for 3 hours			
9	Estimated cost	One reliability engineer for a day at $TBD[a]			
10	The level of confidence in the cost estimate	100% since no numbers are involved.			
11	Reason activity is being done	This is where it comes together			
12	Traceability	49000-290	Updated project plan		
13	Prerequisites		System upgrade components pass subsystem tests successfully		
14	Resources	Roger Lapin: reliability engineer			
15	Person responsible for task	Mark Time: principal systems engineer			
16	Risks	50600-R1	Not ready as scheduled	P = 1	S = 5
		50600-R2	Damage during delivery	P = 1	S = 5
		50600-R3	Damage during installation	P = 1	S = 5
17	Risk mitigation	50600-R1	WP 49000		
		50600-R2	WP 4910		
		50600-R3	WP 3500		
18	Lower level work package IDs (if any)	WP 51610	System installation		
		WP 51620	System integration		
19	Decision points (if any)	Between installation and integration – in case of a problem in installation			
20	Internal key milestones	Installation completed			
21	Assumptions not stated elsewhere	None			

[a] The exact amount is not needed for this exercise; what is needed is the item list to be costed in the real world.

TABLE 7.15

WP 51600 – The Final Acceptance Test

1	Unique WP ID		**WP** 51600		
2	Name of activity		Final acceptance test		
3	Priority		1		
4	Narrative of the activity		The system will be set up and activated. The operation will be monitored for 3 hours, noting that the countdown timers display correctly for pedestrians and traffic. The test will be filmed using two digital cameras. The system will be left in operation following the test. The 3 hours will be either morning or evening rush hour.		
5	Estimated schedule		One day including pre-test set-up and post-test clean-up.		
6	Accuracy of schedule estimate		99.9% since a working day has 8 hours and the test is scheduled for 3 hours.		
7	Products		51600-900	The working upgraded system	
			51600-701	The final acceptance test report	
			51600-702	The completed test procedure	
			51600-703	The discrepancy report (if needed)	
8	Acceptance criteria for products		Countdown timers work correctly and do not interfere with traffic flow Documents signed off by authorized personnel		
9	Estimated cost		One reliability engineer for a day at $TBD[a]		
			Cost of cameras at $TBD		
			Cost of use of the traffic systems equipment (vehicles, etc.) at $TBD		
10	The level of confidence in the cost estimate		100% since no numbers are involved.		
11	Reason activity is being done		Contractual requirement and demonstrate successful completion of the project		
12	Traceability		49000-290	Updated project plan	
13	Prerequisites		50600-710	Successful integration	
			49000-700	Acceptance test review (ATR)	
			41600-291	Validated test procedure	
14	Resources[b]		Roger Lapin: reliability engineer		
15	Person responsible for task		Mark Time: principal systems engineer		
16	Risks	ID 51600-1	Failure of countdown timer interferes with vehicular traffic flow	P = 1	S = 3[c]

[a] The exact amount is not needed for this exercise; what is needed is the item list to be costed in the real world.

[b] Contractor personnel; the Ministry of Transport does not get involved in the actual test.

[c] Severity is 3 because traffic disruptions for 5 minutes per hour are allowed although any disruption is undesirable. '

(Continued)

TABLE 7.15 (*Continued*)
WP 51600 – The Final Acceptance Test

		ID 51600-2	Risk of a traffic accident during the installation, integration and testing.[d]	P = 1	S = 5[e]
17	Risk mitigation	ID 51600-1	Provide a way to restore the system to the condition before the upgrade within 5 minutes[f] (Table 7.11).	Mitigated in WP to be determined (TBD)	
		ID 51600-2	Use of prescribed signage and safety signs including reduction of speed according to the Ministry of Transport regulations	Mitigated in WP TBD	
18	Lower level work package IDs (if any)		None		
19	Decision points (if any)		What to do if the system fails the test. This decision will be made jointly by the Ministry's contract specialist and the contractor and will depend on the type of failure.		
20	Internal key milestones		Set-up of test complete.		
			Completion of test.		
			Completion of post-test clean-up.		
21	Assumptions not stated elsewhere		None		

[d] Same risk as in installation, integration and any other on-site activity.
[e] Severity is 5 because an accident may cause serious damage to people and equipment up to death.
[f] Note how the down time is specified.

- The numbering of the states in the SDP is in accordance with *Systemic and Systematic Project Management* (Kasser 2019b: Section 5.5).
- The risks are shown in the WP in rows 16 and 17 to keep it simple because the requirement was for showing only two risks in each WP. In a real-world situation, the risks would be in a separate table or database and linked to the WP.
- The personnel shown in the WPs are contractor personnel.
- MBE limits are not included.

7.13.1.1.9 *Lessons Learned*

Lessons learned from the exercise include:

- The compliance matrix is very handy for keeping track of what needs to be done and making sure that it is being done.
- The information repeats in the exercise instruction, problem and solution in the problem formulation template. By using the same language, the risk of miscommunication by misinterpretation is eliminated.
- We need to know and understand the SDP (Section 6.1.5.4) before creating the project plan (Section 7.4.2.1) which points out the interdependency between project management and systems engineering.
- Since this project chose to purchase rather than develop the necessary equipment, the standard SDP (Section 6.1.5.4) had to be modified (Section 6.1.5.5).
- It wasn't easy to make everything link together; something which has to be done to minimize the risk of missing a vital task and hence mitigate the risk of project failure.
- The modified SLC for a COTS acquisition is an extended SLC for a developed system since the SDP is embedded in the supplier state.

7.13.2 The Engaporean Maid Reduction Policy

The Ministry of Human Resources is about to release a policy to reduce the number of maids by the amount over a period of time identified in Section 4.10.2.1.3. In addition to meeting the generic requirements in Section 2.7:

1. Format the problem posed by this exercise in the problem formulation template (Section 5.1.7.2).
2. Create an outline project plan for the maid reduction documenting the process listing the steps in reverse from the time the number of maids to the target number of the policy back to the time of creating the policy.
3. State at least two generic risks in each state of the maid reduction process.
4. Show how the risks will be mitigated or prevented.
5. Break the work into at least three WPs, and make reasonable and realistic assumptions for the information in each WP.
6. Show linkages between the WPs.

7.13.3 The Engaporian Drought Relief Policy

The Ministry of Tourism is about to release a policy to ask tourists voluntarily bring at least two bottles of water with them when visiting the country. In addition to meeting the generic requirements in Section 2.7:

1. Format the problem posed by this exercise in the problem formulation template (Section 5.1.7.2).
2. Create an outline project plan for introducing the policy and administering it for at least one year, listing the steps in reverse from the time the target number is achieved back to the initialization of the policy.
3. State at least two generic risks in each state of the plan.

4. Show how the risks will be mitigated or prevented.
5. Break the work into at least three WPs, and make reasonable and realistic assumptions for the information in each WP.
6. Show linkages between the WPs.

7.13.4 THE OFF-WORLD MINING POLICY

The Ministry of Science is about to release its vision for off-world mining. In addition to meeting the generic requirements in Section 2.7:

1. Format the problem posed by this exercise in the problem formulation template (Section 5.1.7.2).
2. Create a very broad outline plan for introducing the vision and administering it until the first five loads have been transferred from space to the Earth, listing the steps in reverse from the landing of the load back to the introduction of the vision.
3. State at least two generic risks in each state of the plan.
4. Show how the risks will be mitigated or prevented.
5. Break the work into at least three WPs, and make reasonable and realistic assumptions for the information in each WP.
6. Show linkages between the WPs.

7.14 SUMMARY

This chapter discussed project management from different perspectives because project management is the heart of effecting any change, caused by the implementation of a policy, a LEP or any type of project. In particular, this chapter provided an overview of the systems approach to projects and project management discussing pure and applied project management, projects, the project lifecycle, the project plan, generic planning, specific planning, the planning process, CRIP charts, ETL charts, MBE, MBO, using prevention to lower project completion risk, stakeholders and stakeholder management and software project risks, and concluded by discussing generic project risks.

REFERENCES

Adams, Scott. 1996. *The Dilbert Principle*. New York: HarperBusiness.
Augustine, Norman R. 1986. *Augustine's Laws*. New York: Viking Penguin Inc.
Bittel, Lester R. 1964. *Management by Exception; Systematizing and Simplifying the Managerial Job*. New York: McGraw-Hill.
Businessdictionary. 2018. *Feasibility study*. WebFinance Inc 2018 [cited November 12, 2018]. Available from http://www.businessdictionary.com/definition/feasibility-study.html.
Chacko, George K. 1989. *The Systems Approach to Problem Solving*. New York, NY: Prager.
CHAOS. 1995. *The Chaos study*. The Standish Group [cited March 19, 1998].
Checkland, Peter, and Jim Scholes. 1990. *Soft Systems Methodology in Action*. New York: John Wiley & Sons.

Chung, Kon Shing Kenneth, and Lynn Crawford. 2015. The role of social networks theory and methodology for project stakeholder management. In *29th World Congress International Project Management Association (IPMA) 2015, IPMA WC2015*. Westin Playa Bonita, Panama.

Cleverism. 2018. *How to conduct a feasibility study the right way*. Cleverism 2018 [cited November 12, 2018]. Available from https://www.cleverism.com/conduct-feasibility-study-right-way/.

Covey, Steven R. 1989. *The Seven Habits of Highly Effective People*. New York: Simon & Schuster.

Crosby, Philip B. 1981. *The Art of Getting Your Own Sweet Way*. 2nd edition. New York: McGraw-Hill Book Company.

Deming, W. Edwards. 1986. *Out of the Crisis*. Cambridge, MA: MIT Center for Advanced Engineering Study.

Deming, W. Edwards. 1993. *The New Economics for Industry, Government, Education*. Cambridge, MA: MIT Center for Advanced Engineering Study.

Dictionary.com. 2013. *Dictionary.com*. Available from http://dictionary.reference.com/.

Donaldson, Scott E., and Stanley G. Siegel. 1997. *Cultivating Successful Software Development a Practitioner's View*. Upper Saddle River, NJ: Prentice Hall PTR.

Drucker, Peter Ferdinand. 1954. *The Practice of Management*. New York: Harper.

Fayol, Henri. 1949. *General and Industrial Management*. London: Sir Isaac Pitman and Sons, Ltd.

Hague, Rod, and Martin Harrop. 2013. *Comparative Government and Politics: An Introduction*. London: Macmillan International Higher Education.

Harrington, H. James. 1995. *Total Improvement Management the Next Generation in Performance Improvement*. New York: McGraw-Hill.

Hofstrand, Don, and Mary Holz-Clause. 2009. *What is a feasibility study?* Iowa State University Extension and Outreach 2009 [cited November 12, 2018]. Available from https://www.extension.iastate.edu/agdm/wholefarm/html/c5-65.html.

Investopedia. 2018. Feasibility study. Investopedia 2018 [cited November 12, 2018]. Available from https://www.investopedia.com/terms/f/feasibility-study.asp.

Irwin, Brian. 2007. Politics, leadership, and the art of relating to your project team. In *PMI® Global Congress 2007-North America*. Atlanta, GA; Newtown Square, PA: Project Management Institute.

Kasser, Joseph Eli. 1995. *Applying Total Quality Management to Systems Engineering*. Boston, MA: Artech House.

Kasser, Joseph Eli. 1999. Using organizational engineering to build defect free systems, on schedule and within budget. In *PICMET*. Portland OR.

Kasser, Joseph Eli. 2000. A framework for requirements engineering in a digital integrated environment (FREDIE). In *The systems engineering, test and evaluation conference (SETE)*. Brisbane, Australia.

Kasser, Joseph Eli. 2018a. *Systems Thinker's Toolbox: Tools for Managing Complexity*. Boca Raton, FL: CRC Press.

Kasser, Joseph Eli. 2018b. Using the systems thinker's toolbox to tackle complexity (complex problems). In *SSSE Presentation at Roche*. Zurich, Switzerland: Swiss Society of Systems Engineering.

Kasser, Joseph Eli. 2019a. *Systemic and Systematic Project Management*. Boca Raton, FL: CRC Press.

Kasser, Joseph Eli. 2019b. *Systems Engineering: A Systemic and Systematic Methodology for Solving Complex Problems*. Boca Raton, FL: CRC Press.

Kasser, Joseph Eli, and Victoria Regina Williams. 1998. What do you mean you can't tell me if my project is in trouble? In the *First European Conference on Software Metrics (FESMA 98)*, at Antwerp, Belgium.

Kast, Fremont E., and James E. Rosenzweig. 1979. *Organization and Management a Systems and Contingency Approach*. 3rd edition. New York: McGraw-Hill Book Company.

Kezsbom, Deborah S., D. L. Schilling, and K. A. Edward. 1989. *Dynamic Project Management. A Practical Guide for Managers and Engineers*. New York: John Wiley & Sons.

Larson, Erik W., and Clifford F. Gray. 2011. *Project Management the Managerial Process*. 5th edition. The McGraw-Hill/Erwin series operations and decision sciences. New York: McGraw-Hill.

Mali, Paul. 1972. *Managing by Objectives*. New York: John Wiley & Sons Inc.

Malotaux, Neils. 2010. *Predictable Projects Delivering the Right Result at the Right Time*. Keo, Japan: N. R. Malotaux Consultancy.

Mar, Ann. 2019. *130 Project risks (list)*. Simplicable 2016 [cited August 30, 2019]. Available from https://management.simplicable.com/management/new/130-project-risks.

Maslow, Abraham Harold. 1968. *Toward a Psychology of Being*. 2nd edition. New York: Van Nostrand.

Mills, Henry Robert. 1953. *Techniques of Technical Training*. London: Cleaver-Hume Press.

OAS. 2013. *Analysis of Alternatives (AoA) Handbook, a Practical Guide to Analyses of Alternatives*. edited by Air Force Materiel Command (AFMC) Office of Aerospace Studies. Kirtland AFB, NM: Air Force Materiel Command (AFMC) OAS/A5.

Palmer, Stephen R., and John M. Felsing. 2002. *A Practical Guide to Feature – Driven Development*. Upper Saddle River, NJ: Prentice Hall.

Pinto, Jeffrey K. 2000. Understanding the role of politics in successful project management. *International Journal of Project Management* no. 18 (2):85–91. doi: 10.1016/S0263-7863(98)00073-8.

PMI. 2013. *A Guide to the Project Management Body of Knowledge*. 5th edition. Newtown Square, PA: Project Management Institute, Inc.

PMI. 2017. *A Guide to the Project Management Body of Knowledge*. 6th edition. Newtown Square, PA: Project Management Institute, Inc.

Reardon, Kathleen K. 2005. *It's All Politics: Winning in a World Where Hard Work and Talent Aren't Enough*. New York, NY: Doubleday.

Royce, Winston W. 1970. Managing the development of large software systems. Paper read at *IEEE WESCON*, at Los Angeles.

Savage, Sam L. 2009. *The Flaw of Averages*. New York: John Wiley & Sons.

Thomas, Kenneth W., and Ralph H. Kilmann. 1974. *Thomas-Kilmann Conflict Mode Instrument*. Tuxedo, NY: XICOM.

Ullman, David G. 2009. *Decisions based on analysis of alternatives (AoA)*. AcqNotes.com [cited October 29, 2018]. Available from www.acqnotes.com/Attachments/Decisions BAsed on AOA.pdf.

VOYAGES. 1996. *Unfinished Voyages, A follow up to the CHAOS Report 1996* [cited January 21, 2002].

8 An Introduction to Managing Risk in LEPs

LEPs, such as airports, urban-transport systems, oil fields, and power systems, constitute one of the most important business sectors in the world. These projects tend to be massive, indivisible and long term artifacts, with investment taking place in waves. Their effects are felt over many years, especially as auxiliary and complementary additions are made.

Miller and Lessard (2000)

This chapter provides an introduction to large engineering projects (LEPs) and managing risks in LEPs. A LEP contains a hierarchy of projects and systems. In particular, the chapter discusses:

1. LEPs in Section 8.1.
2. Risks in LEPs in Section 8.2.
3. The LEP lifecycle in Section 8.3.
4. Risks based on the non-availability of technology in Section 8.4 technology since technology has become ubiquitous in the 21st century.

8.1 LARGE ENGINEERING PROJECTS

LEPs:

- Take place in the higher layers of the risk framework (Section 1.2), and according to the principle of hierarchies (Section 6.3.1) need a different management style to projects in the lower layers.
- Are generally complex, take a long time, will experience changes and need flexible management. For example:

> Of the various elements that combine to make long-duration projects complex, the most significant is the inevitable changes that will occur in the business environment, which will necessitate adjustments to virtually all elements of the project. Knowing this, the successful project leadership team evolves, practicing situational project leadership, adapting and modifying their approach to accommodate the inevitable changes. In addition to adapting to change, the sheer size of the work involved for large projects weighs heavy on the project team. Research has demonstrated that the smaller the project team and the fewer deliverables, the greater the likelihood of project success. Therefore, the project leadership teams need to reduce the size of work packages to 'seem like' many small projects, as opposed to one very large endeavor. As a final point, team

fatigue and burnout lead to complex human interactions and unavoidable staff
turnover, both of which are difficult to predict and manage.

Hass (2019)

Breaking a project or system into sub-projects or subsystems is systems engineering –
the activity (SETA) (Section 6.5.4). Research shows that SETA was recognized as a
contributor to the success of LEPs such as the:

- The Semiautomatic Ground Environment (SAGE) project, a computer and
  radar-based air defence systems created in the US in the 1950s (Hughes
  1998: p. 15). SAGE was a massive networked system of radars, anti-aircraft
  guns and computers.
- The ICBM development of the 1950s where 'systems engineering was the
  methodology used to manage the problem of scheduling and coordinating
  hundreds of contractors developing hundreds – even thousands – of subsystems that eventually would be meshed into a total system' (Hughes 1998:
  p. 118).
- The Public Housing System, industrial development and the air defence
  system (ADS) in Singapore (Lui 2007).

8.2   RISKS IN LEPs

The generic risks faced by a LEP can be sorted into three categories as follows
(Miller and Lessard 2000):

1. Completion risks discussed in Section 8.2.1.
2. Market-related risks discussed in Section 8.2.2.
3. Institutional risks discussed in Section 8.2.3.

8.2.1   COMPLETION RISKS

Perceptions from the *Temporal* HTP noted that completion risks may be grouped as[*]:

- *Construction risks*: basically, the risks associated with actually performing
  the project. These risks can be technical, schedule or people-related.
- *Operational risks*: based on the system produced by the LEP not functioning correctly. These risks can be substantially reduced by using good systems engineering to design redundancy, reliability, robustness, safety and
  security into the system in the early states of the system development process (SDP) (Section 6.1.5.4).
- *Schedule risks*: planning shall be based on identifying risks and ensuring
  that the proposed plan is feasible rather than planning the project down to
  the lowest level of detail during the project planning state (Section 7.4.2).

---

[*] Note the acronym COST for the four types of risks.

Since plans have a high probability of being changed, detailed planning is a waste of time, for example:

- Given the complexity of the organization in a LEP, the probability of stakeholders not meeting their commitments is much higher than in a smaller project which will result in replanning.
- Unexpected environmental conditions may cause the direction of the project to change rendering all prior detailed planning useless. For example, in a tunnel boring project, unexpected underground conditions may require a change in the way the tunnel is bored.

- *Technology risks*: technology has a lifecycle. Accordingly, if the project takes a long time to mature, there is a risk that technology that is current at the start of the project will be obsolete or even unavailable when the project ends. One tool to mitigate this risk is the technology availability window of opportunity (TAWOO) (Section 8.5).

LEPs can take advantage of size and the use of multiple sources to mitigate the risk of sponsors and other stakeholders not meeting commitments by using a system's concept known as no single point of failure. For example, to mitigate the risk of not having any resource that is purchased or provided by stakeholder such as components and funds, the LEP manager can identify more than one supplier. The suppliers can be used:

- In parallel where each supplier provides a part of the resource.
- In series where one supplier is preferred and provides the entire resource until the point is reached where the supplier cannot meet the commitments, and then the next potential supplier of that resource is contacted to provide the balance of the resource required by the LEP.

However, there will be a slight increase in the complexity of the project management due to the need to identify and manage the additional sources.

A further modification is to use a multiple award scenario, where a pool of suppliers is identified and then asked to bid on specific resources for a specific time. This builds robustness and redundancy into the LEP supply system. A further incentive might guarantee potential members of the pool a small percentage of the entire contract as an incentive to join the pool. The multiple award scenario can also be used to meet sovereign requirements for set-aside contracts such as the need to provide a certain percentage of the contract work to 'small and small and disadvantaged businesses' (Kasser 1997). .

8.2.2 Market-Related Risks

Market-related risks:

- Are based on the assumptions about the structure of demand for the product or service provided by the system created by the LEP once in operation.

- Include financial difficulties (financial risks) that any project faces in attracting the finance to complete the project (lenders and investors) given its potential ROI.
- Include the inability to restructure financial arrangements in the event of unexpected changes in cash flows or the sudden need for an infusion of cash due to some external events.
- Are based on the availability or rather the unavailability of material in that the supplier becomes unable to provide at the agreed-upon price or even at all (supply risks).

8.2.3 INSTITUTIONAL RISKS

Institutional risks include:

- *Regulatory risks*: laws and regulations change and can be made to change by relevant stakeholders
- *Sector rules*: pertaining to specific sectors which may be different to the sector rules in another sector.
- *Social risks*: risks that sponsors will meet opposition to the project from local groups, government agencies and other influential pressure groups.
- *Sovereign risks*: the likelihood that a government will decide to renegotiate contracts, award contracts to others that impact the ROI and other issues once the LEP has begun.
- Stakeholder risks: such as:
  - Stakeholders refusing to honour their commitments.
  - Stakeholders failing to understand the linkages between the stakeholders.
  - Stakeholders perceiving the LEP from their perspective (Weltanschauung) and failing to perceive the project from the perspectives of other stakeholders.
  - Stakeholders failing to understand that other stakeholders do not share their perspective.
  - Stakeholders failing to understand the risks in committing to use technology that is still in development when the LEP begins, but is expected to reach maturity in time to be incorporated in the system being developed.
  - Stakeholders failing to understand the environment (laws, regulations, sector issues, etc.).

The key to mitigating and preventing these failures is communications (as in a transfer of knowledge) which often includes the right type of contracts, negotiation and conflict management starting in the project initiation state and continuing through the remaining states. Accordingly:

- Contracts are discussed in Section 8.2.4.
- Negotiation is discussed in Section 8.2.6.
- Conflict management is discussed in Section 7.3.1.2.
- Communications as in communicating with the public is discussed in Section 3.2.2.

8.2.4   CONTRACTS

Since LEPs are typically performed within the framework of contracts. Such contracts may be between buyers and suppliers as well as between stakeholders who might be both a buyer and a supplier at the same time.[*] Effective LEP (and smaller project) sponsors and managers must have an understanding of the contract framework for the project, and the purpose of this section is to provide it. A contract:

- Is a term descriptive of various agreements including ones for procurement of supplies or services (Sherman 1987).
- Is the overt manifestation of an agreement, but it is not the agreement.
- Documents who (supplier) is to supply what (product or service) to whom (customer), the amount of the target price (Section 8.2.4.3.2) and how the payment is to be made.
- Documents any other legal arrangement between the parties.
- May be:
  1. *Fixed price*: supplier and customer agree on a fixed price for the work to be performed (Section 8.2.4.1).
  2. *Cost-plus*: also known as time and materials (T&M). These contracts come in several variations and provide for the payment of reasonable, allowable and allocatable costs incurred by the supplier in the performance of the contract (Section 8.2.4.2).

Uncontrolled escalating costs are a major risk in any project. By understanding the contractual background to projects, the sponsor, project manager and other stakeholders can use the rules in a cost-effective manner. Each type of contract has its advantages and disadvantages, yet the goal of reducing the costs of production applies to each of them.

From the customer's perspective, the choice as to which type of contract to use is determined by the element of risk (the probability of successful completion of the contract) and who pays for it (customer or supplier). Contracts in which there is risk, namely there are known unknowns in the performance of the contract, tend to be cost-plus contracts. Contracts in which there is little or no risk, namely the job is simple or the job has been done before, tend to be fixed price contracts.

A fixed price contract is negotiated based on the customer being willing to pay the supplier more than the cost to produce the product.

- From the supplier's perspective, the price can be based on the:
  - Estimated costs plus a profit factor.
  - Going market rate based on the customer's willingness to pay.
- From the customers' perspective, the price is always based on the amount that the customer is willing to pay.

---

[*] A stakeholder might provide something (be a supplier) while at the same time or even at a later time accept something (be a buyer) from another stakeholder.

In a fixed price contract, cost overruns lead to a smaller profit, excessive cost overruns lead to a loss and really excessive overruns can lead to bankruptcy. In a cost-plus contract, cost overruns are generally covered by the supplier until the supplier cancels the contract.

8.2.4.1 Firm Fixed Price Contracts

Firm fixed price (FFP) contracts are mainly used when everything about the product is known. There is no need for development work, but it is purely a matter of meeting the production schedule. The customer is willing to pay the seller's asking price.

Many commercial contracts are FFP. While accurate estimates and cost-effective implementation are important at all times, in an FFP environment, they are critical. If the cost escalates above the contractual ceiling before completion, everybody is going to be unhappy. The systems approach to contracting always includes penalties for poor performance such as late completion and incentive bonuses for early completion. The thresholds are stipulated such that if performance is nominal, the penalty clauses will never be invoked, and if performance is average, the incentive bonuses will not be paid. Perceptions from the *Generic* HTP note that this type of FFP may be thought of as a FFP with penalties and bonuses by exception contract, somewhat similar to MBE (Section 7.7) in that the action is triggered when the threshold is exceeded.

8.2.4.2 Cost-Plus Contracts

Cost-plus contracts share the following characteristics:

- *Reimbursement of contractor's costs*: During the performance of the contract, the contractor is reimbursed for costs expended on material and paid for the work performed.
- *Fee for performance*: provides the contractor's profit. This is the incentive part (motivation).

Several different types of cost-plus contracts each with slight variations have been developed over years including the:

1. Cost-plus fixed fee (CPFF) discussed in Section 8.2.4.2.1.
2. Cost-plus incentive fee (CPIF) discussed in Section 8.2.4.2.2.
3. Cost-plus award fee (CPAF) discussed in Section 8.2.4.3.

Each of the various types of cost-plus contracts uses a different method of calculating the fee.

8.2.4.2.1 The CPFF

The CPFF is a contract type that reimburses the contractor for all allowable costs and includes a fee which is a percentage of the cost. The fee is expressed as a fixed dollar amount at the time of contract issuance. Since the fee is fixed, the amount of the fee will not change, irrespective of whether the contractor overruns or underruns the target costs. Because of this structure, this contract type represents the lowest level of risk to the contractor and provides little incentive for the contractor to manage costs.

The CPFF was the first cost-type contract. The fee the contractor receives is based on the actual work performed. If the contractor works effectively and produces the product working fewer hours, the contractor gets a smaller fee. This situation tends to encourage ineffective working; for example, the contractor gets paid to produce the defects and then gets paid to fix them.

8.2.4.2.2 The CPIF

The CPIF contract represents an attempt to provide an incentive to the contractor to manage costs. The customer and the contractor negotiate an estimated or target cost of the contract prior to contract start.

The amount of fee earned by the contractor varies depending on how well the actual costs of work performed compare to the estimated target costs. When actual costs are much lower and much higher than the target costs, this contract type acts as a CPFF contract. However, in the range close to the target costs, the fee is a variable percentage of the actual costs. If the contractor manages to conclude the contract at an amount less than the target cost, the fee will be higher than if the contractor concludes the contract above the target cost. The CPIF provides a powerful incentive to 'do it right the first time'. The CPIF seemed like a win–win situation. However, it soon became evident that while CPIF contracts worked to control costs, they were cumbersome to administer and difficult to change.

8.2.4.3 The CPAF

The CPAF contract is the best of the 'cost plus' approaches because it has the capability to build incentives towards high performance into the structure of the contract. CPAF contracts are considered cost-reimbursement contracts A CPAF contract contains the following elements:

1. The statement of work discussed in Section 8.2.4.3.1.
2. The target price discussed in Section 8.2.4.3.2.
3. The minimum fee discussed in Section 8.2.4.3.3.
4. The award fee discussed in Section 8.2.4.3.4.
5. The award fee evaluation criteria discussed in Section 8.2.4.3.5.

The CPAF contract has several advantages. They include:

- *Flexibility in changing requirements*: the CPAF contract can be adapted or changed to reflect new or unplanned events (requirement changes) during the life of a contract more easily than other types of contracts. The relative ease of making changes is due to the increased communication between the customer and the supplier, an integral and necessary part of administration of a CPAF contract.
- *Tendency towards improved quality and lower cost*: this tendency results from judicious selection of appropriate objective award fee determination

criteria as discussed above. Since both the customer and the supplier tend to know what the budget can support, they can work together to provide the maximum capabilities the budget can afford or, as it is commonly called, 'design to cost'.

8.2.4.3.1 The Statement of Work

The statement of work is a narrative describing the work activities to be competed, the funding, and product specifications (Kezsbom, Schilling, and Edward 1989). The statement of work is also used in many FFP contracts, software development contracts and consulting agreements as well as in CPAF contracts. The statement of work must be structured to indicate precisely and clearly what the customer anticipates in the way of results without saying how to achieve them.

The tasks which then evolve from planning the implementation of the work described in the statement of work must be structured so as to present a series of finite short-term objectives that the supplier can aim to achieve. The assumption is that completion of each objective (by the supplier), and the acceptance thereof (by the customer), represents a degree of success along the road to completion of the contract on the part of the supplier as well as allowing for MBO (Section 7.8) by the supplier.

8.2.4.3.2 The Target Price

The estimated or target price may be defined as that price which both the supplier and the customer feel to be the most realistic, the best estimate and the most likely ultimate outcome based on the existing knowledge of the work to be done. The customer needs to estimate costs to determine if the supplier's quote or bid is realistic. The target price may be estimated in two ways:

1. *Price analysis*: a market-based approach. The supplier estimates the costs to produce the product and then estimates the price the customer is willing to pay and sets the target price accordingly. If the supplier can produce the product for less than the target price, the supplier decides to offer the product for sale. In general, if the cost to produce the product is greater than the (price analysis based) target price, the supplier will most probably not offer the product for sale.
2. *Cost analysis*: the supplier estimates the costs to produce the product and adds a percentage for profit, and the result is the target price.

In general, a supplier develops the target price for performing a contract by estimating the sum of the:

1. Cost of the work to be performed based on current labour rates.
2. Cost of the raw material used in manufacturing the product.
3. Cost of the overhead associated with the contractor's way of doing work.
4. Cost of marketing, sales, etc.
5. Amount of profit (fee) desired.

6. A reduction factor, depending on how badly the supplier needs the work. From time to time, a supplier does price something below cost for a specific customer, perhaps to spread the development cost over a number of customers, or to find a first customer to establish a track record to enter a new market. In the past, in US government contracts, contractors who used this approach generally recovered their costs when changes to the requirements were costed.

8.2.4.3.3 The Minimum Fee

CPAF contracts contain a contractually stated base or minimum fee. This fee may range upwards from zero to an agreed amount.

8.2.4.3.4 The Award Fee

The award fee is structured to motivate the supplier to meet the schedule and performance levels and to hold costs to a minimum. In US government CPAFs, the amount of the award fee to be paid is determined (and paid) periodically by the government's judgmental evaluation of the contractor's performance in terms of a set of evaluation criteria agreed upon by the contractor and the government. This determination is made unilaterally by the fee determination official.

The amount of award fee a contractor will receive is based on periodic evaluations made during the life of the contract. The contractor's efforts must only be evaluated on the basis of those factors which are under the contractor's control. For example, failure on the part of another support contractor to deliver equipment as scheduled may be beyond the control of the contractor being evaluated.

8.2.4.3.5 The Award Fee Evaluation Criteria

Evaluation criteria must be both meaningful and achievable. If they are not, inappropriate parameters may be used and the evaluation made on something that does not relate to contract performance. Perhaps the greatest challenge in setting up CPAF contracts lies in determining the evaluation criteria. Evaluations may be:

1. *Subjective*: relating to the inability to spell out the requirements in a given performance situation in a measurable and repeatable manner. Evaluation of contractor performance is made on the basis of a subjective analysis which reflects the evaluator's opinions and impressions as to the level of performance achieved by the contractor.
2. *Objective*: based on well-defined criteria that they are physically or mathematically measurable and are defined. This ensures that different individuals evaluating a contractor in a specific endeavour will assign approximately the same grade in any specific situation.

Unfortunately, there is no universal single set of evaluation criteria for all CPAF contracts. So, when negotiating the criteria for a specific period before the start of the period, both parties must consider the following questions:

1. Are they achievable, realistic, measurable and meaningful?
2. Are they positive motivators?

As stated above, the fee is awarded periodically based on evaluated performance. To optimize performance, the customer and supplier can mutually agree on the evaluation criteria for each period before the period begins. The supplier can then communicate the evaluation criteria to the people doing the work, so they know not only what they have to do, but how they will be graded on their work. When the supplier's employees can see that their effort, or lack of effort, has an effect on the bottom line, they are more conscientious about receiving a high rating (Webb 1992). Perceived from the *Generic* HTP, this practice is in the spirit of the continuous process improvement paradigm of getting all employees involved in the process. Crosby defined Quality as 'compliance to requirements' (Crosby 1979). Assuming the evaluation criteria are based on completing the requirements, perhaps as shown in the CRIP charts (Section 7.5), the total costs will increase as work is performed until the requirements are met. From that point on, any further expenditure represents inefficient use of funds. Tweaking the product is not cost effective, since the product already meets requirements. The advantage of 'doing it right the first time' can also be seen. In this situation, the requirements will be met sooner and at a lower total cost. Here, 'working smart' and 'doing right the first time' can be seen to have a payoff.

8.2.4.4 Disputes and Penalty Clauses

All contracts should include disputes and penalty clauses. The more complex a contract, the greater the probability of a dispute. The purpose of penalty clauses is to:

- Provide an incentive to all parties to comply with a contract.
- Minimize disputes in the event of non-compliance by documenting the exact nature of non-compliance. Perceptions from the *Generic* HTP note the similarity to the need for clear and concise requirements in systems engineering and the need for clear and concisely written policy documents (Section 9.8.1).

The actions that trigger the penalties and the amount of the penalties should be set by the supplier and the customer so that if everything goes well, the penalties are never invoked. Perceptions from the *Generic* HTP note the similarity to MBE (Section 7.7). If a supplier baulks at the inclusion of the penalty clause, then there is a probability (risk) that the supplier knows that there is a risk that it will not be able to comply with the aspect of the contract covered by the penalty clause such as timeliness of delivery and quantity to be delivered. This is all the more reason for the customer to mitigate that risk by insisting on including the penalty clause.

Disputes tend to arise when either of the parties is not satisfied with the performance of the other. There are four basic precautions that mitigate the risk of disputes (Gaitskell 2011: p. 7):

1. Ensure that the parties enter into a finalized contract before work commences.
2. Ensure that the scope and the quality of the work are clearly defined at the precontract stage.

3. Ensure that the contract terms are fair and clear and utilize standard forms of contracts where possible.
4. Ensure that the contract contains provision for the early notification of potential disputes and a well-structured dispute resolution clause that stipulates the type of action to be taken in the event the parties cannot negotiate a settlement to the dispute. Such action includes arbitration, adjudication and mediation as alternatives to court litigation.

8.2.5   UNCONTROLLED CHANGES

Uncontrolled changes accepted informally have a risk of causing cost and schedule escalations, hence the importance of a well-understood change management (CM) process. In an open communications environment in which the customer is part of the development team, it is very easy for engineering personnel used to working in poorly managed contracts to accept informal change requests with little or no thought. In any contract environment, it is critical for the CM process (Section 4.9.2) to determine who pays for a change. The additional costs incurred due to a series of informal changes, if absorbed by the contractor, can make the difference between profit and loss.

8.2.6   NEGOTIATION

Project managers negotiate with customers, suppliers, team members and other stakeholders. Negotiation:

- Takes place in the context of a conflict generally in the form of a dialogue.
- Can only be successful if both parties in a conflict are willing to negotiate.
- The risk in negotiation is the risk of not coming out of the negotiation with the minimum position.

8.2.6.1   Negotiation Positions

When entering into a negotiation, you should always have two positions for each item being negotiated:

1. The starting position which you share with your opponent which states:
   1. What you want.
   2. What you're willing to offer in return.
2. A minimum position which you do not share with your opponent which is:
   1. What you really want which should be less than your starting position.
   2. What you are really willing to offer which should be more than your starting position.

These two criteria should enable you to move away from your starting position in the opponent's favour and still come away from the negotiation with what you really want.

8.2.6.2 Negotiation Outcomes

Perceptions from the *Continuum* HTP note that there are three possible outcomes from a negotiation:

1. *Win–win*: both parties come out of the negotiation with what they really want.
2. *Win–lose*: one party comes out of the negotiation with what it really wants the other does not.
3. *Lose-lose*: neither party comes out of the negotiation with all of what they really want.

If the negotiation covers a number of different items or issues, by recognizing the different parties want different outcomes it is generally possible to achieve a win–win outcome.

8.2.6.3 Negotiating Styles

Different people have different negotiating styles. Perceptions from the *Generic* HTP show that these are similar to the styles adopted in conflict resolution which should not be a surprise since negotiation is one of the ways of resolving conflicts (Section 7.3.1.2). One set of five negotiating styles is (Shell 2006):

1. *Accommodating*: enjoying solving the other party's problems and preserving personal relationships.
2. *Avoiding*: disliking negotiating and not doing it unless warranted.
3. *Collaborating*: enjoying negotiations that involve solving tough problems in creative ways.
4. *Competing*: enjoying negotiations because they present an opportunity to win something.
5. *Compromising*: eager to close the deal by doing what is fair and equal for all parties involved.

8.2.6.4 Negotiating Tips

Here are some tips for negotiating to lower the risk of an unsuccessful negotiation by increasing the probability of a successful outcome from the negotiation:

- Understand what is being negotiated and why.
- Determine both your negotiating positions for all issues and items to be negotiated.
- Understand what your opponent is likely to really want to determine if the negotiation has a chance of being successful.
- If you feel your opponent is going for the lose–lose option, find out why. It may be due to a misunderstanding which you can correct or for an emotional reason which you can do nothing about.
- Understand what your items or issues opponent does not care about.
- Understand the cultural differences between the parties and how they might affect the negotiation.

- Find out the negotiating style of your opponent and adopt the same style or delegate the negotiation to somebody who can adopt the same style.
- Be prepared to walk away.*
- Make sure person who has the final say on the agreement on behalf of both parties is present at the negotiation.† If that person is not present, the negotiations cannot be concluded.
- When negotiating multiple items or issues, don't agree on one item at a time. It leaves no scope for give and take on the last few items, and that is where negotiations can break down.
- Speak your opponent's language or delegate the negotiation to somebody who can.
- Read up on the topic before going into serious negotiations.

Understanding both wants and don't cares can be critical to achieving a successful negotiation. A simple example might be a husband and wife negotiating the type of new car to purchase. The husband wants a sports car and doesn't care about anything else; the wife wants a red car and doesn't care about anything else. Accordingly, a successful win-win outcome of the negotiation is an agreement to purchase a red sports car.

8.3   THE LEP LIFECYCLE

Perceptions from the *Generic* HTP note that the four states of the LEP lifecycle are the same as the four states of the project lifecycle (Section 7.4), namely:

1. The initialization state discussed in Section 8.3.1.
2. The planning state discussed in Section 8.3.2.
3. The performance state discussed in Section 8.3.3.
4. The closeout state discussed in Section 8.3.4.

The systems approach builds risk management into each state of the LEP lifecycle in almost the same way as it builds risk management into the project lifecycle. Part of the approach is to ensure that the core team competency remains steady during the entire LEP although the personnel may change for various reasons including:

1. Expertise used in one state of the lifecycle is no longer needed in the subsequent states.
2. The project takes a long time and people leave to take up other positions due to promotions, retirement, higher salaries in other organizations, etc.

---

* This is a great tactic when shopping in a location where bargaining is expected. I've noticed asking prices dropping markedly the closer I got to the door when walking away.
† Or reachable by telephone from the negotiation site in real time.

8.3.1 THE LEP INITIALIZATION STATE

This is the state in which the core project team is created, the stakeholders are identified and the basic strategy is developed. It is important that the core team consider lessons learned from previous LEPs during the initialization state to take advantages of improvements made in those LEPs and avoid repeating the mistakes made in this LEP.

> Projects fail not because they are complicated, but because they face dynamic complexity. Rising to the challenge of large projects calls for shaping them during a lengthy front-end period.
>
> **Miller and Lessard (2000).**

The risks in this state are those that generally prevent the project sponsor from doing the following eight rights:

1. Deciding on the *right* project discussed in Section 8.3.1.1.
2. Identifying, mitigating and preventing the *right* risks[*] discussed in Section 8.3.1.2.
3. Assembling the *right* mixture of sponsors, bankers and other stakeholders.
4. Creating the *right* organizational structure for the LEP.
5. Creating the *right* core project team.
6. Obtaining the *right* amount of funds, time and other resources from the stakeholders.
7. Identifying, employing and managing the *right* consultants.
8. Developing the *right* lasting political connections.

8.3.1.1 Deciding on the Right Project

Successful project sponsors have the ability to make a rapid decision as to whether a proposed project has value.[†] They can quickly evaluate proposals based on prior experience and knowledge. They may form taskforces consisting of lawyers, consultants and investment bankers and take the time to do a feasibility study (Section 7.4.1.1) on whether the proposal is worth implementing.

8.3.1.2 Identifying, Mitigating and Preventing the Right Risks

Successful sponsors need strong assets and a diversified portfolio of projects producing revenue streams that make it possible to survive a cash crunch on a particular project. Risks originate in markets, technology, finance, logistics, management, type of organization, social responses to the project and governments.

8.3.2 THE LEP PLANNING STATE

The systems approach to project planning (Kasser 2019: Section 5.9) should be used in planning LEPs because change is going to be the norm rather than the exception

---

[*] Most likely to cause the project to fail.
[†] Concept of value is different in government and in industry.

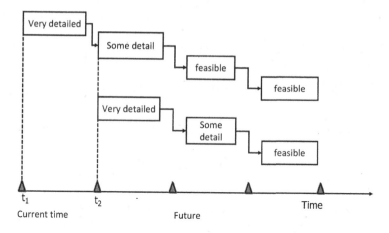

FIGURE 8.1 Planning – level of detail.

as it is in small projects. Accordingly, there is no point in planning a LEP down to the lowest level of detail at the start of the project. The systems approach to project planning is a multiple-iteration problem-solving process based on using the products delivered or handed over at each milestone in the SDP (Section 6.1.5.4.2) working backwards from the delivery readiness review (DRR) to the start of the project using a high-level waterfall chart of the states in the SDP as a planning template (Kasser 2018: Section 14.9). The further into the future, the less feasible it is to plan for (Churchman 1968), and so the lower the amount of detail needed. However, the initial draft work packages (WPs) should be planned in sufficient detail to provide a rough estimate of cost, schedule and risk. As the project progresses, the plan is updated for each milestone and the level of detail is increased as shown in Figure 8.1.

Planning takes time, and if adequate time is not spent on planning, the project is likely to fail. Risks can be reduced by:

- Allowing adequate time for the planning process and some.
- Allowing adequate time for the project performance. Adding a fudge factor of one-third (Augustine 1986) to each WP will increase the accuracy of the scheduling and corresponding cost estimates.
- Stating assumptions and confirming their validity with cognizant personnel.
- Reaffirming stakeholder commitments before incorporating them in the plan.
- Using the appropriate applicable risk mitigation techniques for the project.

8.3.3 THE LEP PERFORMANCE STATE

This is the state in which:

- The validity of the cost, schedule and technical assumptions is shown; under- and overestimating cause changes to designs, often with corresponding delays to rework sections depending on when the assumptions are shown to be incorrect.

- Stakeholders may drop out, not meet commitments or try to renegotiate their arrangements.
- Risks show up in the various sub-projects.
- Unanticipated events occur that may delay or allow early finishing of sub-projects.

Effective risk management can only be performed if the sub-project managers report the status of their projects in a truthful and timely manner. In order to achieve this, the project managers should be encouraged to use categorized requirements in process (CRIP) charts (Section 7.5) and enhanced traffic light (ETL) charts (Section 7.6) to make hiding bad news difficult.* The project managers should be encouraged to not only provide the bad news if applicable (problems), but also to provide plans to recover (solutions) in case those problems impact other projects in the LEP. Punishing the messenger is counterproductive because they will be a tendency to hide bad news until it is too late to do anything about the situation. Perceptions from the *Generic* HTP show that this is very similar to the system subsystem relationship on systems engineering. Here the LEP is the system and the individual subprojects are the subsystems. It is also important to minimize the micromanagement by upper management (MBUM) if the project manager is not going to be replaced.

8.3.4 THE LEP CLOSEOUT STATE

Perceptions from the *Generic* HTP note that the functions in the LEP closeout state are identical to the functions in the project closeout state (Section 7.4.4). Perceptions from the *Continuum* HTP note that the difference is in the scope of the activities and the number of projects being closed out. It is also important to document the lessons learned from the LEP so they can be considered during the initialization state of subsequent LEPs.

8.4 RISKS BASED ON THE NON-AVAILABILITY OF TECHNOLOGY

Technology has a lifecycle; it is developed through research, then moves into widespread use in many products and then becomes obsolete and is replaced. Some projects use technology that already exists; some projects are dependent on technology that is still being developed. Shenhar and Bonen recognized that a single project management methodology would not work for technology in different states of development or existence. They categorized projects into the following four types based on the state of the technology (Shenhar and Bonen 1997):

- *Type A (low-tech projects)*: projects that rely on existing and well-established technologies to which all industry players have equal access.
- *Type B (medium-tech projects)*: projects that rest mainly on existing technologies and incorporate a new technology or a new feature of limited scale.

---

* Preferably by making it a contractual requirement.

- *Type C (high-tech projects)*: projects in which most of the technologies employed are new, but existent – having been developed prior to the project's initiation.
- *Type D (super-high-tech projects)*: projects based primarily on new, not entirely existent, technologies.

The recommended management methodology for improving the probability of success (mitigating the risk of failure) for each of the different types of projects is:

- *Type A (low-tech projects)*: the design cycle in these projects is usually characterized by a single pass through the SDP because the final design of the system is usually frozen prior to the project initiation. The management style is a firm and formal style, in which no changes are allowed or required.
- *Type B (medium-tech projects)*: the design cycle in these projects usually consists of at least one iteration of the SDP. The management style of these projects can be described as moderately firm but more flexible than in Type A projects. The management style requires more communication with the customer (both formal and informal) since more trade-offs and changes are made.
- *Type C (high-tech projects)*: the design cycle in these projects is also iterative and usually entails two or more cycles through the SDP, and the system design freeze takes place at a later stage than in Type B systems. It often occurs as late as the second quarter or the midpoint of the project. The management style of high-tech systems is moderately flexible, since many changes are expected and are a natural part of this type of system development. It involves intensive customer interaction and the use of multiple formal and informal communication channels.
- *Type D (super-high-tech projects)*: the design cycle in these projects requires extensive development of both the new technologies and the actual system being developed by the project. Their development frequently requires building an intermediate, small-scale prototype, on which new technologies are tested and approved before they are installed on the prototype. System requirements are hard to determine; they undergo enormous changes and involve extensive interaction with the customer. Obviously, the system functions are of similar nature – dynamic, complex, and often ambiguous. A super-high-tech system is never completed before at least two, but very often even four, iterations of the SDP are performed, and the final system design freeze is never made before the second or even the third quarter of the project. The management style of these projects is highly flexible to accommodate the long periods of uncertainty and frequent changes. Managers must live with continuous change for a long time; they must extensively increase interaction, be concerned with many risk mitigation activities, and adapt a 'look for problems' mentality.

So, when architecting the project process, the technology risk must be assessed and the appropriate number of iterations of the SDP must be built into the systems engineering management plan (SEMP) (Section 6.5.9.2) and project plan (Section 7.4.2.1).

TABLE 8.1
The TAWOO States and Levels

TAWOO State		Extended TRL Level	Comments
6	Antique	12	Few if any spares available in used equipment market. Phase out products or operate until spares are no longer available.
5	Obsolete	11	Some spares are available; maintenance is feasible.
4	Approaching obsolescence	10	Use in existing products but not in new products. Plan for replacement of products using the technology.
3	Operational	9	Available for use in new products (in general). System 'flight proven' through successful mission operations
2	Development	8	Actual system completed and 'flight qualified' through test and demonstration
		7	System prototype demonstration in an operational environment
		6	System/subsystem model or prototype demonstration in a relevant operational environment
1	Research	5	Component and/or breadboard validation in a relevant operational environment
		4	Component and/or breadboard validation in a laboratory environment
		3	Analytical and experimental critical function and/or characteristic proof-of-concept
		2	Technology concept and/or application formulated
		1	Basic principles observed and reported

© 2016 IEEE. Reprinted, with permission, from Applying Holistic Thinking to the Problem of Determining the Future Availability of Technology, Joseph E. Kasser.

8.5   THE TAWOO

The National Aeronautics and Space Administration (NASA) and the US Department of Defense (DoD) have many projects dependent on technology that is still being developed. NASA developed the technology readiness level (TRL) to minimize the risk of the technology not being available when needed (Mankins 1995). The DoD adopted the TRL with slight modifications. However, the TRL only covers the early stages of the technology lifecycle. Project managers must consider the entire life-cycle because there is no point in incorporating obsolete or almost obsolete technology in a new product. The TAWOO (Kasser 2018: Section 8.12) shown in Table 8.1 (Kasser 2016) was developed to provide policy makers, project managers and systems engineers with such a tool. The TAWOO:

- Allows systems engineers to determine if a technology is mature enough to integrate into the system under development *as well as* estimating if the technology will be available over the operating life of the system once deployed.

Many products use technology represented by sequential 'S' curves inside whale

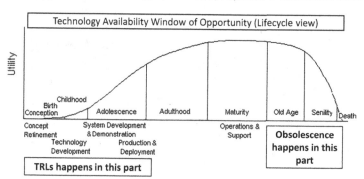

FIGURE 8.2 The TAWOO superimposed on the whale diagram. (© 2016 IEEE. Reprinted, with permission, from Applying Holistic Thinking to the Problem of Determining the Future Availability of Technology, Joseph E. Kasser.)

- When superimposed on the whale diagram (Nolte 2005) as shown in Figure 8.2 (Kasser 2016) provides information about the availability of the technology in the remaining stages of the technology lifecycle effectively extending the TRL to cover the whole product lifecycle including consideration of diminishing manufacturing sources and material shortages (DMSMS) at the end of the technology lifecycle.
- Is described in greater detail in *Systems Thinker's Toolbox: Tools for Managing Complexity* (Kasser 2018: Section 8.12).

Consider the TAWOO from the appropriate progressive and remaining HTPs.

1. *Temporal:* 'Although TRL is commonly used, it is not common for agencies and contractors to archive and make available data on the timeline to transition between TRLs' (Crépin, El-Khoury, and Kenley 2012). Perceptions from the *Temporal* HTP suggest that the data should be archived and used to estimate/predict maturity. If that data were available, one could infer from the *Scientific* HTP that one could use the rate of change of TRL rather than use a single static value at one particular time. For example, Figure 8.3 (Kasser 2016) shows that the technology was conceptualized in 1991 and the development was planned to advance one TRL each year starting in 1993 for production in 1999. However, the development did not go according to plan. The technology did not get to TRL 2 until 1995 advancing to TRL 3 two years later in 1997 and jumping to TRL 6 in 1998. So, can the technology be approved for a project due to go into service in 1999? It depends. If the project can use the technology at TRL 6, then yes. But, if the product using the technology is to go into mass production, the answer cannot be determined because there is insufficient information to predict when the technology will be at TRL 9. The project will have to obtain more information about the factors affecting the rate of change in TRL to make a forecast as to the future.

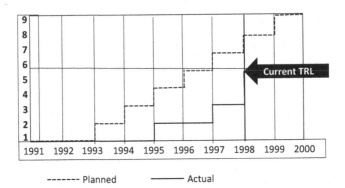

FIGURE 8.3　The TRL 1991–2001. (© 2016 IEEE. Reprinted, with permission, from Applying Holistic Thinking to the Problem of Determining the Future Availability of Technology, Joseph E. Kasser.)

2. *Generic*: perceptions from the *Generic* HTP indicate that projects use earned value analysis (EVA) (Kasser 2018: Section 8.2) and display budgeted/planned and actual cost information in graphs in which future costs are forecast.

3. *Scientific*: combine observations from the *Generic* and *Temporal* HTPs and display the rate of change of the TRL in the form of an EVA financial chart as shown in Figure 8.4 (Kasser 2016). When this is done, one additional significant item of information is obtained. Assuming nothing changes and progress continues at the same rate as in 1997–1998, the technology should reach TRL 9 by 1999. However, the reason for the rate of change between 1996–1997 and 1997–1998 is unknown. This provides the project manager with some initial questions to ask the technology developers before making the decision to adopt the technology. The static single-value TRL has become a dynamic TRL (dTRL) (Kasser and Sen 2013). The dTRL component would

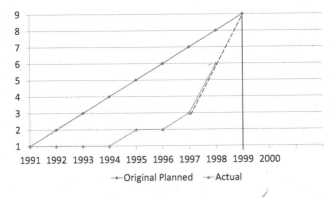

FIGURE 8.4　The dTRL. (© 2016 IEEE. Reprinted, with permission, from Applying Holistic Thinking to the Problem of Determining the Future Availability of Technology, Joseph E. Kasser.)

make adoption choices simpler. Prospective users of the technology could look at their need by date, the planned date for the technology to achieve TRL 9 and the progress through various TRLs. Then the prospective users could make an informed decision based on the graph in their version of the dTRL. If the rate of change projects that the desired TRL will not be achieved when needed and they really needed the technology, they could investigate further and determine if they could help increase the rate of change of TRL.

Perceptions from the *Generic* HTP and insight from the *Scientific* HTP have conceptualized the use of a dTRL to help to predict when a technology will achieve a certain TRL, thus lowering the risk of adopting a technology that would not be available when needed. The need for a dTRL has been recognized in practice, and there has been research into estimating the rate of change of technology maturity (El-Khoury 2012). The dTRL concept was used for quite a few years in the US aerospace and defence industry beginning in the strategic defence initiative era (early 1990s) and took the form of waterfall charts that tracked the TRL (Benjamin 2006).

8.6 EXERCISES

This section provides a number of exercises as opportunities to think about the use of consultants. A partial acceptable solution is provided for the off-world mining policy (Section 8.6.4). It is up to the reader to use the sample as a template when performing the remaining exercises.

8.6.1 The ENU Traffic Light Upgrade

In addition to meeting the generic requirements in Section 2.7:

1. Format the problem posed by this exercise in the problem formulation template (Section 5.1.7.2).
2. Discuss the risks based on the technology aspects of the selected parts of the concept (Section 8.4) and the TAWOO (Section 8.5).
3. Describe at least three risks in hiring consultants for one or more parts of the ENU traffic light upgrade and how they might be mitigated.

8.6.2 The Engaporean Maid Reduction Policy

In addition to meeting the generic requirements in Section 2.7:

1. Format the problem posed by this exercise in the problem formulation template (Section 5.1.7.2).
2. Discuss the risks based on the technology aspects of the selected parts of the concept (Section 8.4) and the TAWOO (Section 8.5).
3. Describe at least three risks in hiring consultants for one or more parts of the Engaporean maid reduction policy and how they might be mitigated.

8.6.3 THE ENGAPOREAN DROUGHT RELIEF POLICY

In addition to meeting the generic requirements in Section 2.7:

1. Format the problem posed by this exercise in the problem formulation template (Section 5.1.7.2).
2. Discuss the risks based on the technology aspects of the selected parts of the concept (Section 8.4) and the TAWOO (Section 8.5).
3. Describe at least three risks in hiring consultants for one or more parts of the Engaporean drought relief policy and how they might be mitigated.

8.6.4 THE OFF-WORLD MINING POLICY

In addition to meeting the generic requirements in Section 2.7:

1. Format the problem posed by this exercise in the problem formulation template (Section 5.1.7.2).
2. Discuss the risks based on the technology aspects of the selected parts of the concept (Section 8.4) and the TAWOO (Section 8.5).
3. Describe at least three risks in hiring consultants for one or more parts of the off-world mining LEP and how they might be mitigated.

8.6.4.1 One Acceptable Solution

A completed exercise could take the form shown in this section. Note it provides one acceptable solution, not the single correct solution as there will be other acceptable solutions (Section 5.1.1.2.2).

8.6.4.1.1 The Assumptions

The assumptions include:

1. The project is achievable within the timeframe.
2. The time frame is approximate. For example, if it takes 55 years instead of 50 years, the Ministry will not complain.

8.6.4.1.2 The Formatted Problem

The problem formatted according to the problem formulation template (Section 5.1.7.2) is:

1. The undesirable situation: is the need to:
   1. Comply with the generic requirements in Section 2.7.
   2. Format the problem posed by this exercise in the problem formulation template (Section 5.1.7.2).
   3. Discuss the risks based on the technology aspects of the selected parts of the concept (Section 8.4) and the TAWOO (Section 8.5).
   4. Describe at least five risks in hiring consultants for one or more parts of the off-world mining LEP and how they might be mitigated.

TABLE 8.2

Compliance Matrix for the Exercise in Section 8.6.4

ID	Instruction	Deliverable	Section	Completed
1	Brainstorming or active brainstorming	No	All	Yes
2	Research	No	All	Yes
3	Use imagination	No	All	Yes
4	Use lists and charts	No	All	Yes
5	Spend less than 60 minutes	No	All	Yes
6	The formatted problem	Yes	8.6.4.1.2	Yes
7	Assumptions	Yes	8.6.4.1.1	Yes
8	Preliminary compliance matrix	No	8.6.4.1.3	Yes
9	Risk based on technology	Yes	8.6.4.1.4	Yes
10	Risk in hiring consultants	Yes	8.6.4.1.5	Yes
11	Lessons learned	Yes	8.6.4.1.6	Yes
12	Updated compliance matrix	Yes	8.6.4.1.3	Yes

2. The assumptions: are:
   1. The exercise can be completed within the 60 minutes.
   2. As listed in Section 8.6.4.1.1.
3. *The FCFDS*: the exercise has been completed and the desired grade has been achieved.
4. *The problem*: how to do the exercise, namely:
   1. Meet the generic requirements in Section 2.7.
   2. Discuss the risks based on the technology aspects of the selected parts of the concept (Section 8.4) and the TAWOO (Section 8.5).
   3. Describe at least three risks in hiring consultants for one or more part of the off-world mining LEP and how they might be mitigated.
   4. Verify that all the requirements for the exercise have been completed and update the compliance matrix accordingly.
5. *The solution*: consists of remedying each of the problems. A partial solution is presented below.

8.6.4.1.3   The Compliance Matrix

The partially completed compliance matrix is shown in Table 8.2.*

8.6.4.1.4   Risks Based on Technology

Risks based on technology (Section 8.2.1) hinder the successful completion of the policy. In this LEP, there will be a number of technologies in different TAWOO

---

* Placing the section number in the completed column not only shows that the requirement has been met but also shows where the response can be found, mitigating the risk of the instructor failing to find and grade the response.

states (Section 8.5) coexisting in different projects. The technologies and risks may be grouped into the following subsystems:

1. *Spaceport*: is expected to use a mixture of technologies that already exist and can be expected to have a long lifetime, namely TAWOO level 9 (Table 8.1). Accordingly, the probability of the technology failing is extremely low although the severity could be high if the failure would cause operations to cease. Perceptions from the *Generic* HTP note that if technology is used in the spaceport in the same way as it is used in airports, then the probability of a technological failure is extremely low.

2. *Launch vehicles*: a few are currently at TAWOO level 9. However, it is expected that new technologies will be adopted and lead to improvements in launch capability. However, since these will not directly concern the spaceport, it is expected that the commercial organizations will manage the change in launch vehicle technology in a reasonable manner.

3. *Low earth orbit (LEO) installations*: the only current operational LEO installation is the International Space Station. It is expected that space station and other facilities to be created during the SDP of the facilities in this subsystem will pass through various TAWOO layers. The Ministry will monitor but not take an active role in managing the risks in developing these facilities.

4. *Deep space vehicles*: at present there aren't any manned deep space vehicles; accordingly, they are at a low TAWOO layer. The Ministry will monitor but not take an active role in managing the risks in developing these vehicles.

5. *Asteroid systems*: are in the same state as deep space vehicles and will be treated in the same way.

6. *Jupiter in-orbit systems*: are in the same state as deep space vehicles and will be treated in the same way.

The Ministry may step in in some way should the necessary technology at a particular milestone not be achievable at a reasonable cost. For example, one milestone is to demonstrate launch capabilities from the spaceport within one year of the award of the contract to construct the spaceport. If at some point in the future this demonstration looks like not being achievable, the Ministry will have to decide to either support some research to achieve the milestone or stretch out the project. It is not reasonable to make that decision at this time.

8.6.4.1.5   The Risks in Hiring Consultants
Risks in hiring consultants include:

1. *The tail will wag the dog*: if the organization hiring the consultants doesn't have any clear goals in mind, the consultants will take over and do what the consultant wants to do while the organization pays the bills. The off-word mining LEP is probably going to require a number of consultants to provide the Ministry with the expertise it needs over the entire LEP lifecycle.

This risk may be mitigated by having a clear concept of the outcome of the activity in which the consultant is providing service. This generally means that while suitable consultants may be identified early in a project lifecycle, they should not be employed until:

- The goals of the project have been approved.
- The objectives for the product or service to be provided by the consultant have also been approved.

2. *Consultants often lack knowledge of in-house procedures*: which can create conflict. This risk may be mitigated by pre-arranging for a pool of consultants during the project initialization state and requiring them to attend an introductory workshop before they are allowed to bid for tasks. The contracting arrangements will be to use task-ordered contracts in which the consultants perform specific tasks.

3. *Consultants fail to perform as expected*: this risk may be mitigated by writing clear and concise task statements stating not only the objectives but also providing a schedule of when those objectives must be met. The contract will include mutually agreed upon penalty clauses for non-compliance to the objectives at the scheduled point in time.

4. *Protection of intellectual property*: the consultant may take intellectual property that belongs to the Ministry or developed while working on the contract and apply that intellectual property elsewhere. This risk should be mitigated by the appropriate non-disclosure clause in the contract. However, since the goal of the LEP is to move humanity into space, the consultant should probably be free to apply that intellectual property in pursuit of the goal. This means the consultant could go work for one of the commercial enterprises and make use of the intellectual property developed while working for the Ministry. Accordingly, the contract should contain the appropriate wording to allow this to happen and also return a small benefit to the Ministry. Perceptions from the *Generic, Quantitative* and *Temporal* HTPS note that NASA developed communication and weather satellites as research projects. Had NASA since been able to receive a small royalty on the information being transferred through commercial communications satellites and the information being provided by the weather satellites, NASA would not have to worry about much of its funding.

8.6.4.1.6 Lessons Learned

- The compliance matrix is very handy for keeping track of what needs to be done and making sure that it is being done.
- It is extremely difficult to keep track of everything going on in a LEP.
- It is important that members of the project team communicate information in a timely manner: bad news as well as good news.
- The five questions from reliability-centered maintenance (Section 3.8.2.1) posed from the *Structural, Operational* and *Functional* HTPs helped to identify risks in the exercise.

8.7 SUMMARY

This chapter provided an introduction to LEPs and managing risks in LEPs because a LEP contains a hierarchy of projects and systems. In particular, this chapter discussed risks in LEPs, the LEP lifecycle and risks based on the non-availability of technology, since technology has become ubiquitous in the 21st century.

REFERENCES

Augustine, Norman R. 1986. *Augustine's Laws*. New York: Viking Penguin Inc.

Benjamin, Daniel. 2006. Technology readiness level: An alternative risk mitigation technique. In *Project Management in Practice: The 2006 Project Risk and Cost Management Conference*. Boston, MA: Boston University Metropolitan College.

Churchman, C. West. 1968. *The Systems Approach*. New York: Dell Publishing Co.

Crépin, Maxime, Bernard El-Khoury, and C. Robert Kenley. 2012. It's all rocket science: On the equivalence of development timelines for aerospace and nuclear technologies. In *the 22nd International Symposium of the INCOSE*. Rome, Italy.

Crosby, Philip B. 1979. Quality Is Free. New York: McGraw-Hill.

El-Khoury, Bernard. 2012. *Analytic framework for TRL-based cost and schedule models*. Engineering Systems Division, Massachusetts Institute of Technology.

Gaitskell, Robert. 2011. Construction Dispute Resolution Handbook. 2nd ed. London: Institute of Civil Engineers Publishing.

Hass, Kathleen B. (Kitty). 2019. Managing *complex projects that are too large, too long and too costly*. PM Times 2019 [cited 3 September 2019]. Available from https://www.projecttimes.com/articles/managing-complex-projects-that-are-too-large-too-long-and-too-costly.html.

Hughes, Thomas P. 1998. *Rescuing Prometheus*. New York: Random House Inc.

Kasser, Joseph Eli. 1997. *The determination and mitigation of factors inhibiting the creation of strategic alliances of small businesses in the government contracting arena*, The Department of Engineering Management, The George Washington University, Washington DC.

Kasser, Joseph Eli. 2016. Applying holistic thinking to the problem of determining the future availability of technology. *The IEEE Transactions on Systems, Man, and Cybernetics: Systems*. no. 46 (3):440–444. doi: 10.1109/TSMC.2015.2438780.

Kasser, Joseph Eli. 2018. *Systems Thinker's Toolbox: Tools for Managing Complexity*. Boca Raton, FL: CRC Press.

Kasser, Joseph Eli. 2019. *Systemic and Systematic Project Management*. Boca Raton, FL: CRC Press.

Kasser, Joseph Eli, and Souvik Sen. 2013. The United States airborne laser test bed program: A case study. In *the 2013 Systems Engineering and Test and Evaluation Conference (SETE 2013)*. Canberra, Australia.

Kezsbom, Deborah S., Donald L. Schilling, and Katherine A. Edward. 1989. *Dynamic Project Management. A Practical Guide for Managers and Engineers*. New York: John Wiley & Sons.

Lui, Pao Chuen. 2007. An example of large scale systems engineering. Presentation at *Keynote presentation at the 1st Asia-Pacific Systems Engineering Conference*, at Singapore.

Mankins, John C. 1995. *Technology readiness levels*. Advanced Concepts Office, Office of Space Access and Technology, NASA.

Miller, Roger, and Donald R. Lessard. 2000. *The Strategic Management of Large Engineering Projects Shaping Institutions, Risks, and Governance*. Cambridge, MA: Massachusetts Institute of Technology.

Nolte, William L. 2005. TRL calculator. In *AFRL at Assessing Technology Readiness and Development Seminar.*

Shell, R.G. 2006. *Bargaining for Advantage.* New York: Penguin Books.

Shenhar, A.J., and Z. Bonen. 1997. The new taxonomy of systems: Toward an adaptive systems engineering framework. *IEEE Transactions on Systems, Man, and Cybernetics – Part A: Systems and Humans* no. 27 (2):137–145.

Sherman, Stanley N. 1987. *Contract Management: Post Award.* Gaithersburg, MD: Wordcrafters Publications.

Webb, Bill. 1992. *How to live with a CPAF contract.* Contract Management, (January):19.

9 Perspectives of Polices

This chapter discusses policies and policy management from the different holistic thinking perspectives (HTPs) and examines policy implementation because many policies are implemented by large engineering projects (LEPs). In particular, it discusses policies as a system, the different types of policies, the generic policy lifecycle, policy documents and ways of mitigating risks in policies.

Perceptions of policies from the HTPs included:

9.1 QUANTITATIVE

Perceptions from the *Quantitative* HTP included:

1. Many problems posed to policy and strategic decision-makers are complex problems discussed in Section 9.1.1.

9.1.1 POLICY PROBLEMS ARE COMPLEX PROBLEMS

Policy problems are complex problems because they contain many coupled controlled and uncontrolled variables (Section 5.1.2.1) and a change in one variable may affect others.

9.2 BIG PICTURE

Perceptions from the *Big Picture* HTP included:

1. Public policies start and end in layer 5 of the risk framework (Section 1.2).
2. Public policies start and end in layer 5 of the Hitchins-Kasser-Massie framework (HKMF) (Section 3.4.4).
3. Policies exist within a system discussed in Section 9.2.1.
4. Policies have reasons for their existence discussed in Section 9.2.2.
5. Policies have stakeholders discussed in Section 9.2.3.
6. Policies are instruments of change and often face resistance discussed in Section 9.2.4.

9.2.1 POLICIES EXIST WITHIN A SYSTEM

The systematic and systemic approach:

- Does not treat a policy in isolation but as a part of a system of policies.
- Considers a policy in the context of policies in existence or under development.
- The system can be represented by a series of context diagrams showing the relationship between the policy and stakeholders.

9.2.2 POLICIES HAVE REASONS FOR THEIR EXISTENCE

A policy is usually created as a result of someone with authority deciding to remove something undesirable from a situation.* The undesirability in the situation may be the result of something that:

1. Is not being done should be done.
2. Is being done should not be done.
3. Is being done is causing the undesirability.
4. Needs to be added to make the situation even more desirable.†

9.2.3 POLICIES HAVE STAKEHOLDERS

Policies have stakeholders who:

- May be direct or indirect (Kasser 2019: Section 10.14).
- Affect or may be affected by the policy.‡
- Help or hinder the creation and validation of the policy in various ways.
- May have different ideas as to the causes of the undesirability and/or different ideas about how the situation should be transformed.
- May be shown in a context diagram which shows relationships between:
  - Policy and direct or indirect stakeholders.
  - Stakeholders and other direct or indirect stakeholders.

9.2.4 POLICIES ARE INSTRUMENTS OF CHANGE

Policies are used to effect changes. These changes can be technical, fiscal, social, etc. Examples include:

- *Technical*: introduction of standards.
- *Fiscal*: encourage or discourage activities via taxes or subsidies.
- *Social*: change behaviours and cultures.

9.3 CONTINUUM

Perceptions from the *Continuum* HTP included:

1. The different categories of policies discussed in Section 9.3.1.
2. Some policies succeed and some policies fail discussed in Section 9.3.2.
3. Policies are only one way of implementing changes discussed in Section 9.3.3.

---

* Perceptions from the *Generic* HTP note that this is the start of an iteration of the traditional generic problem-solving process (Section 5.1.6.1).
† Improve the situation.
‡ The usual definition of a stakeholder is someone who has a stake in the project, but sometimes people just like to get involved. They too can pose risks and provide opportunities.

4. Policies may have unintended consequences discussed in Section 9.3.4.

5. The continuum of care discussed in Section 9.3.5.

9.3.1 THE DIFFERENT CATEGORIES OF POLICIES

Policies may be categorized as:

1. *Steady state*: policies which try to maintain a situation.
2. *Transitional one-shot*: policies which cause a transition from one situation to another.
3. *Ideological*: policies which strive to transition to/maintain an ideological goal. Ideological policies may be steady state or transitional one-shot.

9.3.2 SOME POLICIES SUCCEED AND SOME POLICIES FAIL

Perceptions from the *Generic* HTP note that the reasons for success and failure policies discussed in this section are very similar if not identical to the reasons for success and failure of LEPs in layer 5 of the risk framework (Section 1.2). Accordingly, risk management techniques used in the lower layers of the risk framework may be applicable for managing the risk in policies if appropriate. According to the principle of hierarchies (Section 6.3.1), each layer in the hierarchy needs the appropriate tools and techniques which may not be the same as those in lower layers.

The primary reasons for the failure of new policies include:

1. *Lack of planning*: a lack of planning incorporates the failure to think through the ramifications of the policy. Successful policy implementation is planned and staged in milestones according to a policy implementation process.
2. *Failing to understand the CM process*: failing to understand the CM process is the second primary reason for policy failure because the policy implementation process is a variation of the CM process since policies are instruments of change.
3. *Lack of attention to implementation*: often due to (Edwards III and Sharkansky 1978):
   1. Inexperience in administration.
   2. Lack of time to spend on monitoring implementation.
   3. No incentives for compliance.
   4. No penalties for non-compliance
   5. Lack of independent measurement of compliance.
4. *Unclear communications in instructions and advice*: 'the more accurately policy decisions and implementation orders are transmitted to those who must carry them out, the higher the probability of them being implemented' (Edwards III and Sharkansky 1978).
5. *Lack of resources to carry out the policy*: 'if implementers lack the resources necessary to carry out policies, implementation is likely to be

ineffective' (Edwards III and Sharkansky 1978). For example, insufficient and inadequate staff to:

1. Carry out policy directives.
2. Monitor compliance.
3. Lack of information.
4. Be able to predict the outcomes of a policy, 'Before an agency such as EPA orders a costly change in an industry or its products (such as automobiles), the agency should be able to predict the effects of the change on the economic health of the industry in question. Such information, however, is frequently lacking' (Edwards III and Sharkansky 1978: p. 63).
6. *Blindly following standard operating procedures*: the process. Designed for typical situations, they can be ineffective in unusual circumstances.
7. *Fragmentation*: responsibility is distributed across different organizations

9.3.3 POLICIES ARE ONLY ONE WAY OF IMPLEMENTING CHANGES

A specific policy is but one of a number of ways to shape the situation or achieving a goal. One well-known example was the constitutional ban on the production, importation, transportation and sale of alcoholic beverages from 1920 to 1933 in the US known as Prohibition.

9.3.4 POLICIES MAY HAVE UNINTENDED CONSEQUENCES

Once initiated policies:

- Transform an undesirable situation into a new situation producing consequences or outcomes that can be a combination of:
  - Intended.
  - Unintended.
  - Desired.
  - Undesired.
  - Don't Care.
- Perceptions from the *Generic* HTP note that these are the same as outcomes of actions and decisions. For example, Prohibition designed to reduce alcoholism led to large-scale smuggling, bootlegging and a growth in organized crime, an unintended and undesired outcome.

9.3.5 THE CONTINUUM OF CARE

Many policies fall within a 'zone of indifference'.* These policies will probably be implemented faithfully because implementors do not have strong feelings about them. Other policies, however, may be in direct conflict with the policy views or personal or organizational interests of implementors. When people are asked to

---

* Or the 'Don't Care' situation (Kasser 2018b: Section 3.2.2).

execute orders with which they do not agree, inevitable slippage occurs between political decisions and performance. In such cases, implementors will exercise their discretion, sometimes in subtle ways, to hinder implementation (Edwards III and Sharkansky 1978: p. 90).

9.4 GENERIC

Perceptions from the *Generic* HTP noted that policies

1. *Create change*: actually, 'Most "new" policies simply provide for incremental* changes from existing polices' (Edwards III and Sharkansky 1978: p. 173). Accordingly, policy management needs to incorporate CM.
2. *Are and contain projects*: the entire policy lifecycle may be a project; each state in the lifecycle may contain a number of projects depending on a particular situation.

9.5 OPERATIONAL

Perceptions from the *Operational* HTP included:

1. How generic risks change over the policy lifecycle discussed in Section 9.5.1.

9.5.1 THE GENERIC POLICY LIFECYCLE AND ASSOCIATED RISKS

Implementation of a policy generally initiates a change which moves a system from one state (an undesirable situation) to another (a desirable situation). Once in the second state, a further change is often desired, so a new policy is created and implemented. This is a cyclic situation. The policy lifecycle is the entire policy process loop from conception to verification and has been stated in different ways by different authors. One version of the policy lifecycle contains the following eight steps (Althaus, Bridgman, and Davis 2007):

1. *Issue identification*: the recognition of a problem.[†]
2. *Policy analysis*: gain an *understanding* of the situation, identify the stakeholders and determine what should be done to remove the undesirability.
3. *Policy instrument development*: create the policy that describes how the situation will be changed to remove the undesirability.
4. *Consultation with stakeholders*: (which permeates the entire process) recognize the important role that opposition forces play in making public policy.
5. *Coordination*: with stakeholders.

---

[*] Or adaptive (Section 4.1.2).
[†] An undesirable situation.

6. *Decision*: after due consideration of the benefits and probable resistance, the decision as to implement the policy is taken. If the decision is not to go ahead with the policy, then the cycle ends at this point or loops back to Step 1 to develop an alternative.

7. *Implementation*: the literature tends to gloss over the implementation states of the policy process somewhat taking it for granted. Perceptions from the *Generic* HTP note that just as a LEP (Chapter 8) was implemented in a number of subprojects, a policy is similarly implemented in one or more subprojects, and so the risks in the implementation state are similar to those in the implementation state of a LEP.

8. *Evaluation*: the policy in action is evaluated to determine if the situation is no longer undesirable, and if it is, the policy needs to be rethought and another iteration of the policy loop commences.

Each step in the policy lifecycle contains:

- Its own set of problems and their associated risks and can be a project or a part of a project depending on the complexity of the policy and its implementation.
- The traditional generic problem-solving process shown in Figure 5.3.

Perceptions from the *Generic* HTP note that the four states in the change lifecycle (Section 4.2.1), the project lifecycle (Section 7.4) and the LEP lifecycle (Section 8.3) are the same as the generic policy lifecycle:

1. The policy initialization state discussed in Section 9.5.1.2.
2. The policy planning state discussed in Section 9.5.1.3.
3. The policy performance state discussed in Section 9.5.1.4.
4. The policy closeout state discussed in Section 9.5.1.5.

The policy initialization and planning states remedy a research problem and occur during the first problem-solving process shown in Figure 5.12 and produce the policy implementation plan. The policy performance and closeout states remedy the subsequent intervention problem and take during the second problem-solving process.

9.5.1.1 Generic Risks in the Policy Lifecycle

Generic risks in the policy lifecycle include:

1. Not identifying the appropriate influential and important stakeholders.
2. Not alleviating the fears of the public and other stakeholders who may perceive that downside of the proposed policy by communicating with the public (Section 3.2.2).
3. Policy running open-loop, namely not reporting status back to the policy initiator and manager in a timely manner or at all so that corrective action may be taken to overcome anything preventing the policy from going ahead.
4. Not realizing that the focus of risk changes over the policy lifecycle.

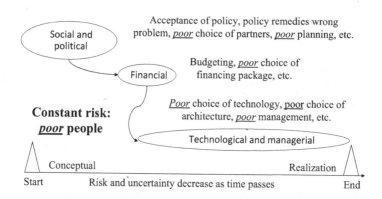

FIGURE 9.1 How risk emphasis changes over time.

The emphasis of risk and uncertainty changes over time as shown in Figure 9.1. At the start of the policy initialization, risk and uncertainty are at a maximum. When the policy is in action and has been determined to be meeting its goals, risk and uncertainty are zero because everything is certain and known. In between, during the policy initiation state, the risks tend to be social and political, including failure to achieve acceptance of the policy, and the policy remedies the wrong problem, poor choice of partners, etc.

When the policy moves into the planning state, the risks tend to be financial, including errors in budgeting, poor choice of financial packages, etc. As the policy moves into the performance state, the risks tend to focus on the technology and other aspects of project risks such as poor choice of technology, poor choice of architecture, poor project management and poor systems engineering.

The common factor is the risk of poor people, one of the quadruple constraints (Section 7.3.2). Because it is people that implement and monitor the policy in action. For example:

- When examining the undesirable situation, the wrong cause may be inferred, which will produce the wrong policy and will not remedy the undesirable situation.
- When implementing the policy, errors of omission and errors of commission (Section 5.1.4.3.1) will cause the policy to fail.
- Failure to communicate the implementation progress leads to not taking timely corrective action and the policy will fail.

9.5.1.2 The Policy Initialization State

Each new policy has to contend with:

1. Factors in its environment.
2. Constraints, such as other existing policies which may conflict, and standard operating procedures.

3. Resources.
4. Time to make the change.
5. Public opinion.
6. Each of the above may help or hinder the policy.

The systems approach to policy making begins with the use of the problem formulation template (Section 5.1.7.2) where:

1. *The undesirable situation*: gives rise to the policy initialization state.
2. *The assumptions*: specific to the situation.
3. *The Feasible Conceptual Future Desirable Situation (FCFDS)*: the undesirable situation with the causes of undesirability removed and without any further undesirability after the policy has been implemented for sufficient time to effect a change.
4. *The problem*: is to create a policy, which when implemented will transform the undesirable situation into the FCFDS.
5. *The solution*: takes place in the four states of the policy lifecycle:
   1. The policy initialization state conceptualizes the FCFDS (what the policy is intended to achieve) and the policy to create the FCFDS.
   2. The policy planning state is when the plans of how to implement the policy are created.
   3. The policy performance state is when the policy is implemented according to the plan.
   4. The policy verification state is when the situation with the policy in action is compared with the FCFDS. If the policy in action has achieved the FCFDS, the policy lifecycle terminates. If the policy in action has not achieved the FCFDS, the policy lifecycle iterates. If some new undesirability is present in the situation, a new policy lifecycle may be initiated.

With reference to the generic policy lifecycle (Section 9.5.1), the policy initialization state is when the following are performed:

1. *Issue identification*: the recognition of one or more undesirable situations and the determination of someone with the authority to do something about removing the undesirability. The selection of which undesirable situation to deal with depends on the agenda of the authority. This activity is characterized by conflict and negotiation between the stakeholders covering factors such as which aspect of undesirability should be dealt with in a situation with insufficient resources to deal with all (e.g. funding or time) or disagreement as to how to remove the undesirability.
2. *Policy analysis*: gain an *understanding* of the undesirable situation, identify the stakeholders and determine what should be done to remove the undesirability. This activity follows the traditional generic problem-solving process (Section 5.1.6.1) to examine a number of options and select the best one for that specific situation at that specific time.

3. *Policy instrument development*: create the policy that describes how the situation will be changed to remove the undesirability. However, if some other method is chosen, the policy lifecycle does not go any further, but the same functions will probably be performed to ensure that the undesirability is removed by the other method.
4. *Consultation with stakeholders*: recognizing the important role that opposition forces play in making public policy. The tools and techniques are the same as those used in consulting with the stakeholders of a LEP and overcoming resistance to change.
5. *Coordination:* with stakeholders.
6. *Decision:* after due consideration of the benefits and probable resistance, the decision on whether to create and implement the policy is taken. If the decision is not to go ahead with the policy, then the cycle ends at this point or loops back to Step 1 to develop an alternative. If the decision is to go ahead with the policy then the policy advances to the next state of the policy lifecycle.

9.5.1.2.1 *Generic Risks in the Policy Initialization State*

Generic risks in the policy initialization state include the risk of not:

1. Selecting the correct problem to address with the policy.
2. Creating the correct policy.

9.5.1.3   The Policy Planning State

The problem posed by the policy planning state may be formatted according to the problem formulation template (Section 5.1.7.2):

1. *The undesirable situation*: the need to plan the implementation of the policy.
2. *The assumptions*: specific to the situation.
3. *The FCFDS:* the plan has been completed and accepted by the important and influential stakeholders.
4. *The problem*: is to create the plan.
5. *The solution*: follow the traditional generic problem-solving process (Section 5.1.6.1) to create the plan.

Perceptions from the *Generic* HTP note that planning a policy is the same as planning a project, namely working backwards from the solution (Kasser 2018a: Section 18.8). Visualize the policy implementation plan being approved, then visualize what happened before the plan was approved (e.g. the plan was presented to the stakeholders), then visualize what happened before that, then visualize what happened before that and so on until you reach the starting point of the policy planning state. Then those activities form the basis of the plan. The contents of the plan are similar to those of a LEP plan and a project plan (Section 7.4.2.1).

9.5.1.3.1 *Generic Risks in the Policy Planning State*

Generic risks in planning a policy are the risks and planning a LEP with a greater focus on the political risks associated with layer 5 of the risk framework (Section 1.2).

9.5.1.4   The Policy Performance State

The problem posed by the policy performance state may be formatted according to the problem formulation template (Section 5.1.7.2):

1. *The undesirable situation*: the need to implement the policy plan.
2. *The assumptions*: specific to the situation.
3. *The FCFDS*: the plan has been implemented; the situation is evaluated to determine if the situation is no longer undesirable.
4. *The problem*: is to implement the plan to put the policy into action and then evaluate the situation.
5. *The solution*: follow the traditional generic problem-solving process (Section 5.1.6.1) to create the plan.

With reference to the items in the generic policy lifecycle (Section 9.5.1), the policy performance state is when the following items are performed:

7. *Implementation: the literature tends to gloss over the implementation states of the policy process somewhat taking it for granted. Perceptions from the *Generic* HTP note that just as a LEP (Chapter 8) was implemented in a number of subprojects, a policy is similarly implemented in one or more subprojects and so the risks in the project performance state are similar to those in the project performance state of a LEP.
8. *Evaluation*: the results produced by the *steady state* policy in action or the results produced by a *transitional one-shot* policy are evaluated to determine if the situation is no longer undesirable. If the situation is
   * *No longer undesirable:* the policy lifecycle advances to the policy closeout state.
   * *Still undesirable:* the policy needs to be rethought and another iteration of the policy loop commences.

9.5.1.4.1   Generic Risks in the Policy Performance State

Since the people who originally determine public policies are usually not the same people who implement them, there is considerable room for misunderstanding and distortion of the decision-maker's intentions.

Edwards III (1980: p. 1)

Most policies require an intricate set of positive actions on the part of many people to be implemented.

Edwards III (1980: p. 1)

Perceptions from the *Generic* HTP note that performing a policy is the same as performing a project. Accordingly, the risks in the LEP performance state (Section 8.3.3) apply in addition to those risks pertaining to polices, which are (Edwards III 1980):

---

* This is item 7 in the referenced list.

1. *Communication*: the major communication risks are instructions that are non-specific, because:
   1. The more levels in the bureaucracy, the greater the probability that the order will be garbled since each person who handles the order can put their own twist on the instruction.
   2. The instructions can be interpreted in ways that favour bureaucratic agendas which may diverge from the intent of the order.
2. *Resources*: inadequate quantity and lack of problem, solution and domain expertise.
3. *Disposition*: documents can be intentionally or otherwise misfiled and lost.
4. *Bureaucratic structure*: the bureaucratic structure favours stability. Policies bring change, and change tends to be resisted (Section 4.5.1). There is generally no incentive for bureaucrats to implement the policy; in fact, most of the time the situation is reversed. The incentives are based on maintaining the status quo.

In addition, a contributor to the failure of policies is the lack of feedback. Policy creators create the policy, issue instructions and assume that the policy will be carried out. Much of the time this assumption is false, and there is no feedback to the policy creator to inform that fact. The risk of running open-loop and not receiving progress reports can be mitigated by requiring periodic progress reports as part of the policy instructions and include the requirement to use categorized requirements in process (CRIP) charts (Section 7.5) and enhanced traffic light (ETL) charts (Section 7.6) in the report.

When implementing the policy, LEPs or smaller projects need to take place. Accordingly, those requirements for progress reports should flow down and be inserted into the contracts. In addition, the project managers need to mitigate the risks appropriate to the project level in the policy risk framework (Section 1.2). If they identify risks at a higher level, then that information needs to be passed along as part of the progress report or even earlier if the situation is urgent. Perceptions from the *Continuum* HTP indicate that importance and urgency may be different (Covey 1989).

At no time should the messenger be shot or otherwise punished unless problems are due to negligence by the messenger. Policy and project managers should be encouraged to provide mitigation approaches as well as notice of risks turning into events.

Perceptions from the *Generic* HTP note that policy instructions are the same as requirements in layers 2 and 3 of the risk framework (Section 1.2) and HKMF (Section 3.4.4). Accordingly, the perennial problem of poor requirements (Section 6.8.3) is similar to that (the perennial problem) of poor policy instructions. The risk of poor policy instructions may be mitigated in the same way that risk of the creation of poor requirements are mitigated by the addition of what the military calls commander's intent as a preface to the policy instruction.* So, for example, a policy instruction would be prefaced by the background and intent to the instruction followed by the instruction written in specific and measurable terms together with a

---

* An out-of-the-box solution in this situation coming from a similar solution in the military box.

specific and measurable metric to know when the instruction has been correctly implemented. In addition, the requirements for writing requirements (Kasser 2019: Section 10.5) in layer 2 of the risk framework apply to writing policy instructions, namely remove the ambiguity and subjective measurements. On the other hand, some flexibility is desirable so the instruction should specify what needs to be done and leave the how it is done to the staff carrying out the policy.

The risk of disputes and legal action should be mitigated as discussed in Section 8.2.4.4.

9.5.1.5 The Policy Closeout State

The problem posed by the policy closeout state may be formatted according to the problem formulation template (Section 5.1.7.2):

1. *The undesirable situation*: the need to document the lessons learned and free up the resources used.
2. *The assumptions*: specific to the situation.
3. *The FCFDS:* the lessons learned have been documented and the resources have been freed and allocated elsewhere.
4. *The problem*: is to realize the FCFDS.
5. *The solution*: document the lessons learned, free the resources and allocate them elsewhere and store the appropriate records following the traditional generic problem-solving process (Section 5.1.6.1).

9.5.1.5.1   Generic Risks in the Policy Closeout State

Perceptions from the *Generic* HTP note that closing down a policy is the same as closing down a LEP (Section 8.3.4) and liable to the same risks. While perceptions from the *Continuum* HTP point out that the situation is different and the effect of those differences needs to be explored.

9.6 FUNCTIONAL

Perceptions from the *Functional* HTP included:

1. Thinking, planning, organizing, directing, controlling, staffing, analysing, problem-solving, influencing, etc. discussed in Section 7.1.
2. Politics in action discussed in Section 9.6.1.
3. Decision-making and problem-solving discussed in Chapter 5.
4. Communicating with stakeholders discussed in Section 3.2.2.

9.6.1 POLITICS

Policies run on politics. Whether it's initiating a policy, drumming up support, overcoming resistance or reporting, there are people involved and people use politics. Politics and their uses for project managers (Section 7.3.1.10) are just as important in the policy lifecycle. There is always a risk of unknowingly upsetting people which motivates them to resist the policy. Many managers with an engineering background may not think that people skills are important; however, in any activity above the system layer in the risk framework (Section 1.2), people skills are an important tool in achieving desired results.

9.7   TEMPORAL

Perceptions from the *Temporal* HTP included:

1. Changes take time discussed in Section 4.2.3.
2. Changes tend to be resisted discussed in Section 4.2.4.
3. The types of risks are different in the different states of the policy lifecycle discussed in Sections 9.5.1.1, 9.5.1.2.1, 9.5.1.3.1, 9.5.1.4.1 and 9.5.1.5.1.

9.8   STRUCTURAL

Perceptions from the *Structural* HTP included:

1. Various definitions including a policy is, 'A course or principle of action adopted or proposed by an organization or individual' (Oxford 2019).
2. Policy documents discussed in Section 9.8.1.

9.8.1   POLICY DOCUMENTS

Policy documents are a major communications tool. Policy information is stored in a set of documents wherein each document should contain the following standard information:

- *Definitions*: clear and unambiguous definitions for the terms and concepts
- *Purpose*: containing:
  1. Why the organization is issuing the policy.
  2. What the desired effect or outcome of the policy should be (the benefit).
- *Applicability and scope*: identifying:
  1. The stakeholders affected by the policy.
  2. The actions that are impacted by the policy.
- *An effective date*: when the policy comes into force.
- *Background*: the reasons, history and intent that led to the creation of the policy.
- *Responsibilities*: identifying the parties and organizations responsible for carrying out individual policy statements.
- *The specific regulations, requirements or modifications to organizational behaviour*: that the policy is creating or changing.

Typical policy documents include:

1. Policy implementation plans discussed in Section 9.8.1.1.
2. Policy briefs discussed in Section 9.8.1.2.

9.8.1.1 Policy Implementation Plans

In addition to the standard information in a policy document, a policy implementation plan should also:

1. Cover each state in the particular policy lifecycle being implemented with greater details in the earlier states.
2. Identify and summarize:
   1. Feasibility.
   2. Activities and associated attributes.
   3. Costs, risks, etc.
   4. Critical actions and dates.

9.8.1.2 Policy Briefs

A policy brief is a summary of the following information (as a minimum) extracted from the policy implementation plan:

1. Problem and its context addressed by the policy.
2. Proposed goals and relevant sub-goals and indicators.
3. Linkages between the proposed policy and national policies.
4. Costs and resources that will be needed during each of the states of the policy implementation cycle.
5. Specific risks and proposed risk management during each of the states of the policy lifecycle.
6. Implications and benefits of the proposed risk management measures.
7. Specific decisions which need to be made, by whom and when.

9.9 SCIENTIFIC

Inferences from the *Scientific* HTP included:

1. A way to create and implement a successful policy discussed in Section 9.9.1.
2. A reason for the confusion between complex problems and wicked problems by policymakers discussed in Section 9.9.2.

9.9.1 A Way to Create and Implement a Successful Policy

To create and implement a successful policy, the policymaker has to determine:

1. What is undesirable about the current situation.*
2. The probable causes of the undesirable aspects of the current situation.
3. The specific changes needed to remove the undesirable aspects of the current situation.
4. The impact of the changes.

---

* There may be nothing undesirable in the current situation, other than something needs to be improved.

5. The stakeholders who will be affected by the changes.
6. The anticipated resistance to the changes.
7. The risks in making the change over the entire policy lifecycle.
8. How to ensure truthful status reports will be generated in a timely manner during the policy performance state.

The policymaker then has to ensure that adequate resources are provided during the policy performance state and the personnel performing the policy report progress or lack of progress in a timely manner. The policymaker deals with the risks at the policy layer in the risk framework (Section 1.2), is aware of the critical risks at the next lower layer but in general allows the project managers in various projects in that layer to manage the risks. The policymaker or person in charge of the during the policy performance state should not use micromanagement by upper management (MBUM).

9.9.2 A REASON FOR THE CONFUSION BETWEEN COMPLEX PROBLEMS AND WICKED PROBLEMS

Policies generally deal with a large number of controlled and uncontrolled variables (Section 5.1.2.1), namely they may be both complex and wicked problems.

9.10 EXERCISES

This section provides a number of exercises as opportunities to think about assumptions, policies and how they relate to project management, systems engineering, politics and risk management using referenced material from the previous exercises[*] and additional material developed for this exercise. A partial acceptable solution is provided for the off-world mining policy (Section 9.10.4). It is up to the reader to use the sample as a template when performing the remaining exercises.

9.10.1 THE ENU TRAFFIC LIGHT UPGRADE

An exercise in this context is not applicable in this chapter because the ENU traffic light upgrade is a relatively simple project.

9.10.2 THE ENGAPOREAN MAID REDUCTION POLICY

In addition to meeting the generic requirements in Section 2.7:

1. Format the problem posed by this exercise in the problem formulation template (Section 5.1.7.2).
2. Assume the policy implementation plan (Section 9.8.1.1) exists, and using reasonable and realistic assumptions, prepare a policy brief (Section 9.8.1.2) to summarize:
   1. The assumptions.
   2. The problem and its context addressed by the policy.

---

[*] There is no need to repeat the work.

3. The proposed goals and relevant sub-goals.
4. At least three linkages between the proposed policy and national policies.
5. At least three specific decisions which need to be made, who makes them and when the decisions are made.

9.10.3   THE ENGAPORIAN DROUGHT RELIEF POLICY

In addition to meeting the generic requirements in Section 2.7:

1. Format the problem posed by this exercise in the problem formulation template (Section 5.1.7.2).
2. Assume the policy implementation plan (Section 9.8.1.1) exists, and using reasonable and realistic assumptions, prepare a policy brief (Section 9.8.1.2) to summarize:
   1. The assumptions.
   2. The problem and its context addressed by the policy.
   3. The proposed goals and relevant sub-goals.
   4. At least three linkages between the proposed policy and national policies.
   5. At least three specific decisions which need to be made, who makes them and when the decisions are made.

9.10.4   THE OFF-WORLD MINING POLICY

In addition to meeting the generic requirements in Section 2.7:

1. Format the problem posed by this exercise in the problem formulation template (Section 5.1.7.2).
2. Assume the policy implementation plan (Section 9.8.1.1) exists, and using reasonable and realistic assumptions, prepare a policy brief (Section 9.8.1.2) to summarize:
   1. The assumptions.
   2. The problem and its context addressed by the policy.
   3. The proposed goals and relevant sub-goals.
   4. At least three linkages between the proposed policy and national policies.
   5. Specific decisions which need to be made, by whom and when.

9.10.4.1   One Acceptable Solution

A completed exercise could take the form shown in this section. Note it provides one acceptable solution, not the single correct solution as there will be other acceptable solutions (Section 5.1.1.2.2). The assumptions made in this example may or may not be entirely realistic due to lack of domain knowledge.

9.10.4.1.1 The Assumptions

The assumptions include:

1. The exercise can be completed within 60 minutes.
2. The response is supposed to cover breadth rather than depth due to the lack of domain knowledge.
3. While the policy implementation plan as summarized in the policy brief (Section 9.8.1.2) will conceptualize a phased implementation using pilot projects to demonstrate that policy can be put into action by the desired target date, there will be no effort to enforce it because the whole enterprise will not be under government control. One reason for this is the perceptions from the *Temporal* HTP that just as in the era of the East India clipper ships, the first ship back to Europe each year generally created a large profit for its owner, the first ship back to earth from the asteroids will also create a large profit for its owner.
4. The part of the question on linkages between policies refers to space policies.

9.10.4.1.2 The Formatted Problem

The problem formatted according to the problem formulation template (Section 5.1.7.2) is:

1. *The undesirable situation*: is the need to:
   1. Comply with the generic requirements in Section 2.7.
   2. Prepare a policy brief summarizing:
      1. The assumptions.
      2. The problem and its context addressed by the policy.
      3. Proposed goals and relevant sub-goals and indicators.
      4. At least three linkages between the proposed policy and national policies.
      5. At least three specific decisions which need to be made, by whom and when.
2. *The assumptions*: are:
   1. The exercise can be completed within the 60 minutes.
   2. As listed in Section 9.10.4.1.1.
3. *The FCFDS*: the exercise has been completed and the desired grade has been achieved.
4. *The problem*: how to do the exercise, namely:
   1. Meet the generic requirements in Section 2.7.
   2. The assumptions.
   3. The problem and its context addressed by the policy.
   4. The proposed goals and relevant sub-goals.
   5. At least three linkages between the proposed policy and national policies.
   6. At least three specific decisions which need to be made, who makes them and when the decisions are made.

TABLE 9.1

Compliance Matrix for the Exercise in Section 9.10.4

ID	Instruction	Deliverable	Section	Completed
1	Brainstorming or active brainstorming	No	All, specially 9.10.4.1.4.1	Yes
2	Research	No	All	Yes
3	Use imagination	No	All	Yes
4	Use lists and charts	No	All	Yes
5	Spend less than 60 minutes	No	All	Yes
6	The assumptions	Yes	9.10.4.1.1	Yes
7	Format the problem	Yes	9.10.4.1.2	Yes
8	Preliminary compliance matrix	No	9.10.4.1.3	Yes
9	The policy brief	Yes	9.10.4.1.4	Yes
10	The assumptions.	Yes	9.10.4.1.1	Yes
11	The problem addressed by the policy	Yes	9.10.4.1.4.1	Yes
12	The context of the problem	Yes	9.10.4.1.4.1	Yes
12	Proposed goals	Yes	9.10.4.1.4.2	Yes
13	The relevant sub-goals	Yes	9.10.4.1.4.2	Yes
14	Linkages between the proposed policy and national policies.	Yes	9.10.4.1.4.3	Yes
15	The specific decisions	Yes	9.10.4.1.4.4	Yes
16	Who makes the specific decisions	Yes	9.10.4.1.4.4	Yes
17	When the specific decisions need to be made	Yes	9.10.4.1.4.4	Yes
18	Lessons learned	Yes	9.10.4.1.5	Yes
19	Updated compliance matrix	Yes	9.10.4.1.3	Yes

7. Verify that all the requirements for the exercise have been completed and update the compliance matrix accordingly.

5. *The solution*: consists of remedying each of the problems. A partial solution is presented below.

9.10.4.1.3   The Compliance Matrix

The partially completed compliance matrix is shown in Table 9.1.[*]

9.10.4.1.4   The Policy Brief

The required information from the policy brief extract is shown in the following sections:

1. The assumptions in Section 9.10.4.1.1.
2. The problem and its context addressed by the policy in Section 9.10.4.1.4.1.

---

[*] Placing the section number in the completed column not only shows that the requirement has been met but also shows where the response can be found, mitigating the risk of the instructor failing to find and grade the response.

3. The proposed goals and relevant sub-goals and indicators in Section 9.10.4.1.4.2.
4. At least three linkages between the proposed policy and national policies in Section 9.10.4.1.4.3.
5. At least three specific decisions which need to be made, by whom and when in Section 9.10.4.1.4.4.

9.10.4.1.4.1   The Problem and Its Context Addressed by the Policy   After a brainstorming and an active brainstorming session,[*] the ideas about the problem and its context addressed by the policy were rearranged into the following HTPs being used as an idea storage template (IST) (Kasser 2018a: Section 14.2):

1. *Big Picture*: perceptions from this HTP include:
   - The purpose of the policy is stated in Section 2.7.4 which introduced the exercise.
   - The Minister's expectation that moving humanity into space is not something that should be left to governments but should be a commercial profit-making enterprise similar to the European colonization companies in the 16th–18th centuries.
   - A whole society inhabiting the asteroids needing all the functions performed in a country including government functions, working, recreation, producing and consuming food and beverages, and sports.
2. *Operational*: perceptions from this HTP include:
   - Terrestrial mining and oil extraction activities.
   - Marketing and selling the material produced.
   - Transporting the material to the markets.
   - Refining the material.
   - Miners and support personnel travelling between the Earth and the asteroids.
   - Miners and support personnel living in closed environmental habitats.
   - The mined materials being transported to the Earth.
   - Ongoing research to lower the cost of the whole policy and action.
   - Tourism in space.
   - Expanded planetary exploration by government, universities and commercial interests fuelled by the hydrocarbons mined Jupiter.
   - Refining of raw materials in space rather than on the Earth.
   - Construction of into planetary spaceships in space rather than on the Earth using the materials mined from the asteroids.
   - Law enforcement.
3. *Functional*: perceptions from this HTP include:
   - Drilling.
   - Tunnelling.
   - Waste disposal.
   - Recycling oxygen and water.

---

[*] This sentence was inserted to give the impression that brainstorming and active brainstorming did take place as part of performing the exercise.

4. *Structural*: perceptions from this HTP include:
   - Oil rigs currently mining for oil.
   - Drills and all the technological equipment used on the oil rigs and in mining and in transporting the desired materials and the waste such as lorries, sheds, containers and computers.
   - Spaceships and the equipment inside them.
   - The hostile vacuum environment of space that will need attention to strict health and safety.

5. *Generic*: perceptions from this HTP include:
   - Many ideas published in the science fiction literature.

6. *Continuum*: perceptions from this HTP include:
   - Current mining and oil extraction activities that while producing the desired materials also produce undesirable materials in the form of waste, some of which is hazardous and toxic.
   - When the European colonization companies in the 16th–18th centuries moved out into Africa, America and Asia, they found the territories were populated and there was the potential for trade. In this instance, the territories will have to be populated so that the inhabitants will be able to mine the asteroids and the atmosphere of Jupiter so that trade can take place.
   - Once miners have established claims, there is the potential for claim jumping, so there will be a need for law enforcement to be established and maintained just like in the American Wild West.
   - Accidents will take place.
   - There are opportunities for sabotage by unscrupulous opposing forces especially in the early years.

7. *Quantitative*: perceptions from this HTP include:
   - All of the costs of mining, drilling, marketing, transportation, etc.
   - Projections of the amount of oil reserves.
   - Projections of the amount of material remaining to be mined.
   - All the statistics associated with mining and drilling.
   - The expected costs of setting up and performing asteroid mining.
   - The anticipated revenues of the asteroid mining.

8. *Temporal*: perceptions from this HTP include:
   - Historical data on the amounts of material mined and drilled.
   - Expectations that as terrestrial resources become harder to find, the costs of mining and drilling and processing will increase.
   - Expected trends in the consumption of terrestrial material being mined and drilled for.
   - The European colonization companies in the 16th–18th centuries moving out into India, Africa, America and Asia.
   - The technology will change as time goes by the same way that aviation technology advanced from the Wright brothers' aeroplane to large commercial airliners during the 20th century.
   - The development of one or more space stations in low Earth orbit to act as way stations between the asteroids and the Earth and serve as

manufacturing facilities for the interplanetary transportation vehicles and asteroid habitats; at least until those manufacturing facilities can be set up in the asteroids.

- In the era of the East India clipper ships, the first ship back to Europe each year generally created a large profit for its owner; the first ship back to the Earth from the asteroids will also create a large profit for its owner.
- The technology to make the policy happen either exists or can be developed in time to meet the 50-year goal.
- Knowledge we don't know we don't know (Section 5.2.4.2) may become available and speed up the process. Such knowledge includes improved propulsion systems resulting in shorter transit times, new ways of mining asteroids, new type of life support and waste management systems of use in space. However, the schedule must be developed as if that knowledge will not become known.

9.10.4.1.4.2   Proposed Goals and Relevant Sub-goals   The goal is for the policy to be in action within 50 years. The sub-goals include:

1. The announcement of the policy during the policy initialization state and raising the awareness for the desirability of the policy by glamourizing the policy in action, pointing out the undesirable effects of the current situation such as pollution and pointing out that with the policy and action there won't be any pollution; the Earth will be greener.
2. Expressions of interest by at least five corporations with a feasible expectation of being able to mine the asteroids within the 50-year expected period.
3. Successful completion of the policy implementation plan at the end of the policy planning state.
4. Three successful pilot projects in LEO. These will include a non-government habitat and a manufacturing facility that has constructed one interplanetary vehicle.
5. One successful mining pilot project returning a commercially viable payload to the Earth.
6. The policy in action with a number of different commercial enterprises cooperating and competing.

9.10.4.1.4.3   At Least Three Linkages between the Proposed Policy and National Policies   Space policy is the political decision-making process for, and application of, public policy of a state (or association of states) regarding spaceflight and uses of outer space, for both civilian (scientific and commercial) and military purposes (Wikipedia contributors 2019).*

---

* In many academic institutions, references to Wikipedia are unacceptable for research papers. This does not mean that Wikipedia material should be used without citation in other works since plagiarism is also unacceptable. It is generally better to consult the sources reference by the Wikipedia article and cite those sources. It is not a good idea to cite the sources found cited in Wikipedia as if you had read the sources, since there is a risk that the sources have been misquoted in the Wikipedia article just like the risk of misquotation in any other secondary source (Kasser 2015: p. 374).

TABLE 9.2
The Three Specific Decisions

	Decision	Decision-Maker	When the Decision Is Made
1	To go public with the policy concept	The Minister	When the Minister decides to do so.
2	To award a contract for the spaceport	The government	One year before the first launch.
3	To create a civilian agency to carry out the policy in the same manner as NASA carries out the US space policy	The government	As part of the policy initialization state so that the effect of the decision can be included in the announcements of the policy concept.

The linkages are to[*]:

1. The US space policy.
2. The space policy of the People's Republic of China.
3. The Russian space policy.

9.10.4.1.4.4 At Least Three Specific Decisions Which Need to Be Made, Who Makes Them and When the Decisions Are Made The decisions, who makes them and when are summarized in Table 9.2.[†]

9.10.4.1.5 Lessons Learned

Lessons learned included:

- Conceptualizing this concept was fun especially the brainstorming and active brainstorming sessions.
- The compliance matrix is very handy for keeping track of what needs to be done and making sure that it is being done.
- It was easy to come up with isolated concepts; developing the linkages was much more difficult.
- Most of the planning had a tendency to focus on the structural, operational and functional perspectives of the planning and implementation states of the policy lifecycle and ignored the risks associated with those perspectives.

---

[*] The question did not ask for further explanation, so none is provided.
[†] Since there is no requirement to justify the decisions, a table makes it easy to see compliance with the instruction. Moreover, the table also acts as a compliance matrix at the same time as providing the information.

9.11 SUMMARY

This chapter discussed policies and policy management from the different HTPs because many policies are implemented by LEPs. In particular, it discussed policies as a system, the different types of policies, the generic policy lifecycle, policy documents and ways of mitigating risks in policies.

REFERENCES

Althaus, Catherine, Peter Bridgman, and Glyn Davis. 2007. *The Australian Policy Handbook* (4th ed.). Sydney: Allen & Unwin.

Covey, Steven R. 1989. *The Seven Habits of Highly Effective People*. New York: Simon & Schuster.

Edwards III, George C. 1980. *Implementing Public Policy*. Washington, DC: Congretional Quarterly Press.

Edwards III, George C., and Ira Sharkansky. 1978. *The Policy Predicament. Making and Implementing Public Policy*. San Francisco: W. H. Freeman and Company.

Kasser, Joseph Eli. 2018a. *Systems Thinker's Toolbox: Tools for Managing Complexity*. Boca Raton, FL: CRC Press.

Kasser, Joseph Eli. 2018b. Using the systems thinker's toolbox to tackle complexity (complex problems). In *SSSE Presentation at Roche*. Zurich, Switzerland: Swiss Society of Systems Engineering.

Kasser, Joseph Eli. 2019. *Systems Engineering: A Systemic and Systematic Methodology for Solving Complex Problems*. Boca Raton, FL: CRC Press.

Kasser, Joseph Eli. 2015. *Holistic Thinking: Creating Innovative Solutions to Complex Problems* (2nd ed.), Vol. 1. Solution Engineering. Charleston, SC: Createspace Ltd.

Oxford. 2019. *Oxford Dictionaries*. Oxford University Press 2019 [cited January 22, 2019]. Available from https://en.oxforddictionaries.com/definition/emergence.

Wikipedia Contributors. 2019. *Space policy*. Wikipedia, The Free Encyclopedia, 27 September 2019 2019 [cited October 4, 2019]. Available from https://en.wikipedia.org/w/index.php?title=Space_policy&oldid=918236880.

10 Afterword

The contents of the book provided a journey of exploration in which each chapter was built on the previous chapters and provided exercises to practice the knowledge explained in the chapter. Namely:

- Chapter 1 introduced the book and the risk framework and recommended a way to read this book if you are not using it as a textbook. The Chapter continued with an introduction to the systems approach and concluded with a summary of the major risks in any activity and some ways to mitigate them.
- Chapter 2 discussed thinking, systems thinking, the nine HTPs and the benefits of going beyond systems thinking. In particular, the Chapter discussed judgement and creativity, thinking, critical thinking, systems thinking and going beyond systems thinking, and concluded with examples of using the HTPs to perceive a camera, a house; a car and a policy.
- Chapter 3 documented thoughts about risks and risk management systemically and systematically from the nine HTPs and suggested generic risk mitigation and prevention activities. Systemic perceptions of risks using the HTPs include the risk management process, the risk rectangle, risk profiles and risk trees.
- Chapter 4 perceived change and change management from the HTPs because processes in which risks occur generally cause change. In particular, the Chapter discussed some change management models, resistance to change and how to overcome it, and some aspects of stakeholder management.
- The risk management process is riddled with changes, problems and solutions because managing risks poses problems which need to be solved or remedied and change is an outcome, mostly desired but sometimes undesired. A risk mitigation strategy is a solution to the problem of managing the risk but poses a problem to the people who will have to implement the strategy. Accordingly, Chapter 5 discussed perceptions of the problem-solving process from a number of HTPs, and then discussed the structure of problems, the levels of difficulty posed by problems and the need to evolve solutions using an iterative approach. After showing that problem-solving is really an iterative causal loop rather than a linear process, the Chapter then discussed complexity and how to use the systems approach to manage complexity. The Chapter then showed how to remedy well-structured problems and how to deal with ill-structured, wicked and complex problems using iterations of a sequential two-part problem-solving process. While discussing managing complexity, the chapter used Miller's rule to specify the minimum number of elements in a system for it to be defined as, and managed as, complex.

Reflecting on this Chapter, it seems that iteration is a common element in remedying any kind of problem other than easy well-structured ones irrespective of their structure.

- Chapter 6 perceived systems and systems engineering from different perspectives because projects are systems and create products which are systems using systems engineering.* Moreover, the project takes place in an environment or context which is a system. The chapter provided an overview of systems, the system lifecycle (SLC), systems engineering and the system development process (SDP) to help the policymaker, implementer and project manager understand the important aspects of risk management in systems and systems engineering. Specifically, the Chapter discussed, the nature of systems, properties of systems, hierarchies of systems, supply chains as systems, an introduction to systems engineering, modeling and simulation, the nine-system model and risks in systems and systems engineering.
- Policies and projects are systems which are realized using systems engineering and project management working interdependently. Chapter 6 provided an overview of systems, the SLC, systems engineering and the SDP to help the policymaker, implementer and project manager understand the important aspects of risk management in systems and systems engineering and the interdependency and overlap between risk management in systems engineering and project management. Accordingly, Chapter 7 followed on by discussing project management from different perspectives because project management is the heart of effecting any change, caused by the implementation of a policy, a large engineering project (LEP) or any type of project. In particular, the Chapter provided an overview of the systems approach to projects and project management discussing the project lifecycle, project plans, project planning, generic and specific planning, the planning process, categorized requirements in process (CRIP) charts, enhanced traffic light (ETL) charts, management by exception (MBE), management by objectives (MBO), using prevention to lower project completion risk, stakeholders and stakeholder management and software project risks, and concluded by discussing generic project risks.
- Chapter 8 provided an introduction to LEPs and managing risks in LEPs because a LEP contains a hierarchy of projects and systems. In particular, the Chapter discussed risks in LEPs, the LEP lifecycle and risks based on the non-availability of technology since technology has become ubiquitous in the 21st century and introduced an FFP with penalties and bonuses by exception contract, somewhat similar to MBE.
- Chapter 9 discussed policies and policy management from different HTPs and examined policy implementation because many policies are implemented by LEPs. In particular, the Chapter discussed policies as a system, the different types of policies, the generic policy lifecycle, policy documents and ways of mitigating risks in policies.

---

* Even if they don't call it systems engineering.

Each chapter also contains a discussion of pertinent risks and cumulative exercises.

Perceptions from the *Generic* HTP show the similarity in the risks in policies, projects and LEPs as well as in daily life. Any action taken contains risk. These risks need to be identified in their context, and their probability of occurrence and severity of impact estimated. All this requires thinking, which is what this book was about. This book:

- Has provided an introduction and overview of the systems approach to risk management and some aspects of the current non-systems paradigm.
- Will not make you an expert in the topics covered in each chapter, it is not intended to. However, it is intended to teach you enough to communicate with the relevant subject matter (domain) experts to:
  - Make an informed decision on the advice (information) they provide.
  - Detect when they are more ignorant than you which will enable you to ignore their advice and find a real expert.
- Has explained the need for applying, and shown how to apply, systems thinking and beyond to risks rather than just adopt lists inherited from other projects.

As this book showed, the systems approach is based on:

- Identifying risks in each area of activity and determining their importance and urgency using domain expertise before managing them.
- Aggregating the risks into categories so that common mitigation and prevention techniques may be applied.
- Classifying the major risks to any endeavour into the following categories:
  1. *Lack of a common vision of the goal**: this situation is represented in Figure 2.1. Without that common vision of the purpose and performance of the solution systems among the stakeholders, time and funds will be expended on non-productive activities. This risk may be mitigated by developing that common vision in the appropriate format for the activity such as in the concept of operations (CONOPS) (Section 6.5.9.1) commonly used in systems engineering.
  2. *People*: incompetent, inexperienced and unmotivated people will not do the job properly. They will waste time and funding doing the wrong thing and introducing errors of commission and omission (Section 5.1.4.3.1) into what they do even if they follow the SDP (Section 6.1.5.4). This risk may be mitigated by employing competent and experienced people, motivating them by showing them 'what's in it for me†?' (Section 7.3.1.4) and providing just-in-time (JIT) training at appropriate

---

* The goal may change over time. Each step in the problem-solving process has its own goal, and different activities have their own goals. Accordingly, there may be more than one goal at any time: the short-term as well as the long-term goal of the project.

† Me being them.

points in the SDP. However, competent people cannot do a proper job with adequate resources and time.

3. *Politics*: people will mess things up for various personal agendas. They can use negative politics to cause failures or modifications that will cause failures. They can withhold resources and fail to meet commitments. They can persuade others to follow their lead with respect to the activity. This risk can often be mitigated by the use of positive politics (Section 7.3.1.5.1).

4. *Funding*: lack of sufficient funding leads to shortcuts, errors and early terminations. This risk can be mitigated by providing adequate funding.

5. *Resources*: insufficient resources lead to failures due to the lack of the resource. This risk can be mitigated by providing adequate resources.

6. *Time*: insufficient time leads to shortcuts and failures. This risk can be mitigated by inserting sufficient time into the schedule.

7. *Technology*: failures happen and need to be minimized and repaired: technology becomes obsolete as in diminishing manufacturing sources and material shortages (DMSMS) resulting is systems becoming unmaintainable and needing replacement. This risk may be mitigated by the use of redundancy, reliability and the correct maintenance procedure.

Finally, the systems approach to planning a project:

1. Ensures that there is no single point of failure in the supply chain.
2. Determines the costs, schedules and resources needed as worst-case values to mitigate the funding, resource, technological and time-related risks. Stakeholders that subsequently reduce those values for political and other reasons reintroduce those risks into the endeavor which can result in a worst-case or even a doomed endeavour.

Author Index

Subject Index

Printed in the United States
by Baker & Taylor Publisher Services